REBELS,

MAVERICKS,

AND HERETICS

IN BIOLOGY

REBELS, MAVERICKS, AND HERETICS IN BIOLOGY

Edited and with an Introduction by

OREN HARMAN and
MICHAEL R. DIETRICH

and with an Epilogue by R. C. Lewontin

Yale University Press

New Haven & London

Published with assistance from the Louis Stern Memorial Fund.

Set in Melior type by Duke & Company, Devon, Pennsylvania.
Printed in the United States of America by Sheridan Books, Ann Arbor, Michigan.

ISBN: 978-0-300-11639-7

A catalogue record for this book is available from the Library of Congress and the British
Library.

10 9 8 7 6 5 4 3 2 1

To Savtale,
who lost a debate all those years ago,
and won our everlasting love
O. H.

To Lydia and Arlena,
who don't take anything too seriously
M. R. D.

Contents

Preface   ix

Acknowledgments   x

1 Introduction: On Rebels, Icons, and the Value of Dissent   1
OREN HARMAN AND MICHAEL R. DIETRICH

2 Alfred Russel Wallace, the Discovery of Natural Selection, and the Origins of Humankind   20
MICHAEL RUSE

3 Rebel With Two Causes: Hans Driesch   37
GARLAND E. ALLEN

4 Wilhelm Johannsen: A Rebel or a Diehard?   65
RAPHAEL FALK

5 Raymond Arthur Dart: The Man Who Unwillingly Ushered in a Revolution in the Evolution of Humankind   84
PHILLIP V. TOBIAS

6 In Weismann's Footsteps: The Cyto-Rebellion of C. D. Darlington   103
OREN HARMAN

7 Striking the Hornet's Nest: Richard Goldschmidt's Rejection of the Particulate Gene   119
MICHAEL R. DIETRICH

8 Rebellion and Iconoclasm in the Life and Science of Barbara McClintock   137
NATHANIEL COMFORT

9 Challenging the Protein Dogma of the Gene: Oswald T. Avery, a Revolutionary Conservative   154
UTE DEICHMANN

10 Roger Sperry and Integrative Action in the Nervous System 174
TIM HORDER

11 Leon Croizat: A Radical Biogeographer 194
DAVID L. HULL

12 Dogma, Heresy, and Conversion: Vero Copner Wynne-Edwards's Crusade and the Levels-of-Selection Debate 213
MARK BORRELLO

13 Peter Mitchell: Changing the Face of Bioenergetics 231
JOHN PREBBLE AND BRUCE WEBER

14 Howard Temin: Rebel of Evidence and Reason 248
DANIEL J. KEVLES

15 Motoo Kimura and the Rise of Neutralism 265
JAMES F. CROW

16 Against the Grain: The Science and Life of William D. Hamilton 282
ULLICA SEGERSTRALE

17 The Iconoclastic Research Program of Carl Woese 302
JAN SAPP

18 Stephen Jay Gould, Darwinian Iconoclast? 321
DAVID SEPKOSKI

19 Culture and Gender Do Not Dissolve into How Scientists "Read" Nature: Thelma Rowell's Heterodoxy 338
VINCIANE DESPRET

20 Bringing Statistical Methods to Community and Evolutionary Ecology: Daniel S. Simberloff 356
WILLIAM DRITSCHILO

Epilogue Legitimation Is the Name of the Game 372
R. C. LEWONTIN

List of Contributors 381

Index 385

Preface

This volume emerged from our mutual interest in iconoclastic biologists. Having devoted our earlier scholarly efforts to elucidating the work of Cyril Darlington (by Oren Harman) and Richard Goldschmidt (by Michael R. Dietrich), we sought to broaden our view beyond these two rebellious biologists. Finding iconoclasts in biology was not as difficult as finding authors who would engage deeply with their rebellion. Although the scientists presented in these pages represent only a selection, the contributors have put their histories to great use and raised important questions about the nature of orthodoxy and dissent, the impact of institutions on innovation, and the power of personality in science.

It is our hope that *Rebels, Mavericks, and Heretics in Biology* will prove to be of interest to historians and philosophers of science, as well as working biologists and scientists. We are also hopeful that a volume dedicated to an analysis of the role of dissent and iconoclasm in science might appeal to a broader audience that finds itself interested in the growth of biology in modern times, in the dynamics of scientific knowledge, or both. There are also great advantages, we feel, to teaching the development of scientific thought through a rigorous examination of the researchers and thinkers who departed from the norm. Such a consideration highlights the need to constantly scrutinize the working assumptions on which "normal" science is based and emphasizes the important role of thinking "outside the box." We feel that at a time in which the biological sciences seem to be moving in new and exciting directions and continually inventing new subdisciplines and languages, these essays will be of considerable interest.

Acknowledgments

We thank the contributors to this book, who bravely withstood the kind of editorial nagging only a rebel (or an unperturbed professional) would swat away. Michael, Gar, Rafi, Phillip, Nathaniel, Dan, Jan, Vinciane, Ute, Bill, David H., David S., Mark, John, Bruce, Jim, Tim, and Ullica: your dedication and fine scholarship helped turn a hunch about the idiosyncrasy of rebellion and the value of its study into a well-supported and broad piece of historical evidence. Thanks, too, to Richard Lewontin, who agreed to write the epilogue to this book, and as usual, enlightened us all.

To Jean Thomson Black, the kindest, most efficient, and most professional of editors, who guided our travels with wisdom and good humor, thanks for your unfailing encouragement and support. And thank you, too, Jane Zanichkowsky, for working with much talent to make this book an infinitely smoother read than it was before coming under your exacting pen.

Last but not least, to all the rebellious minds lurking out there behind the page: *Audentes fortuna iuvat.* We eagerly await your awakening!

REBELS,

MAVERICKS,

AND HERETICS

IN BIOLOGY

CHAPTER ONE

Introduction:
On Rebels, Icons, and the Value of Dissent

OREN HARMAN and MICHAEL R. DIETRICH

The history of science is invariably told through the lives of its heroes, and modern biology is no exception. Men and women who, with
painstaking research and brilliance, have helped advance science to
the heights we call "today" have been our guides. Historians have
studied their lives in order to analyze the growth of the life sciences,
the rise of institutions and disciplines, and the evolution of biologists'
understanding of nature. Insofar as we seek narratives concerning
humankind's quest for knowledge, it is not surprising that some biologists should be cast as heroes. But it is not the whole story.

Seldom is the story of biology told through the tale of its rebels:
men and women who challenged the prevailing picture of life in the
myriad disciplines that, taken together, constitute modern biology.
Some of these researchers were in fact wrong; others, though lam-
basted for their views at the time, will be found—or have already been
found—to deserve a more appreciative treatment. Some have been
called cranks, others gadflies, still others prophetic. Not all have
been heroes. But whether vindicated by history or forgotten, scientific
rebels, as is true for challengers of any kind, may teach the challenged
much about themselves and about the issues that most feel are no
longer in need of scrutiny. Even when such challenges end up being
resisted, the Italian economist Vilfredo Pareto's comment on the importance of dissent is worth remembering: "Give me a fruitful error
any time, full of seeds, bursting with its own corrections. You can
keep your sterile truth to yourself."[1]

In *Rebels, Mavericks, and Heretics in Biology,* we have collected the
stories of leading iconoclastic figures in biology throughout the late
nineteenth and twentieth centuries. The chapters that follow do not
offer full biographical treatments of their subjects but, rather, focus

on particular challenges to particular orthodox assumptions in the life sciences. It is our hope that the collection of these different narratives will provide a fuller understanding of the role of dissent and controversy in science.

The nineteen subjects were chosen because they satisfied two basic requirements of the book, conceived as a kind of alternative history of the growth of modern biological thought. First, they participated in important episodes of dissent from major areas of biology in the twentieth century. Second, they present a broad and varied exposition of the different roles and effects of rebelliousness and iconoclasm in science. The need to strike a balance between both requirements generated a sample that is representative but far from comprehensive.

This is the place to make an important admission: we have not done justice to many rebels who "got away." When canvassing the history of twentieth-century biology, we were amazed at how many rebels we encountered and dismayed that relatively few could be included in our book. In particular, we would have liked to include Max Delbrück's informational challenge to the biochemical understanding of life, Arthur Kornberg's challenge to the belief that the core processes of life, such as replication, are not amenable to thoroughly conventional analytical methodology, Lynn Margulis's endosymbiotic challenge to orthodox neo-Darwinism, John Gurdon's challenge to the icon of cell-loss of pluripotency during development, Tracy Sonneborn's iconoclastic challenge to the nuclear theory of heredity as a champion of the importance of the cytoplasm in heredity and development, Eva Jablonka and Marion Lamb's "Lamarckian" challenge to neo-Darwinian orthodoxy in contemporary evolutionary thought,[2] and still more cases. All we can say, by way of an apology, is that these omissions are due to the kinds of constraints associated with producing books with many authors in a set period of time.

The chapters in this book cover many of the basic problems of the life sciences throughout the twentieth century. Specifically, the nineteen histories emerge from the narratives of human evolution, evolutionary theory, development, systematics, microbiology, biochemistry, physiology, genetics, neurobiology and brain sciences, cytology, virology, and ecology. We began with scientific icons—assumptions held broadly by practitioners in a given field of biology at a given time in the twentieth century—and from there began our search for historical dissenters. Each contributor was asked to focus on a particular scientist and the particular assumption or icon that the figure challenged. We asked the authors to consider the precise dynamics of the challenge and the responses to it. We also urged each contributor to try to assess the impact of this challenge on the given field and to describe the lessons that might be gleaned in view of the present and the future direction of the field.

Ever since the publication of Thomas Kuhn's study *The Structure of Scientific Revolutions,* there has been a great debate about the degree to which consensus among scientists is constitutive of normative (or "normal" or accepted) science.[3] Looking at the fate of biologists who operate outside the consensus of their colleagues, and at the fate of their theories, we feel, will shed light on this debate.

ABOUT THE REBEL

From the analysis of nineteen leading iconoclastic biologists of the twentieth century, a Tolstoyan thesis emerges: although all conventional practitioners in the life sciences may be said to be conventional in the same way, all rebels seem to rebel in their own particular fashion. The category of "rebel" may be elusive and nebulous, but the study of its contrasting forms has value.

Take, for example, the issue of legitimacy. Certain biologists who challenged the mainstream assumptions of their discipline operated within the consensus group of that discipline, and others did not. Leon Croizat, the man who challenged the notion that geographic barriers and biota do not co-evolve, was called a member of the lunatic fringe by the eminent paleontologist G. G. Simpson and generally taken to be an amateur and something of a crank. He lived and published in Venezuela, and, as David L. Hull shows, more often than not the establishment patronizingly referred to him as a kind of wealthy Venezuelan amateur botanist rather than as a legitimate member of the scientific community. On the other hand, Bill Hamilton, who extended evolutionary theory to explain the evolution of altruistic behavior, was a respected member of the Oxford University Zoology Department and a bastion of the natural sciences. He was accorded every accolade a biologist can hope to receive. Likewise, earlier in the century, Wilhelm Johannsen, the man who attacked the notion that the character and the factor behind it were one and the same thing, operated in Sweden with all the accoutrements of a respected establishment scientist; but Alfred Russel Wallace, at about the same time, died very much on the edge of reputability for having objected—for scientific reasons—to the notion that human evolution could be explained naturalistically without recourse to an external guiding force or forces, though in the previous century he had achieved the very same insight that would make Darwin's name recognizable the world over. The Yorkshire-born zoologist Vero Copner Wynne-Edwards was castigated and ridiculed for holding on to the notion that selection works on groups rather than individuals; Mark Borrello paints him as something of a romantic, tragic figure in the tradition of Don Quixote. Howard Temin, on the other hand, as Daniel J. Kevles shows, challenged one of the greatest ideas of modern molecular biology—the central dogma of information flow—but nonetheless was considered a thorough, first-rate scientist operating within the confines of the establishment.

One might object that cranks or fringe elements are labeled so because the challenges they mount are, in fact, wrong-headed and crazy or because the science used to mount such challenges does not satisfy established standards. Wallace, surely, was not being scientific when he argued for supernatural intervention in the evolution of man, however scientific he thought this position to be. Yet Temin, we all know, won the Nobel Prize for discovering reverse-transcriptase, a mechanism that has proved to be of great importance in medical science, and many of Bill Hamilton's equations remain the best means known to biologists of explaining altruistic behavior (and, for that matter, the evolution of sex).

The issue of scientific legitimacy cannot rest on the simple dichotomy between what ends up being "right" and what ends up being "wrong." There is no better example than the figure of Hans Driesch, the German embryologist and philosopher who turned from radical mechanism to radical vitalism but, as Garland E. Allen points out, did so without for one instant losing his reputability and academic stature within a rather rigid and status-conscious German university system. The scientific legitimacy of the rebel no doubt is constructed by the scientific community with reference to the quality of the science that is used to mount the challenge; that quality is a changing standard, however, and it would be a mistake to whiggishly equate it with contemporary notions of right and wrong. The neutral theory of molecular evolution, so forcefully championed by Motoo Kimura, may never be accepted in the terms that Kimura would have preferred. Yet, as James F. Crow demonstrates in his chapter about Kimura, the neutralist challenge to panselectionism has fundamentally changed evolutionary biology.

Perhaps scientific rebels are united by method if not by perceived legitimacy and reputation. Here, too, one fails to find a common thread. The English cytologist Cyril Darlington mounted a challenge to the very methodological workings of his discipline by adopting a theory-driven approach that paid little heed to problematizing exceptions. Thelma Rowell's radically different understanding of primate behavior was rooted in her willingness to simply observe them for longer periods of time than did her peers. Wynne-Edwards and Hans Driesch (in his antimechanist rebellion) also seem to have been driven by theory making rather than an empirical approach. Peter Mitchell, who brought about a revolution in the understanding of cellular bioenergy while proving that doing good science outside of the academy or industry is possible, was also a theoretically inclined researcher. He let primary assumptions and hypotheses rather than experimental observations lead the way, as John Prebble and Bruce Weber show.

Oswald Avery, who in 1944 performed the transforming principle experiment, which showed that DNA and not protein was the transforming agent and therefore most probably the carrier of hereditary information, was anything but a theory-driven investigator. Like Howard Temin and the neurobiologist Roger Sperry, he was a careful experimentalist who let the findings of his experiments speak for themselves rather than search, a priori, for evidence for a well-established theory. Clearly, the thin line between hypothesis-making, deduction, and "naïve" experimentalism is always a problematic one to walk when trying to describe the ways in which scientists operate and develop facts and theories. To the extent that one can speak of leanings or dispositions in scientific styles, however, it is clear that scientific rebels diverge rather than coalesce with respect to this variable, too.

Some rebels operate alone: Darlington, Croizat, and Barbara McClintock are examples. But other cases render this romantic picture of the shunned, castaway individualist fighting against all odds something of a myth. They served as magnets for like-minded challengers to the mainstream and together established alternative schools and traditions. Stephen Jay Gould, as David Sepkoski shows, is one example. So are Driesch and, as William Dritschilo makes clear, the ecologist Dan Simberloff. His group at Florida State University was even given a name: the Tallahassee Mafia.

Some rebels create their own journals and professional instruments of expression, thereby establishing independence from academic censure—Darlington, who co-founded with R. A. Fisher the journal *Heredity,* is an example; others, such as Mitchell and Croizat, escape the usual academic routes by striking out on their own independent and unconstrained paths of discovery. We should mention, however, that consensus opinion in science is sometimes established via independently constructed routes, as well. As Dritschilo shows, one way Robert MacArthur established himself as one of the leaders of mathematical community ecology was by publishing in non-peer-reviewed journals.

If not legitimacy, method, or mode of action, then perhaps a certain disposition or temperament can be said to characterize the scientific rebel. We learn that Darlington wrote in his personal journal, "*Vox populi* is *vox diaboli*," adding, "No subject keeps my interest when I find my own view of it agrees with the accepted or majority view." Indeed, when he felt that his cytological rebellion had run its course, he turned to a further rebellion having to do with the role of the cytoplasm in transmission, and, finally, to outrageously reductive claims about the genetics of man and society. Raymond Dart, as his student Phillip V. Tobias relates in this volume, "had no sense of dedication to a search for human ancestors when coming unwillingly to South Africa early in 1923," the year before he made the momentous discovery of *Australopithecus africanus,* which would place the origins of mankind in Africa. Yet a look at the heterodox views expressed in his earlier work concerning the morphology of the nervous system paints a general picture of a man who, however unwilling, seemed nevertheless always to find himself attached to rebellious and heretical attitudes. This suggests a proclivity.

The iconic geneticist Barbara McClintock was seen as a rebellious personality fighting against a marginalizing scientific community that was not quite able to appreciate the brilliance of her scientific claims. Bill Hamilton, too, seems to have been something of a maverick, taking on theories that others thought mad, such as the theory about the origin of AIDS that led to his untimely death from malaria. In a different vein, but one that also points to a proclivity to resist convention, Hans Driesch spoke vociferously against the Nazis and became one of the few non-Jewish academics to be fired from posts at German universities. Howard Temin risked being caught by the KGB when he helped smuggle illicit materials by Jewish refusniks from the Soviet Union to the United States.

But as Nathaniel Comfort beautifully shows in his chapter about McClintock, rebellion and iconoclasm are not one and the same: rebelliousness is a character trait, whereas iconoclasm is "a sort of professional activity, a demolishing of cherished dogmas or institutions." Clearly, most rebels fail to achieve iconoclasm, and certain icons are sometimes demolished by personalities that are far from rebellious. A case in point is Oswald Avery, who, as Ute Deichmann point out, was a painfully conservative type, far removed from the kind of hand-waving and gate-crashing behavior favored by the truly rebellious. Simberloff did not have the appearance or the manner of a rebel, "just a dry sense of humor and a tenacious opposition to illogic." Rowell, too, as Vinciane Despret shows, did not seek "contrariness" but did not shrink from it, either.

So "scientific rebel" is almost an anticategory, or rather a non-category. At best, it is an heuristic that helps point out the many exceptions and divergences from the intuitive understandings of the term. This would seem to justify the focus of this volume on specific acts of iconoclasm in the history of modern biology rather than on biographies. Still, if there may be said to be some principle corralling scientific rebels into one unifying space, it is this two-pronged generalization: first, the maverick must choose an important and relevant problem; those who tackle irrelevancies are never remembered. What is perceived to meet these criteria is a changing entity, one that even with hindsight is difficult sometimes to establish. Still, rebels would not be perceived as such by their contemporaries unless the latter perceived the iconoclasm to be relevant or important. Second, once a pregnant problematic is defined and attacked, the scientific rebel invariably exhibits what Charles Kingsley called "divine discontent." If one single thing unites all the characters featured in this book, it is that they all exhibited stubbornness and steadfastness in their challenges to orthodox thought. George Bernard Shaw is quoted as saying: "The reasonable man adapts himself to the world. The unreasonable one persists in trying to adapt the world to himself. Therefore, all progress depends on the unreasonable man."[4] This may or may not be true; it would be a stretch to call Temin, Hamilton, Sperry, or Driesch unreasonable, and it is unclear in what way Wallace's or Croizat's iconoclasms benefited mankind. Shaw thought that rebels always pay a personal price, but this is not the case. Some of those featured in this book achieved worldwide fame: Temin, Sperry, Mitchell, and, in the end, McClintock all won Nobel Prizes. Dart, Gould, Hamilton, and, to a lesser degree perhaps, Carl Woese all were acclaimed. Others, however—Croizat, Wynne-Edwards, Goldschmidt, and to a degree Wallace—were ridiculed and put to shame. Success is not the point of

Rebels, Mavericks, and Heretics in Biology. Rather, it is that orthodoxies are sometimes challenged, and when the challenge is substantial,
it invariably can teach us much about the workings of science.

In this respect, scientific iconoclasms may function in much the
same way as historical counterfactuals. In effect, historians have employed two kinds of counterfactuals: imaginary ones that lack an
empirical basis, such as G. M. Travelyan's "If Napoleon had won the
Battle of Waterloo," and empirical attempts at testing hypotheses that
favor calculation over imagination, such as R. W. Fogel's attempt to
compute American economic growth in the absence of the birth of the
railways.[5] The first of these approaches tends to fail by way of making
reductive and hence implausible assumptions and relying too heavily
on hindsight; the second, invariably, fails by way of anachronism. In
his edited volume *Virtual History,* the historian Niall Ferguson argues that despite these heresies of historiography, counterfactuals can
serve as valuable tools for historical investigation.[6] This depends, he
claims, on being able to consider as plausible or probable "only those
alternatives which we can show on the basis of contemporary evidence that contemporaries actually considered."[7] The counterfactual is
meant to serve the larger claim that historians must settle exclusively
for explanations of events in terms of the uncovering of antecedent
events without which the events in question could not have trans-
pired, whereas scientists can use such explanations as hypotheses
to be tested experimentally. Absent the desire—or ability—to invoke
grand covering laws in history, counterfactuals are the only tools
available to the historian with which to test causal hypotheses.

"Rebel," therefore, can serve as a valuable historiographical category because, in a fundamental sense, historical rebels, heretics, mavericks, and iconoclasts represent living counterfactuals to received
wisdom and thought in a given context. Iconoclasms are realized
counterfactuals, a particularly useful historiographical category. On
one hand, they escape most of the forms of criticism leveled at counterfactual history because they actually happened; on the other, they
present the same kind of historiographical opportunities promised
by advocates of counterfactuals in order, among other things, to combat overly simplistic determinisms that seem to trap unimaginative
historians all too often. The rebel can accomplish historical work of
a different kind than the hero, for example, not only because rebels
are not always heroes—nor are all heroes rebels—but because rebels
represent an alternative that allows the intellectual and social his-

torian (in our case, of science) to navigate with more precision and depth the mangrove of causation that is relevant to the area of study. "Why did things turn out the way they did in the history of biology?" or "Was the orthodox view necessary given the past and present state of the field?" are the kinds of questions the use of this category can help address with added rigor.

So although "rebel" is a Tolstoyan category in the sense that all rebels rebel in their own fashion, just as all unhappy families are unhappy in their own way, it nonetheless defies, from the historiographical point of view, Tolstoy's own Newton-inspired deterministic view of history. This view set out, from the pen of one of the greatest painters of the human condition—irony of ironies—to disprove the existence of free will.[8] Rebels are living testaments to the irreverent exercise of free will (and thought) in the face of what might seem, to their more conventional counterparts, necessities or truths in no need of being challenged.

ABOUT THE ICON

The rebel, then, can be useful to the historian. But if rebellion and iconoclasm are not the same, and *rebel* is far from a well-defined term, then one's attention shifts to the meaning of iconoclasm. Before one can begin to contemplate the nature of iconoclastic science, however, one needs to define *icon* and, beyond that, the ways in which an icon is established. A historic analysis of instances of challenge to iconic thought in biology may be of great assistance in answering these questions.

Trying to define the nature of a scientific icon is similar to trying to define the nature of a scientific paradigm. When Thomas Kuhn tried to do this in his classic treatment of scientific revolutions, one commentator objected that the term was so broad that it ended up having little explanatory value.[9] We would like to avoid heading in a similar direction. For that reason, we treat the scientific icon as the central, explicitly stated organizing assumption in a given discipline, an assumption that is held and practiced by a substantial majority of researchers. The best way to establish that an assumption is either held or practiced is to observe that it is not challenged. In a sense, then, the iconic stature of an assumption stems from the very fact that it is not contested. To be sure, there is much noncontested knowledge in a given scientific discipline at any given time. What renders noncontested knowledge specifically iconic is its centrality to the intellectual framework that constrains and directs scientific practices in a given field. Instrumentation and technique often play this role as much as do scientific theories and paradigms;[10] the moves in genetics from breeding experiments to breeding experiments plus cytology to molecular analysis via electrophoresis to polymerase chain reactions to informatics, for example, in turn opened vistas that changed intellectual constraints and frameworks in interesting and important ways. In this book, however, we focus in particular on conceptual assumptions, whether they result from the constraints of instrumentation and method or not. Whatever their etiology, these assumptions function as basic intellectual frameworks around which their fields are conceptualized. At a given point in time, these assumptions were challenged by a historical figure. Here is where we meet them, looking forward and backward to understand what was at stake.

The chapters in history presented in this book make clear that icons can become established in myriad ways. In the case of early twentieth-century cytology, consensus regarding the method of observing chromosomes and describing their behavior was reached, for reasons detailed in the chapter about Darlington, via disciplinary caution. The fear of making unsubstantiated claims ended up solidifying an excessively inductive method of practice for more than a generation, often predetermining the kinds of problems and questions that were thought legitimate to tackle in the first place. Ecology, on the other hand, presents a different kind of consensual etiology. Resurrected as a modern science in the 1960s from its nineteenth-century origins, the new ecology coalesced around a mathematical approach that simply did not consider statistics to be a relevant methodological tool. When Dan Simberloff suggested that ecology needed to become a statistical science, he was resisted. At times, resistance can be due to a prior void, rather than objection to the replacement of something that already exists. In such cases, where there has been no outstanding icon, the chief resistance may be against the erection of another one.

Icons can be established owing to lack of imagination, as well. This seems to have been the case, as Jan Sapp details in his chapter about Carl Woese, with respect to the evolution and taxonomy of bacteria. Before Woese came up with a new, RNA-based genomic method for tracking the evolutionary histories of prokaryotes, no one seemed to believe that such ancient histories could ever be cracked with any certainty. Bacteria were thought to consist of a single kingdom, and that was that. The case of Wilhelm Johannsen presents a similar failure of imagination. As Raphael Falk shows, Johannsen challenged the notion that the character and the factor were identical, suggesting instead that an underlying *genotype* might exist independent of, though obviously related to, the visible *phenotype.* This leap of imagination, as we shall see, actually resulted from harking back to an outdated typological notion of taxonomy championed by Linnaeus, but it would ultimately herald one of the great revolutions in twentieth-century biology.

Sometimes icons are established for pragmatic reasons. The history of classical genetics is an especially stark instance of this kind of dynamic. The materialist, deterministic, atomistic beads-on-a-string model of genetics of the first half of the twentieth century may have been recognized by certain more sophisticated practitioners as somewhat simplistic, but it worked much of the time, affording genetics the great prestige and success that it was to claim. This meant that the kind of problematizing theory presented by rebels such as Darlington, McClintock, Goldschmidt, and, later, Temin simply seemed to many to fly in the face of reason. Daniel Kevles quotes Harry Rubin, otherwise respectful of Temin's scientific credentials, who wrote in a review of the work on Rous sarcoma, "I'll give Howard's idea the amount of time it's worth—none." Goldschmidt was dubbed a heretic for voicing views that challenged the consensual model of the particulate gene. Indeed, as the case of Stephen Jay Gould illustrates, this kind of opposition would seem often to be the result of the difference between focusing on exceptions and focusing on rules. Often the icon is established on the basis of a generalizable rule such as "one gene, one enzyme," "beads on a string," "the autonomous action of genes," or "all traits are adaptive and the result of natural selection." The iconoclast, following William Bateson's exhortation, is he or she who "treasures exceptions."[11]

A related but distinct strain of iconic etiology has to do with orthodoxy. Orthodoxy can be established for pragmatic scientific reasons, as the case of genetics illustrated. It can also, however, be established not so much because of the experimental success of its assumptions (pragmatism) as because its theoretical logic might seem unassailable to practitioners, if devoid of any substantial proof. Before Avery conducted his transforming principle experiment, it was widely assumed that proteins must carry the hereditary information of organisms because DNA was considered to be a simple, even "stupid" molecule comprised of a bit of sugar and phosphate and a mere four nucleic acids. In another instance, the neo-Darwinian assumption that selection works on individuals rather than groups rendered Wynne-Edwards a rebel of sorts, but it, too, was a logical assumption less proved than assumed. Finally, orthodoxies can be political in a general sense: Alfred Russel Wallace's claim that it was necessary to postulate supernatural forces in order to explain human intelligence was resisted owing to a naturalistic, positivistic worldview that had become mainstream through the work and influence of men such as Thomas Henry Huxley and his X-Club friends. No one at the time understood how intelligence had arisen, but a nonnaturalistic explanation was simply not going to be let on the field. Similarly, Raymond Dart's claim that the origins of mankind were in Africa rather than in Asia flew in the face of all political sensibilities at the time. As Phillip Tobias shows, every possible "scientific" argument was marshaled in order to uphold this essentially political piece of wishful thinking.

ABOUT ICONOCLASM

The icon, then, is a central organizing assumption about which there is consensus within a discipline. That consensus can be supported by evidence, theoretical or logical argument, political orthodoxy, or some combination of these. What needs to be asked now is, What does it actually mean that an icon has been challenged? Previous treatments of controversy in science, most recently Machamer, Pera, and Baltas's anthology *Scientific Controversies: Philosophical and Historical Perspectives,*[12] focus on trying to answer the question Why do controversies exist?, striking a middle road between the traditional poles of the logical positivists (controversies are simply due to imperfect knowledge or mistakes) and the social constructivists (controversies are due to the power politics that infuses science just as it does any other human pursuit). The contributors to the anthology show that controversies may involve epistemic values, social environment, ideological commitments, ways of thought, and philosophical or ontological assumptions alongside facts, theories, and experiments. More often than not, rhetoric and persuasion play a crucial role in the acceptance or rejection of a certain argument or claim. An argument is established as convincing in a normative fashion, with style of presentation, language, and dialectic functioning not merely as "verbal ornaments or embellishments" but rather as logical agents of persuasion.[13] The interesting question, of course, is how such persuasion comes to be considered a part of scientific knowledge, that is to say, iconic, consensual science, and how, in turn, it relates to nature. "Once the Paradise of the early moderns and the logical positivists is admitted to be lost because science is recognized as having more dimensions than logic and evidence," the editors write, "and the symmetrical Hell of the social constructivists is also recognized to be inadequate because scientific knowledge needs to be something firmer and more durable than a cultural fashion, the middle way between the two extremes may be alluded to, but has not been found yet. It is not easy to see where to go."[14]

We are happy to accept the claims of Machamer, Pera, and Baltas, but *Rebels, Mavericks, and Heretics in Biology* concerns itself with a different set of questions. We are interested in asking, What are the different natures of challenges to iconic thought? Can they be categorized? Beyond this, are there different "colors" or "hues" of iconoclasm, reflective of different styles of thought of the scientists behind them? Can we, in short, create a kind of taxonomy of iconoclasm that takes into account both the dynamic of the challenge and the underlying cognitive or psychological characteristics of the challenger? We believe that this is possible with the nineteen test cases at hand.

Challenges to iconic thought in the life sciences in the twentieth century were of three general kinds: methodological, conceptual, and experimental. A fourth category, which deserves separate mention though it often fits into any one of the first three, is the iconoclasm that stems from the transgressing of disciplinary boundaries. The first three categories, as we shall see, map onto the further categorization of process, theory, and discovery.

When Hans Driesch castigated his former teacher Ernst Haeckel for his descriptive approach to biological problems, he was representing what Garland E. Allen has called elsewhere a methodological "revolt against morphology."[15] It is true that the result of this revolt was the adoption of an experimental approach, but the primary consideration that led to it was methodological: Driesch and others of the mechanistic bent believed that the truly important questions relating to embryological development simply could not be gotten at by way of a static, morphological approach. Haeckel's own method of comparative analysis of the anatomy of embryos in the service of constructing phylogenetic trees would need to be replaced by an experimental regime based on mechanistic reasoning. As Allen shows, Driesch wanted to ask how an egg differentiated, not what a comparison between chick and human embryology can teach about the evolutionary relationship of the two.

Cyril Darlington, too, was primarily a methodological iconoclast, as were Simberloff and Woese. Adopting August Weismann as his hero, Darlington, as we shall see, enjoyed quoting the nineteenth-century German theoretician: "The time in which men believed that science could be advanced by the mere collection of facts has long passed away." Indeed, although it would become necessary toward mid-century to soften some of the universal generalizations about chromosome mechanics that Darlington theorized in the 1930s, it is clear that the kind of Gordian Knot–cutting associated with his deductive approach played an important role in freeing cytology from the ancillary role it had assumed alongside the much more fecund genetics of the early decades of the century.

In the case of Simberloff, introducing statistical thinking and, in particular, the null hypothesis represented a direct challenge to the methodological underpinning of a science that was just being born anew in its twentieth-century form. As William Dritschilo shows, this was happening in the 1960s, precisely when ecology was being linked to a particular philosophy of life at a time of heightened political polarization. Simberloff's call for the introduction of statistical rigor, therefore, carried higher stakes than those associated with the usual inner battles of a scientific discipline. The pursuits of Woese, on the other hand, may have had fewer apparent political ramifications, but they nevertheless excited much scientific interest. As Jan Sapp shows, Woese's heretical attack on the canonical phylogenetic divide between the prokaryote and the eukaryote was rooted less in microbiology than in his notions of the evolution of the genetic code. Woese's use of ribosomal RNA to establish ancient phylogenies was not "in the air" but rather a singular initiative combining technical opportunity with theoretical preparation. Fundamentally, it represented a methodological revolution against the background of lack of imagination.

Alongside iconoclasms of the methodological kind are those of the conceptual variety. Woese is perhaps an example of the difficulty associated with separating process (method) from concept (theory), for his methodological innovation went hand in hand with his rather unorthodox theoretical views of the evolution of the genetic code. Once again, as with the term "rebel," as long as categorical taxonomies of iconoclasms are understood as heuristics rather than rigid realities, they can be used profitably.

Rather more stark instances of conceptual iconoclasm include Richard Goldschmidt, Bill Hamilton, and Stephen Jay Gould. Goldschmidt likened the chromosome to a string of a violin, with the implication that a gene should not be thought of as a localized entity; rather, systemic regulation of the entire genetic complement played important roles in development and evolution and therefore it, rather than the discrete gene, should be the focus of the attention of biologists. This represented a major departure from the largely American tradition of viewing evolution and transmission in atomistic terms. As Ullica Segerstrale shows in her chapter, Bill Hamilton developed mathematical models of the evolutionary logic of the maintenance of altruism and of sex based on theoretical considerations of genetic relatedness and the benefits and costs that affect fitness. The conceptual shift he pioneered (perhaps following a famous quip of J. B. S. Haldane) was that certain biological phenomena could be explained more satisfactorily if they were analyzed from the point of view of the gene rather than that of the individual organism or the group. Finally, Steve Gould tried to shatter a worldview that he called "adaptationism"—a neo-Darwinian approach that focused on the power of natural selection to mold adaptive traits in organisms over evolutionary time to the exclusion, Gould believed, of the narrative, contingent, historical nature of the evolutionary process.

Some iconoclasms, in the classical tradition of scientific advance via discovery, are due to well-defined experiments that produce previously unknown facts. Barbara McClintock discovered transposable elements —transposons—whether or not her interpretation of them as "controlling elements" was accepted by her peers. Raymond Dart held the Taung baby fossil in his hand before announcing to the world its implications for the understanding of human evolution. Howard Temin, too, was a discoverer in the Magellan-cum-Leeuwenhoek-cum-Amundsen-cum-Flemming sense: he found retroviruses and, eventually, the mechanism that allowed for reverse transcription. Finally, Avery's iconoclasm was first and foremost the result of a set of careful experiments that proved beyond doubt that DNA and not protein was the transforming agent in heredity. All four challenges would carry with them stark conceptual ramifications (though, as Comfort shows, in the case of McClintock they were due to the decoupling of transposition from genetic regulation), but they stemmed from important new discoveries.

Whether methodological, conceptual, experimental, or some combination of the three, what emerges from the historical cases is the relative ubiquity of instances in which iconoclasms result from the transgression of what are seen by researchers to be well-defined disciplinary boundaries.[16] Darlington was a cytologist who thought mainly in evolutionary terms, a fact that allowed him to ask ultimate questions about chromosome mechanics that simply were not on the screen (or rather the microscope slide) of more descriptive, less theory-prone investigators in his field. This led to much resistance to Darlington's ideas, but it also helped ensconce the evolutionary synthesis. Peter Mitchell, too, gained his insight from thinking outside the disciplinary box: as Weber and Prebble show, his perspective came not from the biochemists who were concentrating on understanding oxidative phosphorylation but rather from the field of transport, which at the time was considered to be a part of physiology. Mitchell, therefore, succeeded by coupling biochemistry to physiology—a matchmaking move more conservative practitioners did not even contemplate. Carl Woese was a synthesizer, too, bringing together the fields of microbiology and evolutionary molecular genetics in order to solve a problem that more conservative minds believed did not exist. Jan Sapp perhaps says it best: "Small changes, refinements, occur within disciplines; large-scale changes may result from the sharing of innovations between them."

If we have mapped a heuristic taxonomy of categories of iconoclasm, there remains the color of the iconoclasm—a feature more relevant to the cognitive or psychological approach or disposition of the challenger than to the actual nature of the challenge. An interesting thing emerges: whereas folk intuition would lead one to think of iconoclasms as visionary, or pointing to the future, many are often precisely the reverse, that is to say, reactionary or conservative in nature. Wallace was the co-discoverer of the principle of natural selection and an important naturalist and biogeographer. Yet Wallace's insistence that nonnaturalistic forces played a role in the evolution of mankind was deeply reactionary. Driesch rebelled against a reductive, mechanistic approach to biology, though he had once been a leading figure in the progressive revolt against an old-school biology based on description rather than experiment.

That great scientists should revert to positions deemed anachronistic by their peers may not, after all, be very surprising. What is more interesting is a kind of iconoclasm that, though not strictly reactionary, can be called conservative. Wilhelm Johannsen, for instance, harked back to a Linnaean taxonomy shaped by hereditary forces of a type in the Platonic sense—surely an anachronistic position. Yet, surprisingly, it was this conservatism that, via the concept of the genotype, helped relieve the science of heredity from what Raphael Falk calls "the preformationist hump of the determinist 'unit character.'" Richard Goldschmidt was thought of as heretical for arguing for saltationism, a position that T. H. Huxley had held in the nineteenth century and that was supported by Hugo de Vries and William Bateson. In doing so, Goldschmidt was harking back to a nineteenth-century tradition that failed to see the point of unnaturally decoupling heredity from development and evolution. In this sense Goldschmidt was a conservative rebel. Wynne-Edwards is another example. Although dubbed a heretic, he actually thought of himself as harking back to a distinctly Darwinian tradition of understanding population dynamics in terms of group selection. Gould, too, can be thought of as a conservative iconoclast: critics thought of Gould as departing from the Darwinian camp, but he saw himself as expanding and refining Darwin's original ideas. Sepkoski sees this as the pre-Enlightenment notion of a cyclical return to an original state, rather than a violent upheaval and annihilation of an old regime. But to judge from the other rebels who seem to fit a similar mold, Gould was not such a maverick after all. Rebels, it seems, often look back as much as they look forward.

But there remain, of course, iconoclasts in the classical sense: novel, path-breaking, disruptive. Avery is an example, as are McClintock, Temin, and Sperry. Woese's approach to evolutionary phylogeny also represents a novel iconoclasm, as does Peter Mitchell's synthesis of biochemistry and physiology in the service of figuring out the bioenergetic working of the cell. The first rebellion of Hans Driesch against the stolid morphology of his time consisted in the application of a novel, mechanistic approach to the solution of basic problems in development and embryology.

Between the two extremes of reaction and innovation, an interesting designation emerges. In many cases there is a sense that the rebels are simultaneously behind and ahead of their times. Driesch rejected naïve mechanism, and in this sense was at once reactionary and visionary; Darlington harked back to a theory-based outlook exemplified by nineteenth-century giants such as Weismann that, although considered unscientific by peers, would be picked up again by theoreticians and experimentalists trying to understand the role of dynamic genetic systems in evolution; McClintock and Goldschmidt, too, tied development and heredity to evolution in the nineteenth-century tradition, and, though they were frowned at by contemporaries, actually pointed to problems that would only be dealt with decades later. Raymond Dart harked back to Darwin's own insight about the African origins of humanity, but his theory was not taken on board until a generation after it had been conceived. Finally, as work in the past decade on the intersecting levels of selection in nature increasingly shows, Wynne-Edwards was, in a certain sense, behind and ahead of his time.

Heterodox ideas are difficult to introduce into a system that proceeds by way of grants, peer review, employment and promotion, and professional dependents. Richard Lewontin reminds us in the epilogue that as interesting as idiosyncratic histories of rebellion in science may be, we must always keep in mind that the social structure of science delimits the pitch on which the game is played and that all rebellions must be judged in relation to the way they interact with the known rules of the game. Although some of the cases presented in this book beg to differ with the idea that iconoclastic ideas can only be successfully introduced by prominent people who conform to the accepted processes of scientific legitimation, Lewontin's general notion that science is as much a political game with hard-and-fast rules and binding norms as a free game of the mind and of the elements surely remains the organizing framework within which challenges to orthodoxy must be judged. If we are to look forward to new and exciting heterodox ideas in biology in the twenty-first century, we should be thinking about the structural arrangement of the scientific enterprise as much as looking forward to the arrival of highly original, highly different, and highly rebellious new scientific minds. Tweaks to the grant-provision system or revolutions in on-line publishing or university-industry relations will be as important as the arrival of new geniuses or "out-of-the-box" intellects. Certainly both aspects will, as ever, be interacting with each other.

Thomas Jefferson, in a 1787 letter to James Madison, wrote: "I hold it, that a little rebellion, now and then is a good thing, and as necessary in the political world as storms in the physical."[17] It is our hope that readers will, like us, feel both that rebellion in the physical world is produced as much by men and women as by storms and that Jefferson's comment—applied to science instead of to the political sphere—is everything but an overstatement.

NOTES

1. Quoted in S. J. Gould, *Hen's Teeth and Horse's Toes* (New York: Norton, 1980), 83.

2. On Delbrück see Gunther Stent, "Max Delbrück," in *Phage and the Origins of Molecular Biology,* expanded ed. (Cold Spring Harbor, N.Y.: Cold Spring Harbor Laboratory Press, 1992). On Margulis see Evelyn Fox Keller, "One Woman and Her Theory," *New Scientist* (July 3, 1986), 46–50, and Jan Sapp, *Evolution by Association: A History of Symbiosis* (New York: Oxford University Press, 1994). On Sonneborn see Judy Johns Schloegel, "From Anomaly to Unification: Tracy Sonneborn and the Species Problem in Protozoa, 1954–1957," *Journal of the History of Biology* 32 (1999): 93–132. On Jablonka and Lamb see commentaries in *Behavioral and Brain Sciences* (2007) 30, 353–392.

3. Peter Machamer, Marcello Pera, and Aristides Baltas, eds., *Scientific Controversies: Philosophical and Historical Perspectives* (Oxford: Oxford University Press, 2000).

4. Quoted in Samantha Power, *A Problem from Hell: America in the Age of Genocide* (New York: Basic, 2002), 516.

5. G. M. Travelyan, "If Napoleon Had Won the Battle of Waterloo," in *Clio, a Muse and Other Essays* (London: Longmans, Green, and Co., 1930), 124–135; R. W. Fogel, *Railways and American Economic Growth: Essays in Interpretive Economic History* (Baltimore: Johns Hopkins University Press, 1964).

6. Niall Ferguson, ed., *Virtual History: Alternatives and Counterfactuals* (London: Picador, 1997).

7. Ibid., 86.

8. The analogy is drawn from Leo Tolstoy, *War and Peace* (London, 1978).

9. Dudley Shapere, "The Structure of Scientific Revolutions," *Philosophical Review* 73 (1964): 383–394, 385.

10. Peter Galison, *Image and Logic: A Material Culture of Microphysics* (Chicago: University of Chicago Press, 1997).

11. William Bateson, *The Method and Scope of Genetics: An Inaugural Lecture Delivered 23 October, 1908* (Cambridge: Cambridge University Press, 1908), 22.

12. Machamer, Pera, and Baltas, *Scientific Controversies.*

13. Ibid., 12.

14. Ibid., 14.

15. Garland Allen, *Life Sciences in the Twentieth Century* (Cambridge: Cambridge University Press, 1978).

16. Lindley Darden and Nancy Maull, "Interfield Theories," *Philosophy of Science* 44 (1977): 43–64.

17. Thomas Jefferson, letter to James Madison, January 30, 1787, quoted in *Bartlett's Familiar Quotations,* ed. Justin Kaplan (New York: Little, Brown, 1992), 343.

CHAPTER TWO

Alfred Russel Wallace, the Discovery of Natural Selection, and the Origins of Humankind

MICHAEL RUSE

> Rebel: v. To resist or defy an authority or a generally accepted convention.—*American Heritage Dictionary*

Alfred Russel Wallace discovered natural selection, the driving force of evolution, in 1858. Yet within ten years he was arguing that selection could not possibly account for the evolution of humankind and that only spiritual forces could have done the job. How could this be? Had Wallace, the ultimate man of science, turned into Wallace, the ultimate rebel against science? To answer this question, we need some background.[1]

MID-VICTORIAN THINKING ABOUT THE NATURE OF SCIENCE

The first part of the nineteenth century, particularly the 1830s, was an incredibly active and fertile period for the development of what we call the philosophy of science. This was not a purely intellectual phenomenon, especially not in England. Throughout the first half of the century, with few exceptions, most of the science had to be done by people at Oxford and Cambridge Universities whose main income came from undergraduate teaching in nonscience subjects and who necessarily were ordained members of the Anglican Church. Because they appreciated the great importance of science and technology in an industrial society, they made a major effort to find ways in which science could be supported in a secular setting.[2]

Important examples were the astronomer John F. W. Herschel, graduate of Cambridge and successor to his father as England's leading

Figure 2.1. Alfred Russel Wallace in 1853. From
Alfred Russel Wallace, *My Life: A Record of Events
and Opinions* (London: Chapman and Hall, 1905).

astronomer, and William Whewell, successively professor of mineral-
ogy and professor of moral philosophy at the University of Cambridge
(and inventor of the word "scientist"), and Baden Powell (father of the
scoutmaster) at Oxford University.[3] They were articulating the very
meaning of what it is to be a scientist. Thanks to them and their friends,
new scientific societies were being formed—notably the British Association for the Advancement of Science—and thinkers were turning
their attention to articulating the rules, methods, and standards of
science, especially good science. Isaac Newton was the paradigm. His
work, in particular, his mechanics, was the ideal to which scientists
should aspire. This meant that above all one should explain one's work
using unbroken law, especially deductively connected networks called

axiom systems, or what we today would call hypothetico-deductive systems. One should also strive to be causal, in the sense that one should seek causes to explain phenomena. One should aim for what Newton had called *verae causae,* true causes.

A major topic of discussion was the connection—or, rather, the non-connection—of science and religion.[4] In social and intellectual terms it was becoming important to separate the two. Everyone agreed that science could have implications for religion. Anglicans had always stressed natural theology, and a major reason for promoting science was the way that it proved and glorified God. The 1830s saw the publication of the *Bridgewater Treatises*—a series of works paid for from the legacy of the eighth earl of Bridgewater and glorifying God from the magnificence of his creation (Whewell wrote one)—devoted to the evidence of divine design in nature. Everyone also agreed, however, that it was inappropriate to bring God into one's science. By this time, most of those with pretensions to being real scientists were no longer trying to prove such things as Noah's Flood, and even if they had been, it would have had to be naturally caused. The major problem was the origin of organisms. No one could see how their adaptive nature, their design-like nature, could come through unbroken law. Some, like Herschel, thought that there simply had to be as-yet-undiscovered natural laws. Whewell inclined more to miracles, but he was adamant that any claims of this nature would not be based on science: "The mystery of creation is not within the range of her legitimate territory; she says nothing, but she points upwards."[5]

The efforts of these people—though as they got older they became more conservative—were very successful. By the 1850s one could start to make one's way as a full-time scientist, without need or support of the church. It was not easy, but it could be done. Darwin's great supporter Thomas Henry Huxley got a job in the mid-1850s at the Royal School Mines, a de facto part of the University of London; he always resisted going to Oxford University or Cambridge University, institutions that were introducing science degrees and starting to loosen their ties to the church.

CHARLES DARWIN AND EVOLUTION

Against this background, we turn now to the other discoverer of natural selection, Charles Darwin.[6] Born in 1809, fourteen years before Wallace, Darwin was always interested in science. When at Cambridge he knew the chief scientists (notably the geologist Adam Sedgwick, the botanist John Henslow, and the polymath William Whewell). He spent five years (1831–1836) traveling around the world on *HMS Beagle,* not simply collecting specimens but working hard on the theories involved in understanding the phenomena he saw as well as his specimens (especially those involving geology). He worked nonstop on science when he returned, writing about geology, barnacles (an eight-year study), and many other topics, especially botany (orchids, climbing plants, insectivorous plants, and more). Around 1850, Huxley—who liked to do these sorts of things—rated English biologists and gave Darwin top marks. This was before he knew Darwin and almost a decade before publication of the *Origin.*

In the spring of 1837 Darwin converted to evolution (at the start of the *Beagle* voyage he had believed in the literal truth of the biblical story of creation and then softened to a belief in an unknown natural form of origination for organisms). He discovered natural selection in the fall of 1838, more than two decades before the *Origin* appeared in 1859. Darwin's genius led to his great scientific achievements. But in context, his work is fully understandable and not so very surprising. Paradoxical as it might seem, it would be inappropriate to speak of Darwin as a rebel. Quite apart from the fact that Darwin was not going to alienate important scientists by revealing his thinking about evolution—he did not do this until he was forced into doing so, by which time he had built around himself a group of supporters. His thinking flowed naturally from his background knowledge and training. Darwin made major moves forward, but not as a rebel. He simply did not resist or defy an authority or a generally accepted convention.

For a start, Darwin knew that the origins of organisms constituted the big problem in biology. On the *Beagle* voyage he read Lyell's *Principles of Geology,*[7] a work that insisted that organic origins had to be natural and ongoing and yet dared not grasp the nettle of evolution (or anything else, for that matter). Moreover, Darwin knew all about evolutionary ideas. His grandfather Erasmus Darwin was an evolutionist, and Charles had read the major work (*Zoonomia*).[8] He had talked about evolutionary ideas with Robert Grant when he spent two pre-Cambridge years in Edinburgh in an abortive attempt to become a doctor. In the *Principles,* Lyell wrote about the French evolutionist Lamarck. And Herschel, the leading scientist of the day, had labeled the question of organic origins the "mystery of mysteries."

Having moved to evolution, primarily because that was the only way he could sensibly explain the distribution of the reptiles and birds on the Galapagos Archipelago, Darwin knew that he had to find a cause and be able to present it as a true cause. He fell straight into artificial selection as a means of organic change, thanks to his rural connections (especially the breeding work of Josiah Wedgwood the younger, Darwin's uncle, who would become his father-in-law). For Darwin, artificial selection, which we see and control, is analogous to natural selection—if the former is a cause, then it is reasonable to think that the latter is a cause. Then, after he had hit on the mechanism of natural selection, Darwin knew how to embed his ideas in a well-formed theory, using the mechanism to explain such areas as instinct, paleontology, biogeography, systematics, morphology, and embryology. He sought to bolster the causal pretensions of natural selection and make one think that selection is a true cause. On top of this, of course, Darwin spoke to the design question. Because of his Cambridge training and nonstop diet of natural theology from the works of Archdeacon William Paley, Darwin knew that the defining feature of organisms is their adaptedness—the fact that they exhibit final cause. This is precisely what natural selection addresses.

This last point is noteworthy. When Darwin wrote the *Origin,* he was not yet an agnostic. Although no longer a Christian, he believed in a God who is an unmoved mover. He was a deist. And as such, he believed in God as designer. Shortly after the publication of the *Origin,* he wrote his American friend Asa Gray: "I see no necessity in the belief that the eye was expressly designed. On the other hand I cannot anyhow be contented to view this wonderful universe & especially the nature of man, & to conclude that everything is the result of brute force. I am inclined to look at everything as resulting from designed laws, with the details, whether good or bad, left to the working out of what we may call chance."[9] But not in science! When Gray wanted to introduce directed variations to supplement selection, Darwin simply said that Gray was no longer doing science. The pupil of the philosophers had learnt the message well. There is no place for God in science.

VICTORIAN JONAH

Turn now to Alfred Russel Wallace. He was born to English parents just outside the town of Usk, in Monmouthshire, Wales, on January 8, 1823. He left school early and, with some time out for school teaching, worked for a number of years as a surveyor—an occupation then much in demand given the huge growth of the railways. In 1848, with his friend Henry Walter Bates, he left for the Amazon, where he was to spend about four years collecting specimens (chiefly insects) that were sold to wealthy collectors in England. After a couple of years in the old country, he set off in 1854 for the East and spent eight years collecting specimens on and around the Indonesian Archipelago. After he returned again to England, he married and raised a family. He never held a steady job, making a living from writing, from acting as an examiner for student tests, and eventually from a small government pension. He died on November 7, 1913, and was buried in Broadstone Cemetery in Dorset, England.

If one had to use one word to sum up the life and achievements of Wallace, "unfortunate" would spring at once to mind. His troubles started before his birth, when his father (a solicitor) lost a comfortable inheritance to bad investments and (in his son's words) achieved considerable relief from the fact that his affairs were so bad that they could sink no further. Then there is the missing el at the end of Wallace's middle name, a mistake made at the time of christening and destined to cause scholars infinite troubles ever since. On his return from the Amazon, his ship burned and his collections and papers were all lost. Things were little better when he returned from the East. Wallace thought he had found the perfect spouse until one day the young woman he was courting sent her father to the front door to tell the would-be suitor that he was never again welcome. He was very hurt.

Wallace followed in his father's footsteps by losing the money earned by his collections, first by giving much of it to his brother-in-law's failing photography business and then by investing in lead mines just before the silver mines in the American West discovered that their by-product was huge amounts of cheap lead. He responded to a madman's assertion that the Earth was flat and then, having successfully performed an experiment showing that the Earth is round, discovered that in England wagers are not enforceable by law and had to spend huge amounts of money fending off the crazy flat-Earther. Most unfortunate of all, of course, was that Wallace discovered the mechanism of evolution through natural selection, sent it off to one of his correspondents in England, Charles Darwin, only to discover that Darwin had been sitting on that very idea for twenty years. The following year Darwin published *On the Origin of Species by Means of Natural Selection,*[10] and for ever more the theory has been known as Darwinism rather than Wallacism.[11]

WALLACE AND HIS ENTHUSIASMS

Let us dig a little more into Wallace's nature by comparing him with Darwin. The contrast could not be greater. Wallace was in the middle class but very barely so. For him collecting was a job, whereas for Darwin collecting was a hobby. Darwin did not have to sell his finds to keep himself in funds. Wallace was unfortunate, but all too often his misfortune was of his own making. He may have been generous to relatives, but he threw away the monies earned from his labors in the East. He got mixed up in the question concerning the Earth. No Darwin or Huxley would have soiled his hands in this way. It was no wonder that no one ever wanted to give Wallace a job. Thanks to behind-the-scenes efforts by Darwin, Bates was slotted into the secretaryship of the Royal Geographical Society, and there he became indispensable. One would have had to be as silly as Wallace to have given him the post. It was not that people disliked Wallace. They were happy to pressure Prime Minister William Gladstone to get Wallace a pension. They simply did not trust his judgment.

Above all, there were Wallace's enthusiasms. He probably could have been forgiven for his socialism and his endorsement of nationalization of land, the latter a very popular movement in Britain and America in the last decades of the nineteenth century and supposedly a failsafe method of equalizing the social classes. Darwin was a capitalist, and Huxley admitted that by temperament he was a conservative, but there were others who inclined more to the left with regard to these matters. What did bother people was that in the 1860s Wallace became an ardent spiritualist, believing that there are unseen forces that interfere in the plans and ways of humankind. As a result, Wallace wrote extensively about the subject, he badgered people to go to séances, and he again and again offered public testimonials on behalf of mediums who were accused of fraud.

There was thus something of the crank about Wallace. It will not surprise you to learn that he was anti-vaccination or that he was tempted to vegetarianism—a temptation firmly squelched by his physician, who put him on a diet of raw, finely chopped beef. Then there is Wallace's claim that the future of the human race lay in the hands of young women. Apparently, in the future humanity will be uplifted because young women will choose as mates only the best and finest of young men, those worthy of love and respect. "In such a reformed society the vicious man, the man of degraded taste or of feeble intellect, will have little chance of finding a wife, and his bad qualities will die out with himself. The most perfect and beautiful in body and mind will, on the other hand, be most sought and therefore be most likely to marry early, the less highly endowed later, and the least gifted in any way the latest of all, and this will be the case with both sexes."[12] One can only suppose that if the Wallace children truly behaved that way, they must have been as odd as their father. Not until the rise of American liberal arts colleges and their codes of conduct was the world again to see such a misunderstanding of human nature and of the sexual interests of the young.

WALLACE THE EVOLUTIONIST

If Wallace was so out of touch with reality, if he was so very different from Darwin, how, then, was he able to become an evolutionist and hit on natural selection? Let me stress that there is no mistake about his achievements. He did become an evolutionist, and although there are some differences between the early thoughts of Wallace and Darwin—the former always had a predilection for group selection in a way that was alien to Darwin's thinking—the two men did study exactly the same mechanism.[13] They both saw that there is an ongoing struggle for existence and that randomly appearing features that help in the struggle will tend to be preserved and that this will lead eventually to full-blown evolution. The similarity of the discoveries cannot be denied.

In a way, the important issue is not so much the discovery of natural selection (in 1858) but the move to evolution (around 1845). I do not think that simply using the word "genius" satisfactorily explains the two men's respective discoveries of the process—that would be rather like Molière's joke that a sleeping potion had a dormative virtue—but the fact is that for both Darwin and Wallace there was genius involved. Yet the genius was working in a pattern and with tools at hand. In their work, both men were committed evolutionists and were searching for a cause. Both read Robert Malthus's work about population and realized that organisms are in an ongoing struggle for existence. Both realized that the features of organisms help in the struggle for existence. In other words, there is nothing so very mysterious about the discovery of natural selection, although one can and should stand in awe of the abilities of Darwin and Wallace to put the various pieces together and arrive at a solution.

I have shown that we can start to make sense of Darwin's move to evolution if we take into account the way in which his group was stressing the problem of organic origins, that he knew already about evolution as a putative explanation, that he found things on the *Beagle* voyage that called out for an evolutionary history. I am not pretending that one could have predicted with certainty that Darwin would have become an evolutionist, but that he did is ultimately not surprising. Wallace was different. He had nothing that Darwin had—no training in science, no group of colleagues stressing the interest of the problems, and in 1845 absolutely no first-hand experience of peculiar distributions of organisms in either space or time. Yet he became an ardent evolutionist. Why?

We can find an answer if, as is so often the case with these sorts of things, we take a holistic approach. We should not compartmentalize Wallace the scientist and Wallace the crank. He was, by temperament, always attracted to outrageous positions, particularly if these positions were disliked by the respectable. The move to spiritualism was a classic example of the way his mind worked. Wallace rather liked it precisely because it was laughed at. (The same is true of nationalization of land.) The conversion to evolution fit exactly into this pattern. In 1844 the appearance of an anonymously authored evolutionary tract called *The Vestiges of the Natural History of Creation* scandalized respectable scientists such as Sedgwick and Whewell.[14] Written, as we now know, by the Scottish publisher Robert Chambers, it argued that all of life came from inorganic matter and then progressed steadily upward to become humankind—Queen Victoria, actually. It was an appalling work—scientifically weak, speculative beyond compare, verging on the heretical. In short, if Newton was the filet mignon, Chambers was the fast food of science, and like all fast food was taken up with enthusiasm by the hoi polloi. Half of the fun was precisely that the respectable scientific community loathed it. Wallace read it with great enthusiasm and at once became a convert. It was a minority position, it involved sweeping hypotheses, and it was sneered at by the knowledgeable. There was no way that he could have resisted.[15]

And this is precisely what went into Wallace's move to evolution. It was absolutely right and proper that no serious scientist accepted the evolutionary message of *Vestiges* on the basis of the evidence that it presented. In case you are still worrying about how someone like Darwin could have become an evolutionist, I stress that the evidence of *Vestiges* was not the evidence of Darwin. The older evolutionist had fossils and geographic distributions—the tortoises and birds of the Galapagos—artificial selection, top-quality embryology, and more. *Vestiges* had none of this evidence, or where it had something (such as embryology) it had it wrong. It argued that frost ferns left on windows give evidence of spontaneous generation, that insects come out of electrical experiments, and that birds spontaneously change into mammals. It was not something for the serious scientist. But for a young man who loved outlandish ideas, especially if they were anti-establishment (and Wallace was just then getting very excited about socialism), *Vestiges* was gospel. Had Wallace not been childlike in his trust of the absurd, he never would have become an evolutionist.

What makes Wallace quite exceptional is that when he had an idea or a conviction, he was not about to give it up in the face of opposition from the respectable, and he was prepared to labor for it. He may have been a crank, but he was a dedicated, hard-working crank. He spent years attending séances and supporting spiritualism and spiritualists. He spent more than thirty years as president of the British Land Nationalization Society. Alfred Russel Wallace was no fair-weather friend. In the case of evolution, having become a convert he set about putting his convictions into action. Going to the Amazon was in major part fueled by his felt need to find real evidence for evolution. It was not simply a get-rich project. (It was just as well.) Then he kept thinking and thinking about the issues, trying to stay abreast of ideas back in Britain. In 1855 he published a seminal paper in which he argued that organisms always seem to appear in exactly the places on the globe where similar organisms already exist. If not overtly evolutionary, it was as close as it could be. This led to correspondence with Darwin and explains why, of all people, it was Darwin to whom Wallace sent his paper about selection. (Apparently they may have met earlier in decade, but neither seems to have remembered any such encounter.) Wallace had long known of Darwin's book about the *Beagle* voyage; it was one of his inspirations for going to the Amazon.

HUMAN ORIGINS

I have said that Wallace was a crank. Does this at the same time imply that he was a rebel? Not necessarily. Cranks often become rebels, but one can imagine someone with many ideas that could be thought of as crankish and not think of such a person as a rebel. If one grew up in a family with odd ideas about religion or diet or dress, one might think of such a person as a crank but agree that in context that person was not particularly rebellious. This point applies to Wallace. Had Darwin in the late 1830s come out in favor of evolution, although he might have had good reason to do so—and he did—socially speaking he would have been a rebel. Wallace did not belong to the group of professional scientists. He was part of the great unwashed. He was among the half-informed, the ill-educated, those who would jump on any bandwagon. For this reason, within the context of his class it is not appropriate to think of the Wallace who became an evolutionist in the mid-1840s as particularly rebellious. But now the question arises of the Wallace of the 1860s, who denied that natural causes could explain human origins. Was this a case of rebellion? Let us pick up the story again.

When I say that Wallace was a crank, I am not trying to insinuate that he was stupid. He hit on natural selection when people such as Huxley did not. In the 1860s, he and Darwin had a very sophisticated discussion about the levels of selection. Although in some respects I think Darwin was right and Wallace was wrong, the quality of the debate was not to be equaled until the work of people such as William Hamilton and George C. Williams in the 1960s.[16] And in the 1870s, Wallace did major and lasting work on biogeography. Wallace's Line, which divides Asian and Australasian animals in the Malay Peninsula, is still important.[17] What I do want to say is that Wallace was an untutored, or, rather, a self-taught crank. He had not had the university education of Darwin, nor had he been directed to the books of the philosophers. He had not come to intellectual maturity among a group of men who were trying desperately to articulate the nature and rules of good science. Indeed, his education generally was very spotty. When Wallace tried his hand at school teaching, each night the principal had to help him understand the material for the following day's class. This lack of training and scientific sophistication showed, and nowhere more so than in the case of his explication of human origins in the 1860s. The story is simple. At first, Wallace was eager to offer a natural selection–driven account of human origins.[18] In doing so he pointed out that intelligence was going to be really important in the case of human evolution and that it would be very significant in the struggle for existence. It would be shown to be a major adaptation.

Darwin, for one, was very excited about this thinking, and he wrote to praise Wallace strongly.

Then, as Wallace became enamored of spiritualism, he reversed himself. A natural explanation of human origins was no longer enough. One had to accept that spiritual forces were directing human history. In support of this idea, Wallace instanced a number of features—hairlessness for one, our great intelligence for another—that simply could not have come about by means of selection.[19] Savages have the potential for great thoughts but rarely if ever have them. So the brain cannot have burgeoned in size because it was an adaptation. Some other force must have been responsible. Darwin, of course, was appalled, and he accused Wallace of killing their child. Darwin was also now spurred to write about humankind himself, and in 1871 he published *Descent of Man.*[20] Offering a completely naturalistic account of human origins, Darwin made one concession to Wallace: he agreed that things such as intelligence could not be produced by simple natural selection. For this, Darwin relied on his secondary mechanism of sexual selection, which involves competition within species for mates. Clever people apparently have more offspring. Powerful men have the most attractive wives.

Why did Wallace start appealing to spiritual forces? Let us break this down into two questions. First, why did Wallace accept spiritualism? Second, why did he think he could use it to explain human origins, or rather, why did he think (as he certainly did) that other scientists should accept spiritualism as an explanation of human origins? The answer to the first question is easy. Wallace accepted spiritualism for the usual reasons—he thought that the table-lifting, and the knocks and shrieks, and the rest of the rather weird (and rather pathetic) phenomena could be explained only by invoking the supernatural. In spite of the many cases of fraud, Wallace thought there was enough there to justify belief. He simply refused to accept the criticisms of friends and those who were not so friendly about, for instance, the inability to repeat events in a controlled fashion or to allow observers to sit where they wanted rather than where the medium wanted. In a truly ecumenical fashion, Wallace accepted the miracles at Lourdes and almost took comfort from the fact that such miracles seem to be random and have nothing to do with merit, the gravity of one's disease, or anything else.

Moreover, Wallace refused to balance spiritualism against the great body of evidence that the world simply does not work that way. Sounds produced on an everyday basis are produced by physical objects clashing together and not by forces from another world. Tables that are levitating are being pulled by strings or pushed by fingers, not the dead. John Stuart Mill said flatly to Wallace: "For my own part I have not only never seen any evidence that I think of the slightest weight in favour of spiritualism, but I should also find it very difficult to believe any of it on any evidence whatsoever, and I am in the habit of expressing myself to that effect very freely whenever the subject is mentioned in my presence."[21] Wallace's response to this was that Mill was "very unphilosophical." But, of course, in the eyes of Mill and of others, including Darwin, it was Wallace who was being unphilosophical. It was he who was refusing to abide by the rules of science, including controlled experimentation and so forth, especially for a notion that goes against the huge warehouse of experience. It was not (to put things in a contemporary idiom) that Wallace and only Wallace was a true follower of Karl Popper, prepared to put the most favored hypothesis (that general opinion is wrong) to the test, but that Wallace was no true Popperian, for he would not put his thinking (in favor of spiritualism) to the test.

Why did Wallace behave in this way? In part he did so for the kinds of psychological reasons already discussed. But as significant, if not more so, was the fact that Wallace had never truly internalized the rules of science. He was a man outside the system, and although he had brilliant flashes of intuition, in certain respects his thinking was never disciplined. He did not have Darwin's training nor (to think of people born a few years after Darwin) did he have the social position and responsibilities that Huxley had in the world of science. It is fascinating and surely pertinent to learn that another evolutionist who embraced spiritualism with fervor was Robert Chambers, the still-anonymous author of the *Vestiges*. There were happy exchanges of letters. "I have for many years known that these phenomena are real, as distinguished from impostures; and it is not of yesterday that I concluded that they were calculated to explain much that was doubtful in the past, and when fully accepted, revolutionize the whole frame of human opinion on many important matters."[22] Chambers, like Wallace, was in some respects not a professional scientist but rather was on the fringes of the business and so had not internalized the rules.

The conclusion of Chambers's letter to Wallace (written at about the time that Wallace was going to come out and let the world know about his unorthodox thinking on the subject of humans) gives a clue to the second question posed above: why Wallace plunged ahead and claimed that his position with regard to human origins was scientific. "My idea is that the term 'supernatural' is a gross mistake. We have only to enlarge our conceptions of the natural, and all will be alright."[23] This is the key point. Chambers and Wallace simply were not prepared to separate the natural from the supernatural, the properly scientific from the properly religious. It was not that Wallace wanted to say, as Whewell might have said in the 1830s, that we simply cannot give a scientific explanation of human origins. Instead, Wallace wanted to bring the supernatural into the discussion and claim that he still was being scientific. That was anathema to men such as Darwin. It went against everything they were trying to do as scientists—as people defining their position and role in Victorian society. Wallace had no position and he had no role. He was indifferent to the petty shibboleths of others. If anything, he thought that the rules of science had for many years blocked the way to an appreciation of evolution. And so he had no hesitation in bringing in spiritual forces and thinking that his idea still qualified as science.

So we return to the original question about rebellion. Alfred Russel Wallace was part genius and part crank. But he is not totally mystifying. His greatest work is a function of the same sorts of things that led to his most ludicrous ideas—ludicrous as judged by his contemporaries and as judged by us. The paradox is that he was a brilliant scientist and a man of strong convictions who was prepared to work for them, but he was a man outside science. In part because of inadequate education, in part because of temperament, Wallace rarely played the scientific game as others were then defining it. For this reason he could accept the notion of evolution presented in *Vestiges* when all of the respectable scientists were pulling back from it. But it was also for this reason that twenty years later he could drop natural selection and argue for spiritual forces when it came to the development of humankind. Prima facie he seems the personification of inconsistency, but in his own way he was being completely consistent. The question of rebellion, however, starts to appear more complex because this is something that goes beyond Wallace himself. In the 1840s there was nothing particularly rebellious about Wallace's becoming an evolutionist. He belonged to a group—those who would jump on any bandwagon—for which his moves were almost orthodox. As the correspondence with Chambers shows, that group still existed in the 1860s, so one might want to say that Wallace was still no rebel, and in a way that is true.

But, almost *malgré lui,* by the 1860s Wallace had left the company of the quasi scientists. Thanks to his brilliant discovery of natural selection, in addition to his other work, by right and acknowledgment he qualified as part of the scientific establishment. He ought to have accepted—and in some respects he did accept—the rules of the game. So when Wallace invoked spiritual causes of human development, he was being rebellious. He may not have thought so, but others judged him to be so, and I argue that this is how we, too, should judge him.

ENVOI

Evolutionists know that in order to understand the present, we must understand the past. If we are to understand the twentieth century, we must understand the nineteenth. Alfred Russel Wallace is a fascinating study in his own right, but his entanglement of science and what we would judge to be nonscience shows how science slowly, and at times with difficulty, broke from religion and other topics and became an enterprise or discipline in its own right. The story does not end with Wallace. Still to come, for instance, was the great growth and popularity, at the turn of the twentieth century, of vitalism—a notion that was stuck between the natural and the nonnatural if anything is.[24] But one should think of Wallace's as a story that tells us about the century in which he died as much as about the century in which he was born.

NOTES

1. The chief sources for Wallace's life are Alfred Russel Wallace, *My Life: A Record of Events and Opinions* (London: Chapman and Hall, 1905), and J. Marchant, ed., *Alfred Russel Wallace: Letters and Reminiscences* (London: Cassell, 1916). The spate of recent biographies includes M. Shermer, *In Darwin's Shadow: The Life and Science of Alfred Russel Wallace* (New York: Oxford University Press, 2002); P. Raby, *Alfred Russel Wallace: A Life* (Princeton: Princeton University Press, 2001); and M. Fichman, *An Elusive Victorian: The Evolution of Alfred Russel Wallace* (Chicago: University of Chicago Press, 2004).

2. Consult Michael Ruse, *The Darwinian Revolution: Science Red in Tooth and Claw* (Chicago: University of Chicago Press, 1979), for a more detailed discussion of the philosophy of science in the early nineteenth century.

3. John F. W. Herschel, *Preliminary Discourse on the Study of Natural Philosophy* (London: Longman, Rees, Orme, Brown, Green, and Longman,

1930); William Whewell, *The History of the Inductive Sciences* (London: Parker, 1837); William Whewell, *The Philosophy of the Inductive Sciences* (London: Parker, 1840); Baden Powell, *Essays on the Spirit of the Inductive Philosophy* (London: Longman, Brown, Green, and Longman, 1855).

4. I discuss these issues at length in Michael Ruse, *The Evolution-Creation Struggle* (Cambridge: Harvard University Press, 2005).

5. Whewell, *History,* 3:588.

6. Janet Browne's new biography is definitive. See Browne, *Charles Darwin: A Biography,* vol. 1, *Voyaging* (New York: Knopf, 1995), vol. 2, *The Power of Place* (New York: Knopf, 2002). I have written extensively about Darwin, most fully in Michael Ruse, *The Darwinian Revolution: Monad to Man; The Concept of Progress in Evolutionary Biology* (Cambridge: Harvard University Press, 1996); and Michael Ruse, *Charles Darwin* (Oxford: Blackwell, 2007).

7. Charles Lyell, *Principles of Geology* (London: John Murray, 1830–33).

8. Erasmus Darwin, *Zoonomia* (London: J. Johnson, 1792–94).

9. Charles Darwin to Asa Gray, 22 May 1860, *Collected Correspondence of Charles Darwin,* vol. 8, *1860,* ed. Frederick Burkhardt, Duncan M. Porter, Janet Browne, and Marsha Richmond (Cambridge: Cambridge University Press, 1993), 224.

10. Charles Darwin, *On the Origin of Species by Means of Natural Selection* (London: John Murray, 1859).

11. Wallace himself wrote a book called *Darwinism* (London: Macmillan, 1889).

12. Alfred Russel Wallace, *Studies: Scientific and Social* (London: Macmillan, 1900), 2, 507.

13. I discuss the differences between Darwin and Wallace concerning the levels of selection in Michael Ruse, "Charles Darwin and Group Selection," *Annals of Science* 37 (1980): 615–30.

14. Robert Chambers, *The Vestiges of the Natural History of Creation* (London: Churchill, 1844).

15. For good background concerning the reception of *Vestiges,* see J. Secord, *Victorian Sensation* (Chicago: University of Chicago Press, 2000).

16. For details of the controversy, see Michael Ruse, *Darwinism and Its Discontents* (Cambridge: University of Cambridge Press, 2006).

17. Alfred Russel Wallace, *The Geographical Distribution of Animals* (London: Macmillan, 1876); Alfred Russel Wallace, *Island Life: or, The Phenomenon and Causes of Insular Faunas and Floras, Including a Revision and Attempted Solution of the Problem of Geological Climates* (London: Macmillan, 1880).

18. "The Origin of Human Races and the Antiquity of Man Deduced from the Theory of Natural Selection." *Journal of the Anthropological Society of London* 2 (1864): clvii–clxxxvii.

19. Alfred Russel Wallace, *Contributions to the Theory of Natural Selection* (London: Macmillan, 1870).

20. Charles Darwin, *Descent of Man* (London: John Murray, 1871).

21. John Stuart Mill to Alfred Russel Wallace, March 18, 1868. The letters are printed in Wallace, *My Life.*

22. Robert Chambers to Alfred Russel Wallace, February 10, 1867, in Wallace, *My Life,* 2:285.

23. Ibid., 2:286.

24. See Michael Ruse, *Darwin and Design: Does Evolution Have a Purpose?* (Cambridge: Harvard University Press, 2003).

FURTHER READING

de Beer, G., ed. *Evolution by Natural Selection* (Cambridge: University of Cambridge Press, 1958).

Fichman, Martin. *An Elusive Victorian: The Evolution of Alfred Russel Wallace* (Chicago: University of Chicago Press, 2004).

Morell, Jack, and Arnold Thackray. *Gentlemen of Science: Early Years of the British Association for the Advancement of Science* (Oxford: Oxford University Press, 1981).

Ruse, Michael. *The Darwinian Revolution: Science Red in Tooth and Claw,* 2d ed. (Chicago: University of Chicago Press, 1999).

Wallace, Alfred Russel. *My Life: A Record of Events and Opinions* (London: Chapman and Hall, 1905).

Chapter Three

Rebel With Two Causes: Hans Driesch

GARLAND E. ALLEN

Hans Adolf Eduard Driesch (1867–1941) may never have looked like a rebel, but in his scientific life he was one twice over. A major player in the 1890s movement to make biology in general and embryology in particular experimental and mechanistic sciences, after 1899 he disavowed the mechanistic approach and became one of the leading advocates of a new form of vitalism, the philosophy that claimed that life can never be understood in terms of the principles of physics and chemistry. In both instances, Driesch went against the prevailing tide of opinion within the established biological community. In the 1890s it was the dominant science of morphology, using embryology and other fields for the purpose of reconstructing phylogenetic (evolutionary) histories, against which Driesch rebelled. In its place he advocated a science that sought to study the processes of embryonic development—for example, the question of how cells differentiate—for their own sake and not as an adjunct of evolutionary theory. In the early decades of the twentieth century it was that same approach against which he came to rebel. The machine analogy that motivated so much of the mechanistic approach, Driesch now argued, was incapable of understanding the most fundamental properties of living organisms: their ability to self-regulate. Although he is often characterized as a mystic, Driesch employed the methods of analytical philosophy in an attempt to rigorously defend his antimechanistic, vitalistic position. Well respected by his colleagues in both biology and philosophy, he served on the philosophical faculties of three different universities from 1912 until his forced retirement by the Nazis in 1933. Yet despite his acceptance into the established academic community of his day, Driesch was indeed a rebel at heart.

BACKGROUND

Driesch had been fascinated in the life sciences since his childhood, at least in part as a result of his mother's interest in collecting exotic birds and other animals.[1] After graduating from the Gelehrtenschule der Johanneums (Gymnasium) in Hamburg in 1886, he decided he wanted to become a zoologist. He spent one year (1886–1887) studying at the University of Freiburg with August Weismann (1834–1914), and in 1887 he matriculated at the University of Jena to pursue his doctorate under Ernst Haeckel (1834–1919). It was at Jena that he met another of Haeckel's students, Curt Herbst (1866–1946). Driesch and Herbst became close friends and collaborators. Herbst, indeed, had more direct influence on Driesch's scientific work than his mentor, Haeckel. For his doctoral dissertation Driesch studied the factors governing the growth of hydroid colonies (published in 1890 and 1891), work that was more experimental than descriptive and, most significant, it had nothing to do with phylogeny. Driesch's first rebellion was to join with other younger biologists to promote experimentalism and remove their field from the domination of Haeckel's phylogenetic program.

After receiving his doctoral degree in 1889, Driesch worked for nearly a decade off and on at the Stazione Zoologica in Naples, where Herbst was also located. Established in 1874 by the German morphologist Anton Dohrn (1840–1909), the Naples Station (as it was referred to in English) had become one of the most important marine laboratories in the world. It was at Naples that Driesch met the young American biologist Thomas Hunt Morgan (1866–1945), resulting in a life-long friendship and collaboration on joint papers on development of individual blastomeres of ctenophores and on the development of unfertilized eggs with protoplasmic defects.[2] Driesch, Herbst, and many of their generation were leaders in the movement to make biology, especially embryology, more experimental, and the Stazione was their mecca. It was at the Naples Station that Driesch carried out most of his best-known experiments, often in collaboration with or inspired by Herbst.

An aspect of Driesch's maverick nature in the early years was the fact that he, along with Herbst, had independent means and thus did not settle immediately into the academic path followed by so many German zoologists after receiving their doctorates. From roughly 1889 to 1899 Driesch worked on various aspects of invertebrate embryology at several marine stations, including those at Naples and Trieste. At

Figure 3.1. Hans Driesch in about 1890.
Reprinted with permission from the
Ernst-Haeckel-Haus in Jena, Germany.

the same time, he and Herbst would travel extensively; for example, in 1889–1890 they traveled in the Mediterranean, Ceylon, Java and India, and later Algeria, Tunis, Palestine, Syria, and Greece. Europe, including Scandinavia, was also in their travel plans later in the decade. Many of these trips were combined with meeting other biologists or conducting experiments on marine organisms.

In 1899 Driesch married Margarete Reifferscheidt and moved to Heidelberg, where Herbst had taken a position at the Zoological Institute. Driesch was not officially a faculty member, and with sufficient means from his family, he was able to live the life of an independent scholar. Driesch did eventually take a regular faculty position, habilitating[3] at Heidelberg as extraordinary professor of philosophy in 1912. Although he continued to carry out experimental work until 1908, Driesch began to lose faith that the very approach he had championed would ever solve the real problems of self-regulation in organisms. In place of a highly mechanistic view, Driesch advocated a vitalistic philosophy that attempted to understand life as a holistic, self-regulating system directed by other processes than those encompassed by the laws of physics and chemistry.

In 1919 Driesch accepted the ordinary professorship of philosophy at the University of Cologne and in 1921 he accepted the professorship of philosophy at the University of Leipzig, where he remained for the rest of his career. On three different occasions during this time he served as a visiting professor of philosophy: in China (1922–1923 at the universities of Nanking and Beijing), the United States (1926–1927, at the University of Wisconsin), and Brazil (in Buenos Aires, 1928). In 1933, with the ascent of the National Socialists in Germany, Driesch was forced into "early retirement."

DRIESCH'S FIRST REBELLION: EMBRACING MECHANISM

Although introduced to formal biological work by the popular writings of Haeckel, soon after entering the Zoological Institute at Jena Driesch found himself out of sympathy with his mentor's evolutionarily based research program. One of the best-known naturalists of the later nineteenth century, Haeckel was particularly known for promoting a newer version of Karl Ernst von Baer's recapitulation theory, the idea that the more similar the sequence of stages through which an embryo passes in its individual development (ontogeny), the more closely related species are to each other. After reading Darwin's *On the Origin of Species* in 1860, Haeckel put this idea into an evolutionary framework with his "biogenetic law," the claim that the embryonic development of the individual (ontogeny) repeats or recapitulates the adult stages of the ancestors in its evolutionary lineage (phylogeny). Combined with comparative anatomy, physiology, and the study of life histories (an area in which Haeckel was a pioneer), the recapitulation theory served as the basis for his broadly based research program, known generally as morphology. In order to understand better what Driesch's first rebellion was about, it will help to provide a brief discussion of what Haeckel's morphological program meant to biologists at the time.

Haeckel had been trained as a descriptive morphologist specializing in the study of marine invertebrates, in particular, the sponges and radiolarians. Well-regarded as a taxonomist, Haeckel published a major work about the radiolarians in 1862 and described the taxonomy of several invertebrate groups from the H.M.S. *Challenger* expedition,[4] all of which made him one of the leading naturalists of the late nineteenth century. Yet it was as an evolutionary theorist, inspired by Darwin's work, that he ultimately became best known throughout the world and initially attracted many students to work with him at Jena. Haeckel's evolutionary research program involved using comparative anatomy of both adults and embryos to reconstruct the phylogenetic history of life. By studying the intricate details of early embryonic development—for example, the pattern of cleavage in early cell divisions of the embryo—it was possible to get a glimpse into the deep evolutionary history of the *Stammbaum,* or major lineages of life. This enterprise provided research problems for a whole generation of biologists. Followers of Haeckel and other morphologists investigated various phylogenies, working out in painstaking detail such issues as the specific nature of mesoderm formation, the formation of specific structures such as the eye or the gills, or comparison of larval types as a means of discovering more specific relationships. Although Haeckel's biogenetic law was shown later to be highly oversimplified (embryonic processes also evolve, so that earlier stages in the development of the individual do not represent ancestral adults in any strict sense), it did provide a way to reconstruct phylogenies in the absence of clear fossil evidence.

The problem with Haeckel's phylogenies was that they were largely speculative. Embryonic clues were often not clear enough to provide a firm connection between two groups. Numerous alternative interpretations were plausible, and there was often no clear way to distinguish between them.[5] For students of Driesch's generation, the morphological program increasingly seemed to be an exercise in futility. It was in this context that Driesch's first rebellion occurred. Although Haeckel's student, he chose for his thesis a nonmorphological, experimentally based problem, namely, the laws governing the growth of hydroid colonies.[6] This was, however, only the start of his revolt against the descriptive and speculative nature of morphology. He pursued this course by becoming a major advocate of what was known at the time as the "mechanistic interpretation of life."[7] Driesch, along with others of his generation, helped lead the rebellion against Haeckel's morphology.

THE NEW EMBRYOLOGY: WILHELM ROUX AND THE *ENTWICKLUNGSMECHANIK* PROGRAM

Although he distanced himself from the morphological program in writing his doctoral dissertation, Driesch did not develop his own alternative approach until a year after receiving his doctorate at Jena. In 1888 a fellow German, Wilhelm Roux (1859–1924), published a highly suggestive paper that initiated a new era in the study of ontogeny.[8]

Also a student of Haeckel and a rebel against phylogeny, Roux saw the embryo as presenting critical research problems in its own right. The embryo was much more than a vehicle for tracing out evolutionary histories. A major problem of embryology was discovering how a single fertilized egg cell differentiated via ontogeny into the myriad of specialized cells that made up the adult organism. For Roux this problem could be approached by experimental analysis and thus did not have to rely on mere speculation. In his 1888 paper, Roux proposed a hypothesis that could explain the differentiation process by a simple mechanical process. When the fertilized egg divided, the hereditary particles that determined all the various parts of the organism (head, tail, right and left sides, specific organs, and so on) were qualitatively parceled out into the daughter cells. With each successive cleavage, the range of determinants was restricted, so that eventually each cell of the adult would contain determinants only for the characteristics of cells of that type. This was what became known as the mosaic theory of development. A similar view had been put forward by August Weismann in the 1870s; after 1888 it became known as the Roux-Weismann hypothesis.

In order to test this hypothesis, Roux carried out a series of experiments on the frog's egg in which, using a hot needle, he killed one of the first two blastomeres (formed after the first division of the egg) and then cultured the altered embryo through later divisions. If the mosaic theory was correct, Roux reasoned, he should get only half-embryos because the remaining live blastomere now contained only half the determinants for the organism's traits. Although large numbers of the embryos died, some survived as far as the stage of gastrulation, producing only partial embryos (figure 3.2a). Roux claimed a victory for the mosaic theory and from this approach promulgated a whole new research program that he named *Entwicklungsmechanik*. The term was chosen carefully to indicate the dominant, mechanistic materialist underpinning that Roux wanted to instill into embryological research in particular and biology in general. Mechanistic materialism was a philosophical position of particular interest to biologists in the nineteenth and early twentieth centuries (versions of it had also appeared in the seventeenth and eighteenth centuries) that claimed that all living processes could be understood in terms of the known laws of physics and chemistry. There was no special vital or life force that differentiated living from nonliving matter: it was simply a matter of degrees of complexity. All processes in the universe—living as well as nonliving—could be understood and explained in terms of matter in motion. Roux's hypothesis epitomized the mechanistic approach by postulating a purely mechanical process for the complex problem of differentiation, the differential parceling out of hereditary determinants with each cell division of the embryo.

In addition to the highly mechanistic nature of the Entwicklungs-mechanik program, Roux also emphasized the importance of experimentation in solving biological problems. Experiments allowed the biologist to intervene in a complex process in precise ways, asking one question at a time and controlling all variables except the one under investigation. It was possible with experiments to discover the immediate, or proximate, causes (as opposed to Haeckel's historical, or ultimate, causes) of particular processes such as differentiation (an approach that became known as *causal analysis*). For Roux, the cause of differentiation in the developing embryo could be attributed to the differential parceling out of hereditary determinants, a process that was subject to experimental investigation. By contrast, for Haeckel, the cause of a particular sequence of stages in ontogeny lay in the historical evolution of the species, where new, adaptive stages were added onto the end of ancestral stages. There was no way to test such a claim, however, which left it in the realm of speculation. Although Roux's mosaic hypothesis was also speculative, it was testable, and, as Roux's 1888 paper showed, appeared to be consistent with the results (that is, the development of half-embryos).

Driesch was immediately attracted to Roux's work and the opportunity it provided for young biologists to emancipate their field from the domination of speculative morphology. While at the Naples Station in 1891, Driesch decided to repeat Roux's experiment, using the sea urchin, which was in plentiful supply in the Mediterranean. Instead of killing one of the first two blastomeres, Driesch proceeded to separate the first two blastomeres by shaking them in sea water, a method he adopted from the earlier work of Oskar and Richard Hertwig and Theodor Boveri (he later modified the technique by using calcium-free sea water, as suggested by Herbst). In this process neither cell was injured, and to Driesch's surprise, each proceeded to develop into a complete but slightly smaller-than-normal embryo (see figure 3.2b). He repeated the experiment with four- and eight-cell embryos, with similar results. Driesch concluded that the mosaic theory could not be correct and that each blastomere must contain *all* the determinants for the development of the complete organism. This interpretation was

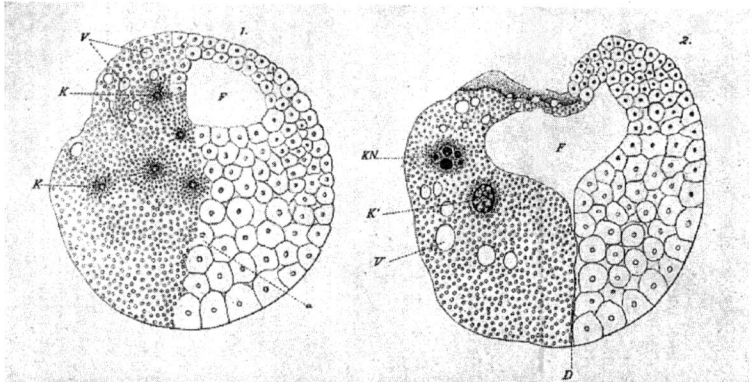

Plates II and III

All these figures represent frog embryos (*Rana fusca* and *esculenta*).

F	Segmentation cavity	Ch	Chorda dorsalis (notochord)
Ec	Ectoderm (external germ layer)	Md	Neural fold
En	Endoderm (internal germ layer)	U	Gastrocoele; in Fig. 12 blastopore
Ms	Mesoderm (middle germ layer)	D	Yolk cells
		V	Vacuoles

Figure 3.2a. Roux' experiment. Roux killed (with a hot needle) one of the first two blastomeres of the frog embryo. The other blastomere divided normally, producing a half-blastula state (the right-hand side of the embryo); the heat-killed blastomere gave rise to an amorphous mass of protoplasmic material and cell nuclei (the left-hand side of the embryo). Roux interpreted these results as verification of his mosaic hypothesis, which claimed that there was a qualitative parceling out of determinants during embryonic cell division. From Wilhelm Roux, "Beiträge zur Entwicklungsmechanik des Embryo: Über die künstliche Hervorbringung halber Embryonen durch Zerstörung einer der beiden ersten Furchtungskugeln, sowie über die Nachentwickelung (Postgeneration) der fehlenden Körperhälfte. *Circhow's Archiv für pathologische Anatomie und Physiologie und klinische Medezin* 114 (1888): 113–153.

also consistent with the recent cytological studies of cell division, which showed that all cells of the developing embryo, as well as of the adult, contained the full complement of chromosomes characteristic of the species. At least at the chromosomal level, then, there appeared to be no visible parceling out of nuclear components during differentiation.[9]

Driesch's experiments displayed his rebellious spirit in a number of ways. First and foremost, by adopting the new Entwicklungsmechanik program, he was rebelling openly and enthusiastically against the descriptive, historically based work of his mentor Haeckel. Second,

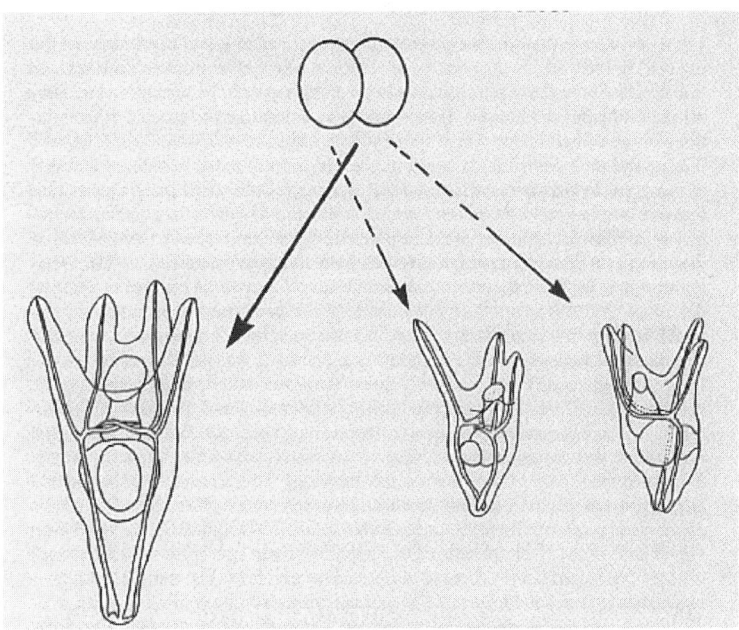

Figure 3.2b. Driesch's experiment. Driesch separated (by shaking) the first two blastomeres of the sea urchin egg, leading to the production of two smaller-sized but complete plutei larvae (right-hand side, dotted arrows). If the two cells remained together they produced one large pluteus (left-hand side, solid arrow). From Viktor Hamburger, *The Heritage of Experimental Embryology* (New York: Oxford University Press, 1988), 10.

by also challenging the mosaic theory, he was rebelling against Roux's simplistic view of development as a purely mechanical process. He concluded that the fate of embryonic cells was not fixed at the two-cell stage or even later at the blastula stage and, indeed, that each cell has the full *prospective potency* characteristic of the fertilized egg to form a complete organism. If left in place, each blastomere would give rise to differentiated parts of the embryo, fulfilling what Driesch called their *prospective significance.* The fate of each cell in the developing embryo was a result of its position in the whole. The process of development could not be broken down into a series of mechanical or chemical processes that functioned automatically or independently. Driesch's experiments had shown that the embryo could adjust itself to altered circumstances: it was what he would refer to later as a "harmonious equipotential system." For Driesch, one of the marvelous properties of living systems was this ability to adjust, to respond, and to act epigenetically. Cells of the embryo were not predestined to become only one type of adult tissue or organ. The whole affected the destiny of the parts just as the parts affected the destiny of the whole.

Published in the prestigious journal *Zeitschrift für wissenschaftliche Zoologie* in 1891,[10] Driesch's experiments initiated a controversy that raged for a decade or more, with Roux vigorously resisting the challenge to the mosaic hypothesis, primarily in the pages of his own journal, the *Archiv für Entwicklungsmechanik.* The controversy was eventually resolved by repeating Roux's experiment but separating the heat-killed from the living blastomere; under these conditions the normal blastomere developed into a complete frog larva. The differences in outcome were due to subtle but important differences in experimental technique.

Between 1891 and 1909 Driesch continued to perform a variety of experiments that demonstrated the ability of embryos to adjust to altered conditions. He squeezed four-cell embryos between cover slips (altering the cytoplasmic context in which the nuclei came to reside after the next division), fused two embryos at the blastula stage (producing a giant blastula), exposed cells to sea water with different ionic concentrations, and centrifuged eggs to alter the polarity of their cytosplasmic components, among other things. In all cases the embryos developed into normal, functional larvae. What was becoming clear from Driesch's experimental work was his understanding and appreciation of the embryo *as a whole,* as a functioning entity in which the parts affected the whole as much as the whole affected the parts. As Oppenheimer has observed, "these concepts were widely disseminated and extremely influential in their day."[11]

DRIESCH'S *ANALYTISCHE THEORIE DER ORGANISCHEN ENTWICKLUNG*

The most complete statement of Driesch's mechanistic views can be found in his *Analytische Theorie der organischen Entwicklung* of 1894.[12] It is worth examining closely Driesch's approach to analyzing development from a mechanistic point of view, because it shows clearly the sophistication of his attempt to correlate mechanistically based experiments with broad theoretical considerations. Driesch was no naïve mechanist who saw the embryo as only a collection of individual machine-like parts. In the nearly two hundred pages of the *Analytische Theorie,* Driesch raised the critical question of *how* events in morphogenesis occur when and where they do (displaying both temporal and spatial determination). With the sea urchin as his ever-useful example, he noted that the mesenchyme (the first organ-forming event in the developing embryo) always originated at the vegetal pole near the end of the blastula stage. Because mesenchyme formation does not occur at the animal pole, there must be a difference between the two regions of the blastula. As Driesch reasoned: "A priori, when there is a difference, we inquire into its cause. There must be a cause for mesenchyme formation, and since it is restricted locally, there must be a cause acting at its location. Since light, gravity, etc. have no influence on the localization of mesenchyme formation, this cause must be a differential, intrinsic in the blastula. . . . [And since the nuclei are all equivalent,] the only differential in the blastula which has direction [i.e., polarity] is the peculiar structure of the protoplasm. . . . We have to assume this, even in cases where its existence has not yet been demonstrated, because a homogeneous structure cannot transform itself into inequalities; such an assumption would overthrow our entire natural science."[13]

Driesch thus argued that the polarity of the egg cytoplasm sets the stage for the development of the polarity (the anterior-posterior axis) of the whole organism. In this context, polarity is conceived in terms of a gradient of chemical substances in the cytoplasm that ultimately is transformed into the polarity of the organism as a whole. This means that for Driesch, the nuclei of the developing embryo were equivalent but the cytoplasm exhibited polarity. The nuclei of all cells of the embryo (and adult, for that matter) retain their prospective potency, while demonstrating their prospective significance as a result of the influence of different cytoplasms.

But what is it that causes the changes that are actualized during differentiation? For Driesch, there are two factors at work: "the capacity [of the nucleus] to respond in a specific way, and the capacity to respond to a specific cause" or stimulus. He continues by exploring the relationship between nucleus and cytoplasm as an interactive system: "In other words, the organ affected by a stimulus must be capable of *perception* of the stimulus, and capable of a specific *response.* We localize the capacity for *response* to a stimulus in the *nucleus,* and the capacity for *reception* of the stimulus in the *cytoplasm,* which is chemically specific in each elementary organ. Hence, the cytoplasm is the mediator (the 'zone of perception') between the releasing cause and the nucleus (the 'zone of action')" [italics in original]. Driesch placed the stimulus-response process of embryonic development in a dialectical context: "Each cell in ontogenesis is the carrier of totality of all *Anlagen* [determiners], insofar as it is in possession of a nucleus. But insofar as it possesses a specific cytoplasm, it is capable of receiving only certain stimuli."[14]

Driesch recognized that the response of cells during differentiation is a function of two components: the responsive capacity of the nucleus—that is, its constituent Anlagen—and the capacity of the cytoplasm to receive the stimulus that ultimately triggers a given nuclear Anlagen's response. The capacity of the cytoplasm to receive a given stimulus is a function of its position within the whole embryo, in turn a function of differential cytoplasmic composition deriving originally from the polarity of the cytoplasm of the egg. For Driesch, the differences in cytoplasm between cells in different regions of the blastula (and later gastrula) served as "releasing factors" (*Auglösungen*) for specific Anlagen in the cell nuclei. Releasing factors were chemical in nature, a point Driesch emphasized in detail in subsequent sections of the book. The Anlagen responded chemically as did "ferments," the name at the time for enzymes, catalyzing new changes in the cytoplasm. Extending this analysis, he suggested that embryogenesis could be broken down into a series of releasing events in a cascade effect. Once Anlagen were activated by their specific cytoplasmic releasers, they altered the cytoplasm of the cell by their action, leading to further feedback effects from that cytoplasm on other Anlagen in the same cell. In addition, as cells differentiated they had effects on other cells nearby, on germ layers, or even on other developing organs, providing a reciprocal cascade of influences that led development progressively toward completion in an orderly and predictable way. As Fred Churchill has phrased it: "The releases were viewed by Driesch as chemical events which in turn released secondary chemical changes of far greater magnitude than the release of the trigger itself. The released reactions could in themselves play the role of trigger for a third generation of chemical reactions, and so ontogeny, becoming an ever-expanding constellation of stimuli and responses, continued to progress."[15]

Without being anachronistic, it is possible to see in Driesch's mechanism an early version of Hans Spemann's concept of embryonic induction. Spemann's ideas were more fully formulated than Driesch's, but the core mechanism—a series of chains of inductive processes, all chemically mediated, with one induced change becoming in turn the inducer for the next—was the same. Although Driesch was not alone in emphasizing the influence of different cytoplasms on the cell nucleus, he was significant in the later nineteenth century in pointing out the reciprocal influence of cytoplasm on the nucleus and vice versa. Although he recognized that all cell nuclei were totipotent, differentiation could still be viewed as an epigenetic (that is, developmental) process mediated by specific chemical agents. In other words, the problem was one in which as development proceeded and cell division produced qualitatively more restricted cytoplasm, the cells became increasingly restricted in the stimuli to which they could respond.

Yet even with this general scheme in hand, Driesch was dissatisfied with the extent to which he could apply a purely causal-analytical approach. In the *Analytische Theorie* he had uncovered a series of individual events (interpreted as successive releasings) but no overall indication of how the process was guided to the same outcome in every case. There was a harmony in the process that analysis of the individual components failed to reveal. That harmony was ever-present in normal development but was even more apparent in experimentally altered cases. As Driesch explained his dilemma, it was as if he had entered a shipyard where he saw structures lying all about him but could make no sense of the construction process until he knew the ultimate purpose of each piece—that is, how it functioned with respect to the completed ship. It was only by introducing a "teleological principle" into embryogenesis—analogous to the ship designer's overall architectural plan—that he was able to understand the individual releasing steps.[16] Teleology for Driesch was not at this moment a denial of cause and effect or the introduction of something mystical into his thinking, but rather a heuristic device that helped bring the parts of the embryonic process into a discernible relation to the whole. Something more than purely blind, mechanical processes was at work: "Because the viable whole is given as a clear recognizable end of the totality of all the processes on ontogeny, we judge on the ground of an objective necessity, therefore, these processes to be as though they were fixed according to quality and order by an intelligence. By these words we give the really adequate expression to the critical teleological standpoint."[17]

Where teleology was of greatest heuristic value was in understanding the initial polarity of the egg cytoplasm, something fixed in place by an "intelligence" or teleological principle that was beyond mechanical analysis. Once cleavage began, the ensuing events were understandable by analytical means, but the original structure of the egg was not. This was the grand architectural plan analogous to the ship's blueprint, something that could never have come into being on its own. Driesch introduced the term *Bildungstrieb* for this teleological guiding force, a term that was at once a nonmystical "force" but also a catch-all for the organizing process in development that defied mechanical analysis.[18]

THE EMBRYO AS A HARMONIOUS EQUIPOTENTIAL SYSTEM

The Bildungstrieb could account for the orderly progression of development under normal circumstances, but what could account for the ability of the embryo to reorganize itself and proceed forward after it was disturbed? In 1894–1895, after publishing the *Analytische Theorie,* Driesch had carried out a series of remarkable experiments in which he cut sea urchin gastrulae near the equator so that each half carried with it both ectoderm and endoderm. Not only did both halves repair the wound, but each developed into a full-scale but smaller larva (as in the earlier separated blastomere experiments). This time, however, Driesch noted that in addition to regenerating the parts themselves, the regenerating gastrulae maintained the original geometric proportions of all parts to each other and to the whole. In the normally developing gastrula, the regulation of proportions of organs could be explained as the effect of diffusible stimuli and successive inductions. This explanation seemed woefully incomplete when applied to a process as complex as regeneration of a whole larva from a half-gastrula. These and similar results led Driesch in 1899 to introduce the concept of the embryo as a "harmonious equipotential system," emphasizing the plastic, self-regulatory dimensions of embryonic development.[19] In the harmonious equipotential system not only did any given embryonic part (cell, germ layer, or organ) contribute to the development of the whole, but the path a particular cell took was also a function of all the new influences deriving from the particular position it occupied within the whole embryo. The balance of physico-chemical causes and effects, Bildungstrieb (teleology), and the harmonious equipotentiality of the system as a whole allowed development not only to proceed along predictable lines when undisturbed but also to respond to disturbances imposed from the outside.

As the historian Jane Oppenheimer has pointed out, Driesch's significance in these early years of his career as an experimental embryologist lies especially in the new vistas he opened up for experimental
analysis based on a more subtle and holistic approach to the problem
of development than the Roux-Weismann mosaic theory. Such a holistic view always informed his thinking, and it became the basis for
his ultimate rejection of any physico-chemical or mechanical understanding of embryonic development. Although in the mid-1890s he
still retained his general adherence to the Entwicklungsmechanik
approach, by 1899 he had explicitly moved into a more abstruse position characterized by an explicit acceptance of vitalism as a necessary
philosophical basis for biological research. This represents Driesch's
second rebellion.

THE NECESSITY OF VITALISM AND THE "ENTELECHY"

Several factors led Driesch to abandon the search for a mechanistic
basis for embryonic development. One was the fact that he did not find
in other organisms (Tubularians, Ascidians, Ctenophores) the same
degree of self-regulation he had observed in the sea urchin. These
observations suggested that there might be no generalizable physically
or chemically based laws operating across animal phyla. Although
he could describe the process, Driesch recognized that this brought
him no closer to understanding the cause of vital processes. As the
historian Anne Harrington put it, "Unclear how to proceed and frustrated by his inability to expand his empirical base, [Driesch] turned to
philosophy for direction, studying the works of Kant, Schopenhauer,
Eduard von Hartmann, Locke, Hume."[20] By 1900 Driesch had more
or less given up his experimental work, turning his attention instead
to developing an antimechanistic philosophy that emphasized the
vital nature of life processes, properties that could not be reduced
to known laws of chemistry and physics. In particular, Driesch set
out to show why the machine analogy, so popular with many of his
contemporaries, was not only insufficient to explain biological (and
specifically developmental) phenomena but also could not account
for human social—moral and ethical—behavior.

On the embryological front, Driesch concluded that development
occurred with such regularity, and in the case of the sea urchin proceeded toward an appointed end despite all sorts of introduced disturbances, that it must be guided by a teleological force that had no

counterpart in the mechanical or physico-chemical world. This force he called *entelechy,* a term borrowed from Aristotle and referring to an "active principle of converting possibility into actuality."[21] Entelechy was for Aristotle not a principle of mere potentiality but a directionality of development built in from the outset. For Driesch, entelechy had a more modest function, as Hilde Hein reminds us: rather than being the primary cause or architect of actualization, it was a regulator governing which of the various potentialities resident in a material system is to be realized and which is to be restrained.[22] Entelechy was thus not the blueprint of an organism's organization, nor the creative agent that brings it about, but a kind of a mediator, similar to a homeostatic governor, that protects the tendency of the system from being disrupted by extraneous factors (including the embryologist's experiments).

While conceding that entelechy might seem like a metaphysical concept, Driesch pointed out that it was not necessarily more metaphysical than the physicists' concept of energy or energy states of matter. Energy, he argued, was "nothing but a measurement of causality in space,"[23] as it was for entelechy, which does not depend on substance for its effect. Entelechy did not violate the physical laws of nature. It was not a means of creating energy outside the laws of thermodynamics. Rather, it was a nonmaterial "force," perhaps a form of energy, that kept embryonic development moving toward its specific end point. Directionality differentiated Driesch's entelechy from the random effects of energy as characterized by the classical kinetic theory of gases.

From the principle of teleology and its specific embryological form, entelechy, Driesch moved on to advocate an openly vitalistic philosophy of biology. It was first made manifest in Driesch's Gifford Lectures, delivered at the University of Aberdeen in 1907–1908 and published as *The Science and Philosophy of the Organism* in 1908.[24] A distillation of these ideas published six years later as *The Problem of Individuality* (1914)[25] will serve as the framework for explicating Driesch's line of argument in favor of vitalism.

Driesch begins his discussion by pointing out that it is not logically possible to "prove" vitalism but only to demonstrate its likelihood by exclusion of its alternative, that is, by showing that mechanistic principles as exemplified by the machine analogy cannot be at the foundation of living systems. To reveal the inadequacies of the machine analogy, which had been most prominently articulated in the works of Jacques Loeb in the 1890s and early 1900s,[26] Driesch used a number of examples, or what he called "proofs." Three will suffice. First, referring back to his own experiments on separating the two blastomeres of the sea urchin embryo, he asked, What machine could be divided in half, repair itself, and proceed to function as two machines? A machine can function only when it is whole, and no machine is known that can repair itself when it is altered by external circumstances: "The embryonic 'machine,' then, that is supposed to exist in the normal system, would be obliged to be present in its completeness in one part of the system also, and also in another such part. . . . For we know that any part of the [embryonic] system, contingent as to its size and as to its position in the original system, can give rise to a complete being. . . . In light of these facts the machine theory as an embryological theory becomes an absurdity. These facts contradict the concept of a machine; for a machine is a specific arrangement of parts, and it does not remain what it was if you remove from it any portion you like."[27] The animal ovary provides the second example. Ovarian cells (oocytes) multiply many times to produce thousands of eggs, again a process that has no counterpart in the behavior of a machine. If the egg is not produced by a machine-like process, then it cannot itself be a machine. And if the egg is not a machine, how is it possible to regard the organism that develops from it as a machine?[28] The very premises of mechanistic analysis would reject the idea of a machine autonomously emerging from a nonmachine.

For the third example Driesch turns to the process of memory.[29] Comparing human memory to the playing of a record on a phonograph, he noted certain similarities: (1) Both display actions based on historical input (the human from his or her experience and the phonograph from the information that was put into the record). (2) The "behavior" of each is initiated by some causal event; for example, the phonograph is turned on or the person's memory is triggered by some verbal or other clue. (3) Both involve some aspect of direct recall in which the historically based input is repeated: the phonograph replicates what has been recorded on the record, and the human activates specific synaptic connections in the brain established by experience.

Despite these basically superficial similarities, for Driesch the differences between the two examples were profound. The phonograph is restricted to or playing as sound only what is fed into it from the record—and this it does with exact replication. It is incapable of any adaptive modification of its own. The human being, on the other hand, never responds to stimuli in exact replication of previous behaviors. Humans are capable of infinite degrees of modification of their response to a stimulus, and hence their behavior bears no real relation to that of the phonograph or any other machine.

Driesch's logic in providing various rejections of the machine analogy is based on the principle of exclusion, a sort of proto-Popperian sleight of hand in which elimination of one alternative automatically strengthens another. He wrote, "[T]he machine theory was the *only* possible form of a mechanistic theory that might *a priori* seem to be applicable to the phenomena of morphogenesis." To "dismiss the machine theory, therefore, is the same as to give up the attempt of a mechanical theory of these phenomena altogether."[30] The distinction between a machine and a living organism, then, resided in the ability of the latter to readjust and reproduce itself according to external changes—in other words, to function as a harmonious equipotential system. This ability is the vitalistic element that qualitatively distinguishes living from nonliving matter.

For Driesch, the machine analogy also had troubling social and biological implications. Like many of his contemporaries, Driesch was concerned with one of the great social and philosophical legacies of the nineteenth century, the question of free will and determinism. Machines such as the phonograph exercise no free will. The philosophy of vitalism, by contrast, offers the possibility of true freedom in the sense of "*freedom* of becoming"[31]—that is, the embryo or the consciously acting human is free to become itself, to actualize its potentiality despite a variety of changing inputs. That freedom in not rigidly programmed but functions in the manner of a harmonious equipotential system acting as an integrated whole. Driesch saw vitalism as an expression of the holistic view of nature that had gained a visible group of adherents, especially in Germany in the interwar period.[32] In his view, holistic philosophy was capable of restoring humans to a more peaceful life, alienation from which had occurred as a consequence of mechanistic thinking. Although not a Marxist, Driesch would have agreed with the Marxist psychoanalyst Wilhelm Reich (1897–1957), who wrote in 1933 with an explicitly antifascist intent: "What is called civilized man is in fact angular, machine-like, without spontaneity; it has developed into an automaton and a 'brain machine.' Man not only believes that he functions like a machine, he does in fact function like a machine."[33] Driesch thought that the machine was the very antithesis of life.

REACTIONS TO DRIESCH'S VITALISM

Driesch's philosophy was always grounded in empiricism and was presented with analytical logic combined with straightforward prose. It therefore was as clear as it could be, given the metaphysical framework within which it was cast. His philosophical ideas were circulated widely while he was still doing experimental work. Before he was habilitated in philosophy in the Faculty of Natural Sciences at the University of Heidelberg (1909), he had been invited to give the Gifford Lectures. Clearly, Driesch was taken seriously not only by his followers and proponents but by his opponents as well. Among his most avid followers were German holistic biologists such as Baron Jakob von Üxeküll (1864–1944), a prominent physiologist who was perhaps best known at the time for his concept of the *Umwelt,* an approach to understanding animal behavior by viewing the animal in conjunction with its environment as a single, nondivisible entity that must be studied as a whole.[34] In 1913 Üxeküll was a top contender for directorship of the Kaiser-Wilhelm Institute for Biology in Berlin-Dahlem. (Because of concerns about his metaphysical leanings and his tendency to engage in what was considered excessive theorizing, he was ultimately passed over in favor of Theodor Boveri.)

Üxeküll had developed his own sort of teleological principle, the *Bauplan,* much like Driesch's entelechy, that directed morphogenesis and adaptation. He was thus poised to see in Driesch's vitalism the beginnings of a whole new renaissance in German biology. He found Driesch's experiments "the starting point for a new approach to an 'exact biology' emancipated from the tyranny of outmoded linear-causal thinking."[35] The philosopher Martin Heidegger (1889–1976) was also enamored of Driesch's vitalism (and Üxeküll's Umwelt). In the work of both men Heidegger recognized the importance of the holistic character of the organism and the integration of the organism with its environment.[36] Such views meshed well with Heidegger's own concept of *Dasein,* the understanding of any phenomenon or process in the world from the inside as a part of the process itself and not from the viewpoint of a detached or dispassionate observer.[37]

Anne Harrington has argued that Driesch's vitalism thus struck a chord with a variety of intellectuals who were distressed by the turn toward materialism and mechanism of modern industrial capitalism.[38] In Germany, a prominent holistic movement developed of which Driesch was the acknowledged leader—one whose prominence was

enhanced by his strong scientific reputation. That much of this movement was tinged with various mystical ideas, including vitalism and
Volkish unity with nature, only reflects the disenchantment with contemporary society that many people felt in the interwar period. Vitalism, as the name implies, involved a reassertion of the importance of
life itself in relation to the machine and other impersonal forces that
were felt to be consuming humanity. The holistic movement was a
rebellion against the machine, and Driesch was in the lead. The irony,
of course, is that this was a revolt against the very principles of mechanistic science that Driesch had at one time promoted so strongly.

But not everyone was so enamored with Driesch's introduction of
full-blown vitalism into biology. One of Haeckel's last students, Julius
Schaxel (1887–1943), a socialist and Marxist geneticist who was not
inherently hostile to holism, attacked what he called the "categorical
vitalism" of Driesch as "definitely not the right view."[39] Jacques Loeb
attacked both Driesch and Üxeküll for their vitalism and teleology.
Seeing them as "brilliant biologists," he found their influence all the
more pernicious because they had the ability to "sway scientific and
public opinion in false directions."[40] Loeb wrote his book *The Organism as a Whole* (1916), which provides a mechanistic alternative
to the issue of how the organism functioned as an integrated entity
and not simply a mosaic of independent parts, largely in response
to Driesch's and Üxeküll's work.[41] Thomas Hunt Morgan, despite
his strong personal friendship with Driesch, admired the attempt
to deal with embryonic development holistically but found vitalism
and entelechy problematic. After reading *The Science and Philosophy
of the Organism,* Morgan wrote to Driesch that although his was the
best presentation of the case for vitalism that he had seen, "when
it comes to the critical point and the entelechy steps in to alter the
physical series of events without itself entering the energy chain, I fail
to be convinced."[42] In the same letter Morgan predicted that Driesch
would receive "abundant abuse from the scientists" and confessed,
"I would abuse it somewhat if I had the opportunity." Such scientists included the physiologist John Scott Haldane (1860–1936), who
found the argument that a machine cannot be divided and subdivided
and still function as a whole unacceptable, though he was willing to
accept the qualitative difference between a phonograph and human
memory.[43] Haldane's son, the geneticist J. B. S. Haldane (1892–1964),
rejected entelechy as a mystical phenomenon that provided no basis
for further research, although he admitted that "from the standpoint

of mechanistic principles we can form no idea of how it is that the capacity of reproducing an elaborate organism is handed down from cell to cell."[44] So influential had Driesch's ideas become in biological circles, however, that as late as 1942 the embryologist Joseph Needham felt compelled to insert a whole section (2.16) in his *Biochemistry and Morphogenesis* to attacking the doctrine of vitalism.[45]

Philosophers also found Driesch problematical. At the Prague International Congress of Philosophy in 1934, at one of his last public presentations, Driesch was attacked by the Viennese logical positivists Rudolf Carnap, Hans Reichenbach, and Moritz Schlick. They found the concepts of holism, vitalism, and entelechy too vague and lacking in the law-like character they felt was necessary for scientific discourse. Schlick complained that Driesch's entire system was nothing more than a linguistic exercise.[46]

What is particularly frustrating from a modern perspective is that while Driesch was pointing to an important philosophical issue— wholeness, or the idea that the whole is greater than the sum of its parts—that was recognized by numerous biologists (especially embryologists) of the day, he did not have the intellectual framework within which to analyze the issue more successfully. In the 1920s and 1930s some biologists, confronting the issues of holism, self-regulation, and emergent properties, had adopted the philosophy that after 1909 was known as dialectical materialism. A view associated with Karl Marx and Friedrich Engels, dialectical materialism had been taken up by J. B. S. Haldane, Joseph Needham, and J. D. Bernal in England and by Marcel Prenant in France, among others. Driesch could have availed himself of this viewpoint, which encompasses issues that were of considerable relevance to his concerns: the whole being greater than the sum of its parts (the principle of emergent properties), the coming into being of entities from historical processes in past entities, a strong emphasis on dynamic change and opposition to the mechanistic view. Yet all of these were grounded in a staunch materialism. A dialectical perspective would have allowed Driesch to retain a materialist understanding of biological processes while still dealing with the problems that most concerned him. But of course the advocates of dialectical materialism at the time were mostly Marxists and therefore were regarded with even more suspicion than was vitalism. We cannot rewrite history, but it would be interesting to know whether Driesch, the rebel, knew of dialectical materialism as an alternative philosophical system to vitalism and whether he would have been rebellious enough to adopt it in the face of the general condemnation such views faced at the time.

In supporting vitalism, Driesch was clearly going against the rising tide of mechanistic thinking and experimentation that had been the dominant methodology in biology throughout the first four decades of the twentieth century. As we have seen, he was not a lone rebel, but he was a rebel nonetheless, challenging the machine analogy and the simplistic assertion that life could be completely understood in terms of the known laws of physics and chemistry. He never argued that physics and chemistry could not answer *some* questions about living processes, but he became convinced that they could not answer the most important question of all: how organization and interrelatedness of parts became established in such a regular and adaptive way during morphogenesis.

DRIESCH'S LATER CAREER AND THE RISE OF NATIONAL SOCIALISM

The holistic science movement was taken up in various ways by the Nazis, but as with much else in their philosophy, it showed considerable inconsistencies. They used holistic biology as the basis for reviving old German cultural myths of the Volkish unity of man and nature, capturing and building opportunistically on the same sense of dénouement that pervaded Germany in the aftermath of the Treaty of Versailles. Driesch's vitalistic philosophy was appropriated by the Nazis, for a while at least, to confer legitimacy on their holistic views of everything from medicine to health foods and the youth-based *Wandervogel* (hiking) movement. Anne Harrington has described this appropriation of Driesch's views by the Third Reich quite well.[47]

What made this situation particularly ironic is that Driesch was an ardent anti-Nazi, a pacifist, and a liberal humanitarian who publicly positioned himself against war, Prussian militarism (in World Wars I and II), fascism, and nationalism. He had been a member of the pacifist pan-European Human Rights League during World War I, and in the early 1930s he repeatedly spoke out against hypernationalism in scholarly writings and popular newspapers. He pointed out the irrational and mystical elements that had crept into popular Nazi cults of holism and sought to turn the language of biological holism toward more peaceful, humanitarian ends. Vitalism, he argued, was the very affirmation of life and was opposed to war and nationalist aggression. Entelechy, he wrote, recognized no state boundaries. The only whole to which a person can rightfully (meaning biologically) belong is humanity.[48]

In his personal actions Driesch was consistent in his political allegiance to democracy and opposition to nationalism and fascism. He took part in political rallies against Nazi candidates, defended junior colleagues who were attacked by Volkisch student groups, and made his view of the Hitler regime public. For both his history and his persistent anti-Nazi activity after 1933, Driesch became one of the first non-Jewish professors to be forcibly retired from his academic post. He was not allowed to accept speaking engagements in Germany, though he did travel abroad to meetings during the following two years. In 1935 this privilege was also withdrawn, and Driesch spent his remaining years in semi-isolation.

CONCLUSION

What can we learn from the rebellious career of Hans Driesch? What does the trajectory of his changing and challenging ideas about the nature of the organism and of how to study it tell us about biology in general and about rebels in science in particular?

One answer is that rebels often point to real issues in their field. They advance important viewpoints that are neglected, even consciously sidelined, by the mainstream, paradigm-bound practitioners within the scientific community. In both his rebellions Driesch challenged prevailing ideas and especially methodologies. And in both cases he offered his own alternatives. When he challenged the descriptive and speculative use of embryology to reconstruct phylogenies in the 1890s, he simultaneously championed the innovative methodology of experimental embryologists such as Roux, who sought to discover proximate causes for such embryonic processes as cell differentiation. Embryos had been neglected as objects of study in their own right, and the problem of differentiation had been a key issue for embryologists since Aristotle. As a subrebellion within experimental embryology, Driesch soon came to challenge the mosaic theory of two giants in the field, Roux and Weismann, articulating in its place his own concept of the embryo as a nonmosaic, totipotent, and self-regulating system. After the turn of the century, in his second rebellion, Driesch disavowed the very mechanistic viewpoint that had motivated his first revolt when he concluded that the methods he and others had pioneered could not determine what made development a self-regulating and dynamic process (an issue that continues to tantalize developmental biologists to this day).

The second point that emerges from Driesch's case is that rebels in science are often rebels in spirit as well as in ideas; they are not content to accept the status quo in any area and are almost driven to reject orthodoxy. They are imbued with what Charles Kingsley (1819–1975) identified as that "divine discontent"[49] which can be a source of both iconoclasm and creativity. It was characteristic of Driesch: as soon as the mechanistic science he had advocated in his youth became established and fashionable, Driesch revolted again, challenging the new orthodoxy to the point of abandoning experimental biology altogether. At another level, he was clearly courageous, if not rebellious in the usual sense, in refusing to accept the tenets of National Socialism, even when it cost him his job. To some extent, perhaps, all rebels must have the personal courage (or fearless indifference to opposition) to buck the prevailing attitudes and entrenched views of their times.

Driesch's case also illustrates the point that in science, at least, not all mavericks are forced to function outside the standard academic framework. Driesch, like many others profiled in this volume, worked in prestigious academic institutions. This was in part a result of his analytical abilities. His defense of vitalism was argued from the point of analytical philosophy. It was not without its fatal flaws, but his presentation was clearly within the vein of acceptable philosophical argument. He may have been an idealist, but he was not a mystic. He was also an empiricist to the extent that he never lost sight of the biological phenomena that his vitalism was meant to explain. He was a considerate colleague and friend who challenged ideas, not individuals.[50] He thus retained the respect of his biological and philosophical colleagues even when they disagreed vehemently with his conclusions.

NOTES

1. Biographical information about Driesch in English is sparse. Much of the brief material presented in this essay is drawn from Jane Oppenheimer, "Driesch, Hans," in *Dictionary of Scientific Biography,* ed. Charles Gillespie (New York: Scribners, 1971), 4:186–189. More scientific detail can be found in Curt Herbst, "Hans Driesch als experimenteller und theoretischer Biologe," *Wilhelm Roux Archiv für Entwicklungsmechanik der Organismen* 141 (1941–1942): 111–153. More personal information is contained in Margarete

Driesch, "Das Leben von Hans Driesch," in *Hans Driesch: Persönlichkeit und Bedeutung für Biologie und Philosophie von Heute,* ed. A. Wenazel (Basel: Ernst Reinhardt, 1951), 1–20.

2. Hans Driesch and T. H. Morgan, "Zur Analysis der ersten Entwicklungsstadien des Ctenophoreneies, I. Von der Entwicklung einzelner Ctenophorenblastomeren. II. Von der Entwicklung ungefurtchter Eier mit Protoplasmadefekten," *Wilhelm Roux Archiv für Entwicklungsmechanik der Organismen* 2 (1895): 204–215; 216–224.

3. In Germany in the nineteenth and early twentieth centuries the process of formally joining a university faculty was known as *Habilitation* and involved, among other activities, a formal lecture, the *Habilitationsrede,* before the assembled faculty.

4. Ernst Haeckel, *Die Radiolarien (Rhizopoda radiaria): Eine Monographie,* 3 vols. (Berlin: Georg Reimer, 1862–1868); Ernst Haeckel, *Generelle Morphologie der Organismen: Allgemeine Grundzüge der organischen Formen-Wissenschaft, mechanische begrünndet durch die von Charles Darwin reformirte Descendenz-Theorie,* 2 vols. (Berlin: Georg Reimer, 1866); for the *Challenger* reports, see, e.g., "Report on the Deep-Sea Keratosa: The Scientific Results of the Voyage of H.M.S. Challenger During the Years 1873–1876," *Zoology* 32 (1889): 1–92. In all, Haeckel wrote four reports based on material from the *Challenger* expedition.

5. For example, there had been a long-standing debate as to whether a group of arthropods known as the sea-spiders (Pycnogonids) were more closely related to the Crustacea (lobsters, crabs) or to the Arachnids (true spiders). Various morphologists such as Anton Dohrn, founder of the Naples Zoological Station, had struggled with this issue with no clear-cut conclusions.

6. Hans Driesch, "Tektonische Studien and Hydroidpolypen," *Jenische Zeitschrift für Wissenschaft* 24 (1890): 189–226.

7. The phrase comes originally from an 1858 essay by Rudolf Virchhow, but it was made more relevant for the twentieth century by German-born U.S. émigré Jacques Loeb in his book *The Mechanistic Conception of Life* (Chicago: University of Chicago Press, 1912).

8. Wilhelm Roux, "Beiträge zur Entiwcklungsmechanik des Embryo: Über die künstliche Hervorbringung halber Embryonen durch Zerstörung einer der beiden ersten Furchtungskugeln, sowie über die Nachentwickelung (Postgeneration) der fehlenden Körperhälfte," *Virchow's Archiv für pathologische Anatomie und Physiologie und klinische Medezin* 114 (1888): 113–153, translated by Hans Laufer and reprinted in Benjamin Willier and Jane Oppenheimer, eds., *Foundations of Experimental Embryology,* 2d ed. (New York: Hafner, 1974), 2–37.

9. Although it was not completely clear in the 1890s that chromosomes were the material bearers of hereditary particles, that view was gaining ground among embryologists and cytologists alike. The so-called chromo-

some theory of heredity would garner increasing support through the first decade of the twentieth century by the work of Theodor Boveri (1901), Nettie M. Stevens and Edmund Beecher Wilson (1905), and T. H. Morgan and his group from 1910 onward. See Garland E. Allen, *Thomas Hunt Morgan: The Man and His Science* (Princeton: Princeton University Press, 1978): 129–140.

10. Hans Driesch, "Entwicklungsmechanische Studien. I. Der Werth der beiden ersten Furchungszellen in der Echinodermementwickluung: Experimentalle Erzeugung von Theil-und Doppelbildungen," *Zeitschrift für wissenschaftliche Zoologie* 53 (1891): 160–178. All told, Driesch published ten papers under the title of "Entwicklungsmechanische Studien" between 1891 and 1893.

11. Oppenheimer, "Driesch, Hans," 187.

12. Hans Driesch, *Analytische Theorie der organischen Entwicklung* (Leipzig: Wilhelm Engelmann, 1894).

13. Ibid., 32. (Translation of portions by Viktor Hamburger; unless otherwise noted, all translations of quotations from the *Analytisch Theorie* are from this source.)

14. Ibid., 79–84.

15. Frederick Churchill, "From Machine-Theory to Entelechy: Two Studies in Developmental Teleology," *Journal of the History of Biology* (1969): 165–185. The quotation appears on pp. 169–170.

16. Driesch, *Analytische Theorie,* 129; quoted in Churchill, 174.

17. Driesch, *Analytische Theorie,* 31; quoted in Churchill, 174.

18. Note that Driesch, like many of his contemporaries (for example, T. H. Morgan) also opposed the Darwinian theory of natural selection on similar grounds—namely, its reliance on chance events to produce an array of wonderfully adaptive characteristics in a species. With all its well-coordinated adult and embryonic traits, a species could not possibly turn into another equally well-adapted species by passing through a series of imperfect stages. Imperfect eggs could not even reach maturity, and hence some more overarching guidance was required.

19. Hans Driesch, "Die Lokalisation morphogenetischer Vergänge: Ein Beweis vitalischen Geschehens," *Roux Archiv für Entwicklungsmechanik der Organismen* 8 (1899): 35–111. See also Driesch, "Zur Analysis der Potenzen embryonaler Organisation," *Roux Archiv für Entwicklungsmechanik der Organismen* 2 (1895–1896): 169–203.

20. Anne Harrington. *Reenchanted Science: Holism in German Culture from Wilhelm II to Hitler* (Princeton: Princeton University Press, 1996), 51.

21. *A Dictionary of Philosophy,* ed. M. Rosenthal and P. Yudin (Moscow: Progress, 1967), 142.

22. Hilde Hein, "The Endurance of the Mechanism-Vitalism Controversy," *Journal of the History of Biology* 5 (1972): 159–188; see especially p. 170.

23. Hans Driesch, *The Problem of Individuality* (London: Macmillan, 1914), 35.

24. Hans Driesch, *The Science and Philosophy of the Organism* (London: Adam and Charles Black, 1908), 2 vols.

25. Driesch, *Problem of Individuality.*

26. Loeb had claimed, for example, that phototropic insects were "photometric machines" enslaved to the light. See Jacques Loeb, *The Mechanistic Conception of Life* (Cambridge: Harvard University Press, 1964 [1912]), Chapter 2, "The Significance of Tropisms for Psychology," 41.

27. Driesch, *Problem of Individuality,* 18.

28. Ibid., 19.

29. Ibid., 25*ff.*

30. Ibid., 19.

31. Ibid., 80.

32. Harrington, *Reenchanted Science,* chap. 2.

33. Quoted in Daniel Pick, *War Machine: The Rationalization of Slaughter in the Modern Age* (New Haven: Yale University Press, 1993), 213. Pick's discussion of Reich was noted in Harrington, *Reenchanted Science,* 189.

34. Harrington, *Reenchanted Science,* 41–44.

35. Ibid., 52.

36. Ibid., 53.

37. Andy Clark, *Being There: Putting Brain, Body and World Together* (Cambridge: MIT Press, 1997), 171.

38. Harrington, *Reenchanted Science,* chap. 1.

39. Ibid., 53.

40. Ibid.

41. Jacques Loeb, *The Organism as a Whole* (New York: Putnam, 1916).

42. Morgan to Driesch, January 30, 1909, quoted at length in Allen, *Thomas Hunt Morgan,* 323.

43. Driesch himself cited Haldane's objection in the second lecture of *Problem of Individuality,* 21. See Haldane's *Mechanism, Life and Personality* (New York: Dutton, 1913).

44. J. B. S. Haldane, *The Philosophy of a Biologist* (Oxford: Oxford University Press, 1935), 38.

45. Joseph Needham, *Biochemistry and Morphogenesis* (Cambridge: Cambridge University Press, 1942), 119–124.

46. Harrington, *Reenchanted Science,* 191–192.

47. Ibid., 188–193.

48. Ibid.

49. Charles Kingsley, *Health and Education* (London: W. Isbister, 1874).

50. In his memoirs Driesch recounts a meeting with Roux at the 1901 International Congress of Zoologists in Berlin, where they "often laughed at [their] 'battles' and the peace terms that ended it; a token that the most sharp

theoretical differences do not have to spoil human relations, as long as they derive from honest convictions." Driesch, *Lebenserrinerungen: Aufzeichnungen eines Forschers und Denkers in entscheidender Zeit* (Munich: E. Reinhardt, 1951), 97 (author's translation).

FURTHER READING

Churchill, Frederick. "From Machine-Theory to Entelechy: Two Studies in Developmental Teleology," *Journal of the History of Biology* 2 (1969): 165–185.

Driesch, Hans. *The Science and Philosophy of the Organism: The Gifford Lectures Delivered Before the University of Aberdeen in the Years 1907–8,* 2 vols. (London: Adams and Charles Black, 1908).

———. *The Problem of Individuality: Lectures Delivered Before the University of London in October, 1913* (London: Macmillan, 1914).

Hein, Hilde. "The Endurance of the Mechanism-Vitalism Controversy," *Journal of the History of Biology* 5 (1972): 159–188.

Mocek, Reinhard. *Wilhelm Roux—Hans Driesch: Zur Geschichte der Entwicklungsphysiologie der Tiere* (Jena: Fischer, 1974).

———. *Der Werdende Form: Eine Geschichte der Kausalen Morphologie* (Marburg an der Lahn: Basilisken-Presse, 1998).

Oppenheimer, Jane. "Driesch, Hans," *Dictionary of Scientific Biography* (New York: Charles Scribner's Sons, 1971), 4:186–189.

CHAPTER FOUR

Wilhelm Johannsen:
A Rebel or a Diehard?

RAPHAEL FALK

> Is the growth of science essentially so slow and so continuous that
> our attention is attracted only by the sudden showy change, which,
> like the bursting of a chrysalis, is merely the sequel to something of
> more importance which went before? Or, does a particular piece of
> work . . . have a value per se which transcends the others completely?
> Probably both questions should have affirmative answers.

Four major concepts shaped the science of genetics in the past one
hundred and fifty years: Mendel's notion of discrete pairs of *Faktoren* for characters, which segregate independent of other pairs in
the gametes and meet upon fertilization; Johannsen's discrimination
between the level of appearance and the level of inheritance of traits;
Morgan and his students' chromosomal theory of inheritance; and
Watson and Crick's physicochemical model of the architecture of DNA
molecules.

Wilhelm Johannsen's analysis of the notion of inheritance challenged the perspective that had been expounded in the first decade of
the twentieth century, following the so-called rediscovery of Mendel's
work. Johannsen's insight allowed him to reformulate the relation
between empirical, observed traits and the inferred Mendelian *Faktoren*. According to Johanssen, unit characters, if properly chosen, were
nothing but good empirical "markers" for the factors. Thus the ground
was laid for the bottom-up experimental approach to the study of the
mechanics of heredity as articulated in the chromosomal theory of
inheritance and the study of population genetics that culminated in
the New Synthesis, on one hand. On the other, the foundation was also
provided for a top-down approach to the analysis of the hereditary
input to development and behavior. In short, Johannsen led the way in

both relieving genetics from its early tendency toward paradigmatic reductionism and, paradoxically, in the later adoption of precisely such an extreme reductionist "genocentricity."

Wilhelm Johannsen of Denmark (1857–1927), a pharmacologist by training, was appointed lecturer in plant physiology at the Royal Veterinary and Agricultural University in Copenhagen in 1892 and for the last twenty-five years of his life served—without any formal university degree—as professor of plant physiology at the University of Copenhagen.[1] Johannsen's work is primarily experimental, and he was deeply committed to mathematical analysis of data. Unlike the early Mendelians, he was interested in the plants' quantitatively varying properties rather than in well-defined binary unit characters. He was influenced in particular by Galton's rule of ancestral inheritance, as revealed by the effect of selection in normally distributed populations. Empirically speaking, the mean of the progeny of selected parents regressed toward the mean of the population, rather than corresponding to that of the selected parents. Galton's regression coefficient, which quantitatively expressed these observations, corresponded to Darwin's notion of evolution by slow, gradual, and continuous change of a population's inheritance that accumulated over the generations. This contrasted with notions of the need for abrupt, stepwise evolutionary progress by sports, independent of ancestral inheritance or, as formulated by Hugo de Vries, by mutations.

Johannsen was also aware of the breeding programs operating in the Svalöf Experimental Station in nearby Sweden. The station's head, Nils Hjalmar Nilsson, juxtaposed already in the early 1890s the two prevalent strategies of selection for the breeding of crops: the painstaking step-by-step mass selection program of the best plots year after year, as indicated by the Darwinian notion of evolution and that of picking up outstanding individual plants—sports—and breeding from them. The latter practice was based on the pedigree method, elaborated by the French breeder Louis Vilmorin, and on de Vries's notion of evolution by mutational jumps. Nilsson, who believed in the constancy of biological types, a principle according to which "continued one-sided selection of variants does *not* lead to a gradual replacement of the type," voted for the latter strategy.[2] Consequently, Johannsen was caught between Nilsson's belief in the Linnaean notion of essential types, which may be changed by discontinuous saltations, and Galton's law of ancestral inheritance, which suggested that progeny inherit half of their characteristics from each of their parents, a

Figure 4.1. Wilhelm Johannsen. Courtesy of the Royal
Library, Copenhagen.

quarter from each of their grandparents, an eighth from each of their
great-grandparents, and so on. His solution was that of conceiving of
two levels of variation in populations: the observable, empiric level
and the deduced, conceptual level. Continuous changes at the superficial level did not refute the notion of the constancy of types at the
deeper level.

TYPES OR TRAITS?

Like Mendel, Johannsen was interested in a quantitative, numerical
analysis of the problem of inheritance, and both persons relied on
breeders' experience. But unlike Mendel, who was concerned with the
inheritance of specific, individual characters, Johannsen was primarily

interested in the general biological aspect of inheritance of the species as a type.[3] He maintained that the varieties and subspecies of the Linnaean species, which show typical characters from generation to generation, were "systematic units," or "constant form-types." Darwinism, like neo-Lamarckism, in Johannsen's view, suffered from the idea of continuous variation in heredity. He rejected the Darwinian conception of species as continuously changing entities and "reconciled evolution and unchangeable (stable) species by letting the elementary species change discontinuously. Through such sudden changes, mutations, new elementary species appear spontaneously."[4]

Johannsen started his critical experiments in 1900 with the purchase of eight kilograms of bean seeds. He followed the efficiency of selection of two quantitative characters: seed weight and seed circumference. Beans reproduce by self-fertilization; consequently, any changes achieved via selection for a character in successive generations would indicate the breeding potential of the examined sample, rather than the outcome of some possible hybridization with foreign types. To start with, both characters showed normal distribution about a mean. When he selected the twenty-five heaviest and the twenty-five lightest beans as seeds for the next generation, their progeny in 1901 showed, as expected, partial regression to the mean of the population. The progeny of nineteen of the lightest seeds (and in a parallel manner, the progeny of the heaviest seeds) were individually tracked further in the following years. After two generations of self-fertilization there was no more regression to the mean of the original population, but there was practically full regression to the mean of the individually selected seeds: "pure lines" were thus established, each being characterized by its mean, which is maintained, in spite of continued selection, with complete regression to the mean of the line.

These experimental results convinced Johannsen that the Linnaean species concept should now be conceived more specifically, that is, in terms of a "geno-species" (*Antlægsart*). A geno-species, being a type that includes all individuals with the same hereditary makeup, is a *geno-type*. Ordinarily, such types could not be easily identified or maintained because organisms interbreed. They were, however, identified by inbreeding among self-fertilizing beans. The establishment of pure lines provided Johannsen with the experimental evidence for his notion of essential types. He concluded that the empirical, statistical mean of a population was not necessarily identical with the notion of the biological type.[5] The observed mean is a superficial appearance statistic, a *phenotypic* variable that must be conceived as distinct from *something* inherent in the biological type, or the genotype.[6]

This was a conceptual breakthrough. For a decade the young science of genetics had developed under the spell of de Vries's notion of the unit character,[7] as introduced in the opening sentences of de Vries's rediscovery paper: "According to pangenesis the total character of a plant is built up of distinct units. These so-called elements of the species, or its elementary characters, are conceived of as tied to bearers of matter, a special form of material bearer corresponding to each individual character."[8] In one stroke Johannsen severed the Gordian knot tying the character to its hereditary factor.[9] Concurrently, by deriving the term *gene* from his genotype, Johannsen actually provided new legitimization to heredity of particulate entities, rather than to the eighteenth- and nineteenth-century belief in heredity as a force (of a certain kind of formative matter).[10] As the historian of biology Nils Roll-Hansen noted, however, "[a]s is the case with many innovators in science, some of Johannsen's ideas are surprisingly archaic. . . . Johannsen's explicit endorsement of an essentialist Linnaean species concept is especially striking."[11]

Mendel's experimental design was based on the logic of physical and chemical reductionist or "bottom up" research; Johannsen's conception was that of "top down" biologists who considered living organisms essentially different from nonliving entities. Mendel judiciously selected individual unit characters that had proved in preliminary experiments to each provide two distinct alternative appearances: yellow or green, smooth or wrinkled, tall or short—a procedure that inadvertently allowed Mendel to avoid the need to distinguish between genotype and phenotype. Johannsen judiciously analyzed quantitative characters that do not assume distinct or discrete unit characters—a procedure that forced him to discern hereditary inputs from nonhereditary inputs. Although no rules other than those applying to the nonliving world were needed, the organisms were the entities of Johannsen's reference. Actually, he retained a conception of an equilibrium theory similar to embryologists' notions of form and function at the time, according to which the great variability following hybridization was due to "a destruction of the equilibrium in the inner constitution" of the progeny.[12]

When Johannsen divided a population's variance into a continuous and a discontinuous component, his top-down conception indicated that the genotype is the cause (an Aristotelian formal cause rather than an efficient cause) of the phenotype of individual organisms. In this sense his conception retained the spirit of Galton's "stirp," which determines the specific development of the individual and is transmitted unchanged from generation to generation: "The genotype is a theoretical entity somewhat like the ideal Aristotelian form, belonging to the organism as a whole."[13]

A direct consequence of this organismic conception was Johannsen's rejection of the chromosomal theory of inheritance. More significant to us, this approach made the unit characters superfluous. Johannsen consistently rejected the reductionist unit character, and to the end of his career he continued to talk of genotypical (and phenotypical) variation and remained reserved about the meaning of the concept of the gene.[14]

PHENOTYPES AND GENOTYPES

The historian of biology Fred Churchill referred to the conceptual distinction between phenotypes and genotypes, fashioned in 1909 by Johannsen, as "one of the major accomplishments of the history of biology." He quoted the geneticist Leslie C. Dunn as stating that "Johannsen's place in the history of biology may come to be seen as a bridge over which nineteenth-century ideas of heredity and evolution passed to be incorporated, after critical purging, into modern genetics and evolutionary biology."[15] We may gain an appreciation of how difficult it was for Johannsen's input to become established from Jacques Loeb's introduction to *The Organism as a Whole,* describing heredity from the standpoint of a physiologist: "When, however, the biologist is confronted with the fact that in the organism the parts are so adapted to each other as to give rise to a harmonious whole . . . doubts as to the adequacy of a purely physico-chemical viewpoint in biology may arise. The difficulties besetting the biologist in this problem have been rather increased than diminished by *the discovery of Mendelian heredity, according to which each character is transmitted independently of any other character.*"[16] In 1916 Loeb had not yet assimilated Johannsen's insight that the characters themselves are not transmitted.

In his historic review Dunn commented that the failure to appreciate that the phenotype depends on the interaction of many genes with each other and the environment "had led to the retention of the notion of unit characters which plagued genetics for two decades and doubtless delayed the clarification of some of its basic concepts."[17]

Nonetheless, eventually, three major aspects of the science of genetics were profoundly overhauled with the introduction of the conceptions of the genotype and the phenotype: that of the stability of Mendelian factors, that of continuous Darwinian evolution, and that of Weismann's preformationism.

STABILITY OF MENDELIAN FACTORS

Of the three claims, that for the nonuniversality of Mendel's laws of the independence of the hereditary *Faktoren* in hybrids was the easiest to dispose of. This point is best demonstrated in the American biologist William Castle's experiments with selection of the unit character for hooded in the progeny of rat hybrids, in which the hooded and the nonhooded color patterns did not maintain their segregation and independence of function.[18]

As Castle wrote in 1919: "My own experimental studies of heredity, begun in 1902, early led me to observe characters which were unmistakably *changed* by crosses and so I have for many years advocated the view that the gametes are not pure . . . and I was in consequence led to adopt the hypothesis that unit-characters are 'inconstant' in varying degrees."[19] As late as 1914 Castle had maintained that the chief genetic factor concerned may be undergoing quantitative variation. This view had been vehemently rejected by Hermann J. Muller "on the ground that this explanation is not 'in harmony with the results of Johannsen and other investigators.'"[20] Castle did not understand how "the experiments of Johannsen have any direct bearing on the case since no single *Mendelizing unit-factor* was demonstrated in that connection." As he saw it, Muller "might with propriety cite the bean work as bearing on the interpretation of the inheritance of body size in animals . . . since both involve blending inheritance. But neither of these cases has any direct bearing on the question of unit-character constancy, since in neither case has a unit-character, either constant or inconstant, been shown to exist."[21]

Eventually Castle's colleague at Harvard University's Bussy Institute, Edward M. East, convinced him that Johannsen's insight administered the final blow to the concept of unit character.[22] Unit character is an efficient hereditary unit as long as it is not modifiable. But once we accept that not only must environmental conditions be standardized and kept constant in order for the unit character to be discretely definable but also that other internal conditions and traits affect it, the notion of the unit character as defining a hereditary factor loses its meaning. East wrote:

> I believe that we may describe our results simply and accurately by holding that unit factors produce identical ontogenetic expressions under identical or similar conditions. If under identical conditions the expression *is* different, then a new standard, a new unit, must be assumed; . . . To be sure there are numerous changes of expression of characters when external and internal conditions are not so uniform . . . these changes can all be described adequately and simply by ascribing them to modifying conditions both external and internal. When external we recognize their usual effect in what we called non-inherited fluctuations, when internal we recognize their cause in other gametic factors inherited independently of the primary factor but modifying its reaction during development.[23]

Remember that Johannsen's approach was a top-down one, and to the extent that he—or, for that matter, East—viewed characters as units, these implicitly embraced external and internal effecting factors. As East pointed out with respect to that quotation: "This is a physiological conception of heredity, as it recognizes the great cooperation between factors during development."[24] Castle's conclusions with respect to hooded (and piebald) rats made sense only as long as he regarded the traits from a bottom-up perspective, as unit characters. But because they existed not only in an external environment but also in the internal environment of the organism, the unit character beyond its instrumental level became meaningless.

East summarized his analysis of Castle's argument:

> Taking into consideration all the facts, no one can deny that they are well described by terminology which requires hypothetical descriptive segregating units as represented by the term factors. What then is the object of having the units vary at will? There is then no value to the unit, the unit itself being only an assumption. It is the expressed character that is seen to vary; and if one can describe these facts by the use of hypothetical units theoretically fixed but influenced by environment and by other units, simplicity of description is gained. If, however, one creates a hypothetical unit by which to describe phenomena and this unit varies, he really has no basis for description.[25]

East advanced a strictly instrumental approach to the concept of the Mendelian factors: "a factor, not being a biological reality but a descriptive term, must be fixed and unchangeable." He went on: "Expressed in Johannsen's words, the basis of the modern conception of heredity is: 'Personal qualities are the *reactions of the gametes joining to form a zygote;* but the nature of the gametes is not determined by the personal qualities of the parents or ancestors in question.'"[26]

CONTINUOUS DARWINIAN EVOLUTION

The most vociferous dispute to which Johannsen contributed was that concerning the quality of the variation available for natural selection, or the role of continuous and discontinuous variation in evolution. Although the British statistician Udny Yule had shown in 1902 that the two were compatible,[27] the biometricians, notably Karl Pearson and W. F. Raphael Weldon, did not accept the notion of discontinuous heredity. Thus, according to the historian of biology Garland Allen, "[t]he most significant effect of Johannsen's 1903 work was to reinforce what the Mendelians and other critics of Darwin had long claimed, that natural selection was powerless to produce new species."[28] In spite of his indebtedness to Galton, Johannsen soon found himself criticizing the biometric school precisely because he fully assimilated the statistical meaning of their conceptions. Whereas Weldon stated that "[i]t cannot be too strongly urged that the problem of animal evolution is essentially a statistical problem," and Pearson insisted that "the solutions to these problems are in the first place statistical, and in the second place statistical, and only in the third place biological,"[29] Johannsen stated at the outset of his *Elemente der Exakten Erblichkeitslehre* that "we must pursue the science of heredity *with* but not *as* mathematics."[30]

Johannsen showed that the question of whether selection acts on continuous or discontinuous variation was misguided because conceptually the variation that Pearson was interested in was not that which the Mendelians were interested in.[31] Although phenotypic variation was the only variation that could be empirically studied, it was merely one observable expression of a deeper, conceptually significant genotypic variation. Variation of type did occur even in pure lines owing to environmentally induced "fluctuations" as well as inadequacies in measuring techniques. These fluctuations, indeed, follow the Quetelet-Galton law of binomial distribution. Mutations, on the other hand, are defined as the "suddenly occurring larger or smaller deviations" from the parent type and are not predictable by the rules of binomial distribution.[32] Thus Johannsen differentiated

(1) a variation of the progeny from the parents' "type" that does not undergo the Galtonian regression to the mean from (2) fluctuating variations in the parents' type that regress (completely) toward the mean. In *pure lines* all variation is fluctuating, and regression to the mean is complete. Variation about the mean of a *population* may be due also to variability of types, or genotypes, in the population. It is only this latter variation that is inherited and thus was amenable to evolutionarily significant selection.

The historian Roll-Hansen suggests that until 1902 Johannsen believed that "maybe the borderline between mutation and individual variation is not quite as sharp as Bateson and de Vries assume[d]" and that Johannsen, too, accepted the inheritance of continuous variations. But by 1903 he realized that reproduction by self-fertilization provided a very special case to follow the individual rather than the species as the essential type—the *Formtypus*—over the course of successive generations. Thus, the law of regression correctly described the behavior of many populations because they contained a mixture of hereditary types.[33] Consequently, in 1905 he introduced the Danish terms *Livs-type* (life-type) and *Tal-type* (number-type), which in 1909 he translated into *genotype* and *phenotype* in the first edition of the *Elemente*.[34] He wrote, "The 'type' in Quetelet's sense is only a phenomenon of superficial nature which can be deceptive. . . . Therefore I have designated a statistical, i.e., purely descriptively established type, as an 'appearance type' [*Erscheinungstypus,*] a phenotype. . . . Through the term phenotype the necessary reservation is made, that the appearance itself permits no further conclusion to be drawn. A given phenotype may be the expression of a biological unit, but it does not need to be."[35] The stable type of the pure line provided a stable equilibrium. It is significant that, although Johannsen's insight actually laid the foundation for the notion of the concealed genetic variation of populations, Johannsen did not consider contributions to variation that were due to segregation of factors besides that of mutations. He considered only the rare sporadic "sports" that were due to new mutations in existing factors to provide raw material for evolution, and he ignored possible contributions of existing variation of factors in populations to evolution. It appears that in the early 1900s his understanding of the Mendelian segregation of characters was inadequate; it was also largely irrelevant to his top-down perspective.[36]

Diagrams showing five different pure lines of beans and a "population" formed by their union are shown in figure 4.2. In each case the beans enclosed in glass tubes are marshaled in equidistant classes of length; identical classes are superposed. The pure lines show transgressive fluctuation: it is mostly impossible to state by simple inspection of any individual bean the line to which it belongs. The fluctuations about the average length (their phenotype) within the pure lines as well as in the mixed population show no characteristic difference.

The strong emotions that accompanied the dispute about the legitimacy of analyzing the observed statistical variation into conceived biological components led to radicalization of positions. Garland Allen suggested that there was continuity between Johannsen's distinction between genotype and phenotype and August Weismann's hypothesis of distinct germ plasm and somatoplasm, which emphasized the disjunction of the impacts of heredity and environment, thus placing "one more nail . . . in the coffin of neo-Lamarckism."[37] Another historian of science, Jan Sapp, on the other hand, believed that "the distinction between the genotype and the phenotype . . . served as a polemic against descriptive, speculative, and morphological approaches to the study of heredity—categories within which Weismann's theory itself proliferated."[38] Johannsen, however, was quite explicit in rejecting attempts to relate his conceptions to those of Galton and Weismann: the conception of the genotype, even if initiated by Galton and Weismann, was completely revised, and "[o]f all the Weismannian armory of notions and categories it may use nothing."[39] A decade later Johannsen was yet more explicit in distancing himself from attempts to reduce his top-down notion of the genotype into genes, though he admitted, "originally I was somewhat possessed with the antiquated morphological spirit in Galton's, Weismann's and Mendel's viewpoints."[40]

THE CONCEPTION OF TRANSMISSION VERSUS THE CONCEPTION OF GENOTYPE

Perhaps the most profound aspect of Johannsen's insight was the physiological and embryological one. Johannsen's segregation of genotype and phenotype made the distinction between preformationists and epigenesists irrelevant. Heredity could now be conceived as a process, a *production:* it was competent to investigate the causes of development and function, rather than merely those of transmission of factors that unfold to become traits.

Preformationism was prevalent among geneticists of the period. One of the avid opponents of preformationism for whom Johannsen's notion provided the way out was Thomas Hunt Morgan:

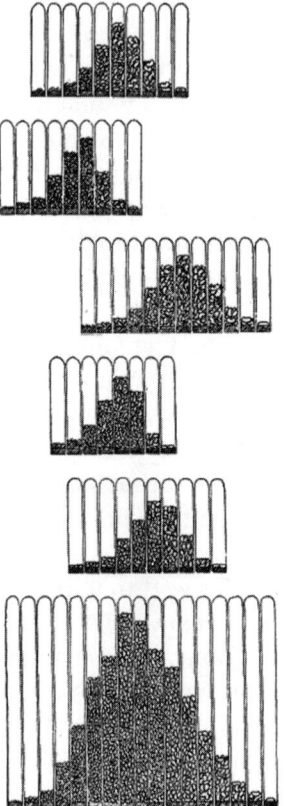

Figure 4.2. Johannsen's presentation of nonhereditary fluctuations within genotypically different "pure lines" and the phenotypic variation obtained in the combined "population." From Wilhelm Johannsen, "The Genotype Conception of Heredity," *American Naturalist* 45, no. 531 (1911): 129–159, 136.

[W]e now realize that it is not the characters that are transmitted to the child from the body of the parent. . . . The modern literature of development and heredity is permeated through and through by two contending and contrasting views as to how the germ produces the characters of the individual. One school looks upon the eggs and sperm as containing *samples* or *particles* of all the characters of the species, race, line or even the individual. This view I shall speak of as the *particulate theory of development.*

The other school interprets the egg or sperm as a kind of material capable of progressing in definite ways as it presses through a series of stages that we call its development. I shall call this view the *theory of physico-chemical reaction,* or briefly the reaction theory. . . .

The *modern* theory of particulate inheritance goes back no further than the discovery that the sperm transmits equally with the egg the characters of the race; . . . Around these simple statements the whole edifice has been erected. We owe to Weismann more than to any other biologist, the peculiar trend that this speculation has followed.[41]

Morgan felt a "distinct disinclination to reduce the problem of development to the action of specific particles in the chromosome," at least in the sense of the Roux-Weismann assumption, which argued for "nicely separating at each division the different kinds of materials of which the chromosomes are composed."[42] Like Johannsen, Morgan was skeptical with regard to the meaning of the Mendelian factors: "[T]hose not engaged in the immediate work itself have, I believe, often been misled in regard to the meaning attached to the term factor, and by the assumed relation between a factor and a unit character. The confusion is due to a tendency . . . to speak of a unit character as the product of a particular unit factor acting alone."[43] Morgan ascribed this confusion to the "attempt to impute to the factorial hypothesis the same interpretation that Weismann made use of in his theory of determinants." Weismann "identified each character of the organism as the product of a special determinant. The factorial hypothesis assumes only that the cell in one case is different from the cell in the other, the difference relating, it is true, to some part, but the character produced may be the result of the whole or much of the cell, and not of one part alone."[44]

In his interpretation of the dominance and recessivity of Mendelian characters as the presence and absence of the corresponding *Faktoren,* the English geneticist William Bateson explicitly referred to them as preformed determinants. With the introduction of Johannsen's distinction between the phenotype and the genotype, Morgan conceived that "[t]he presence and absence *system of nomenclature . . .* has till the present time justified itself, when properly interpreted, by its usefulness." But once Johannsen's distinction was adopted, it became "unwise to commit ourselves any longer to a view that a recessive character is necessarily the result of a loss from the germ-cell."[45]

Nonetheless, by 1910 Johannsen's claim about the stability of the genotype in pure lines was commonly accepted by geneticists. But as Roll-Hansen noted, "by then a new rival to the genotype theory was gathering momentum, the chromosome theory. In this conflict Johannsen and his genotype theory became the losing party."[46]

When Morgan adopted and elaborated the chromosomal theory of inheritance, he tried to maintain a top-down approach to it and to resist the tendency of his co-workers, notably Muller, to conceive of genes as its material atomic components.[47] As a matter of fact, Johannsen warned Morgan against the temptation of the "morphological spirit" to conceive of genes as particles with a certain structure.[48] But it was to no avail.

For Johannsen, the question of chromosomes as the presumed "bearers of hereditary qualities" seemed to be an idle one: "I am unable to see any reason for localizing 'the factors of heredity' (i.e., the genotypical constitution) in the nuclei. The organism is in its totality penetrated and stamped by its genotype constitution. All living parts of the individual are potentially equivalent as to genotype constitution."[49] Methodological constraints, however, as well as the empirical results of detailed linkage maps upheld by cytogenetics observations, tipped the balance in favor of the bottom-up notion of chromosomes as carriers of discrete genetic determinants.[50] Whether instrumental or material, the genes' role as stable, autonomous determinants juxtaposed them against environmental factors. Thus Johannsen's notion of the organismal genotype as only *one* of the contributors to the phenotype was passed over and the foundations of genocentricity, more in line with what was conceived to be Weismann's opposition of germ and soma and Galton's opposition of nature and nurture, were consolidated. Johannsen wrote:

> My term "gene" was introduced and generally accepted as a short and unprejudiced word for unit-factors. . . . From a physiological or chemico-biological standpoint . . . *there are no unit characters at all!* . . . We may in some way "dissect" the organism descriptively, using all the tricks of terminology as we please. But that is not allowed in genetical explanation. Here, in the present state of research, we have especially to do with such genotypical units as are separable, be it independently or in a more or less mutual linkage.[51]

Only in the third edition of his book, published in 1926, "after reviewing some of the arguments against regarding the chromosomes as primary structures, [did] Johannsen reluctantly confess that 'the be-

havior of the chromosomes . . . shows in many cases such an astonishing parallel with certain hereditary phenomena after hybridization, that one is well persuaded to ascribe to them a wholly special
significance.'"[52]

THE PARADOX OF THE GENOTYPE CONCEPTION

By the 1920s Johannsen's "genie" was out of the bottle. His distinction
between genotype and phenotype was well established. The chromosomal theory of inheritance of Morgan's school overcame preformationism.[53] Likewise, evolution was conceived as "inheritance *of*
variation" rather than "inheritance and variation" amenable to Fisher's
Fundamental Theorem of Natural Selection of variation of discrete,
discontinuous entities as the motor of Darwinian evolution.[54] Finally,
the constant stability of hereditary factors rather than that of their
products was formulated in Beadle and Tatum's "one gene, one enzyme" formulation.[55]

  It is, however, a major paradox that by the very elimination of the
unit character and the introduction of the genotype as the "something"
of inheritance behind phenotypic appearance, Johannsen provided a
framework for opposing genetics to environment, thus inadvertently
upholding the notion of nature versus nurture and spawning "genocentricity." In methodological terms this reductionism was most re-
warding. It culminated in Watson and Crick's model of the deoxyribonucleic acid molecule, the possible hereditary consequences of which
did not escape the notice of its authors.[56] It gave birth to maps showing
of intragenic linkage in minute detail, and such maps were found to be
colinear with the genes' protein products.[57] The climax of all this was
the determinist "Central Dogma" of gene action.[58] Yet in conceptual
terms this genocentric construct was bound to falter. By the end of
the twentieth century, the results of the Human Genome Project, with
its detailed mapping of genomes and their corresponding phenotypic
proteomes, finally overcame the notion of autonomous discrete structural "atoms of heredity," leading us back to the nineteenth-century
notion of epigenetic forces of the genotype as components of a system
rather than the notion of particulate genes.[59]

  In retrospect, from the ashes of Johannsen's diehard notion of
extending the Linnaean hierarchical classification down to the essence of the individual organism shaped by hereditary forces emerged
the phoenix that revolutionized the young science of genetics. The

introduction of the concept of the genotype as distinct from the phenotype relieved the science of heredity of the determinist unit character and allowed genetics to envisage its own entities of reference.
Though for a long time it was problematically reductionistic, genetics
became an independent discipline that could address from its specific
perspective problems of evolution, physiology, development, and behavior. Thus, near the centennial of Johannsen's *Elemente der Exakten Erblichkeitslehre* we may again fully appreciate the fundamental
transformation introduced by this diehard rebel.

NOTES

Epigraph: E. M. East, "Mendel and His Contemporaries," *Scientific Monthly*
6 (1923): 225–237.

1. F. B. Churchill, "William Johannsen and the Genotype Concept," *Jour-
nal of the History of Biology* 7 (1974): 5–30; N. Roll-Hansen, "The Genotype
Theory of Wilhelm Johannsen and Its Relation to Plant Breeding and the
Study of Evolution," *Centaurus* 22 (1978): 201–235.

2. Roll-Hansen, "Genotype Theory," 204–205.

3. C. Stern and E. Sherwood, *The Origin of Genetics: A Mendel Sourcebook*
(San Francisco: Freeman, 1966), 5: "Thus the study breaks up into just as many
experiments as there are constantly differing traits in the experimental plants."

4. Roll-Hansen, "Genotype Theory," 222.

5. W. Johannsen, "Concerning Heredity in Populations (*Über Erblichkeit
in Populationen und in reinen Linien.* 1903. Gustav Fischer, Jena)," in *Selected
Readings in Biology for Natural Sciences,* ed. the Staff of Natural Sciences 3
(Chicago: University of Chicago Press, 1955), 172–215.

6. W. Johannsen, *Elemente der exakten Erblichkeitslehre* (Jena: Gustav
Fischer, 1909), 113–128.

7. Quoted in L. C. Dunn, *A Short History of Genetics* (New York: McGraw
Hill, 1965), 99.

8. H. de Vries, "Das Spaltungsgesetz der Bastarde," *Berrichte der deut-
schen botanischen Gesellschaft* 18 (1900): 83–90, as quoted in Stern and Sherwood, *Origin of Genetics,* 107.

9. R. Falk, "The Gene—A Concept in Tension," in *The Concept of the
Gene in Development and Evolution: Historical and Epistemological Perspectives,* ed. Peter J. Beurton, Raphael Falk, and Hans-Jörg Rheinberger, Cambridge Studies in Philosophy and Biology (Cambridge: Cambridge University
Press, 2000), 317–348, 320.

10. See Jean Gayon, "From Measurement to Organization: A Philosophical
Scheme for the History of the Concept of Heredity," in *The Concept of the Gene
in Development and Evolution: Historical and Epistemological Perspectives,*

ed. Peter J. Beurton, Raphael Falk, and Hans-Jörg Rheinberger, Cambridge Studies in Philosophy and Biology (Cambridge: Cambridge University Press, 2000), 69–90.

11. Roll-Hansen, "Genotype Theory," 221.

12. E.g., E. S. Russell, *Form and Function* (Chicago: University of Chicago Press, 1916 [1982]); Roll-Hansen, "Genotype Theory," 207.

13. Roll-Hansen, "Genotype Theory," 224.

14. W. Johannsen, "Some Remarks About Units in Heredity," *Hereditas* 4 (1923): 133–141, 136–137. See also R. Falk, "What Is a Gene?" *Studies in the History and Philosophy of Science* 17 (1986): 133–173, 135–141.

15. Churchill, "William Johannsen and the Genotype Concept," 6.

16. Quoted in J. Sapp, "The Struggle for Authority in the Field of Heredity, 1900–1932: New Perspectives on the Rise of Genetics," *Journal of the History of Biology* 16 (1983): 311–342, 317, italics added.

17. Dunn, *Short History of Genetics,* 99.

18. W. E. Castle, *Heredity* (New York: Appleton, 1913), Chapter VII, 106–127.

19. W. E. Castle, "Piebald Rats and the Theory of Genes," *Proceedings of the National Academy of Science, Washington* 5 (1919): 126–130, 126.

20. W. E. Castle, "Mr. Muller on the Constancy of Mendelian Factors," *American Naturalist* 49 (1915): 37–42, 37, quoting H. J. Muller, "The Bearing of the Selection Experiments of Castle and Philips on the Variability of Genes," *American Naturalist* 48 (1914): 567–576.

21. Castle, "Mr. Muller," 37, italics added.

22. E. M. East, "The Mendelian Notation as a Description of Physiological Facts," *American Naturalist* 46 (1912): 633–655.

23. Ibid., 648–649.

24. Ibid., 649.

25. Ibid., 651. This passage was italicized in the original.

26. Ibid., 633–634.

27. U. Yule, "Mendel's Laws and Their Probable Relations to Intra-Racial Heredity," *New Phytologist* 1, nos. 9, 10 (1902): 193–207, 222–238.

28. Garland E. Allen, "Naturalists and Experimentalists: The Genotype and the Phenotype," in *Studies in the History of Biology,* ed. W. Coleman and C. Limoges (Baltimore: Johns Hopkins University Press, 1979), 179–209, 197.

29. W. B. Provine, *The Origins of Theoretical Population Genetics* (Chicago: University of Chicago Press, 191), 31, 51.

30. Johannsen, *Elemente der Exakten Erblichkeitslehre,* 2, as quoted in Churchill, "William Johannsen and the Genotype Concept," 8.

31. Allen, "Naturalists and Experimentalists."

32. Roll-Hansen, "Genotype Theory," 210.

33. Ibid., 211–212.

34. Ibid., 213.

35. Johannsen, *Elemente der exakten Erblichkeitslehre,* 124, (not pages 162–163, as quoted by Dunn, *Short History of Genetics,* 91–92).

36. Roll-Hansen, "Genotype Theory," 216.

37. Allen, "Naturalists and Experimentalists," 205.

38. Sapp, "Struggle for Authority," 326.

39. W. Johannsen, "The Genotype Conception of Heredity," *American Naturalist* 45, no. 531 (1911): 129–159, 132.

40. Johannsen, "Some Remarks About Units in Heredity," 136.

41. T. H. Morgan, "Chromosomes and Heredity," *American Naturalist* 44 (1910): 449–498, 449–452.

42. Ibid., 453.

43. T. H. Morgan, "Factors and Unit Characters in Mendelian Heredity," *American Naturalist* 47, no. 553 (1913): 5–16, 5.

44. Ibid., 9.

45. Ibid., 11.

46. Roll-Hansen, "Genotype Theory," 221.

47. See the agenda for his research project: H. J. Muller, "Variation Due to Change in the Individual Gene," *American Naturalist* 56 (1922): 32–50.

48. Johannsen, "Some Remarks About Units in Heredity"; Roll-Hansen, "Genotype Theory," 226.

49. Johannsen, "Genotype Conception of Heredity," 154.

50. See R. Falk and S. Schwartz, "Morgan's Hypothesis of the Genetic Control of Development," *Genetics* 134, no. 3 (1993): 671–674.

51. Johannsen, "Some Remarks About Units in Heredity," 136–137.

52. See Churchill, "William Johannsen and the Genotype Concept."

53. Morgan, "Chromosomes and Heredity," esp. 452–453.

54. See Muller, "Variation Due to Change in the Individual Gene," 35; R. A. Fisher, "The Correlation Between Relatives on the Supposition of Mendelian Inheritance," *Transactions of the Royal Society, Edinburgh* 52 (1918): 399–433: R. A. Fisher, *The Genetical Theory of Natural Selection* (Oxford: Clarendon, 1930).

55. G. W. Beadle and E. L. Tatum, "Genetic Control of Biochemical Reaction in *Neurospora,*" *Proceedings of the National Academy of Science, Washington* 27 (1941): 499–506.

56. J. D. Watson and F. H. C. Crick, "Molecular Structure of Nucleic Acids," *Nature* 171 (1953): 737–738.

57. S. Benzer, "The Elementary Units of Heredity," in *The Chemical Basis of Heredity,* ed. B. Glass and W. D. McElroy (Baltimore: Johns Hopkins University Press, 1957), 70–93; D. R. Helinski and C. Yanofsky, "Correspondence Between Genetic Data and the Position of Amino Acid Alteration in a Protein," *Proceedings of the National Academy of Science, Washington* 48 (1962): 173–183.

58. F. H. C. Crick, "On Protein Synthesis," in *Symposium of the Soci-*

ety for Experimental Biology, *The Biological Replication of Macromolecules* (Cambridge: Cambridge University Press, 1958), 138–163.

59. H. Pearson, "What Is a Gene?" *Nature* 441 (2006): 399–401; P. E. Griffiths and K. Stotz, "Genes in the Postgenomic Era," *Theoretical Medicine and Bioethics* 27 (2006): 499–521.

FURTHER READING

Allen, G. "Naturalists and Experimentalists: The Genotype and the Phenotype." In *Studies in the History of Biology,* ed. W. Coleman and C. Limoges (Baltimore: Johns Hopkins University Press, 1979), 3:179–209.

Lewontin, R. C. "Genotype and Phenotype." In *Keywords in Evolutionary Biology,* ed. E. F. Keller and E. A. Lloyd (Cambridge: Harvard University Press, 1992), 137–144.

CHAPTER FIVE

Raymond Arthur Dart:
The Man Who Unwillingly Ushered in a
Revolution in the Evolution of Humankind

PHILLIP V. TOBIAS

Unlike Eugene Dubois, who went to Java in 1888 with the avowed aim of finding a "missing link," Raymond Dart "had no sense of dedication to a search for human ancestors when coming unwillingly to South Africa early in 1923." Indeed, he went so far as to write in 1940: "I have striven ever since I was given my first piece of anatomical research work as a student in Sydney a quarter of a century ago to avoid both bones and mathematics. Circumstances thrust anthropology upon me after I had chosen to follow even more useless trails as a neurological embryologist."[1]

Although Dart's greatest heresies—which justify his inclusion in a work about iconoclastic biologists—were in the field of paleoanthropology, before he bent his footsteps in this direction he had expressed unorthodox opinions about other matters, especially the morphology of the nervous system. In the first paper that he presented to the Anatomical Society in London, in 1920, he disconcerted the company by his untoward ideas about the half-forgotten Froriep's ganglion. Within the year, Dart, with his fellow Australian Joseph Shellshear, had the temerity to challenge Wilhelm His's nineteenth-century doctrine of the cell to the effect that the neural tube and neural crest gave origin to all the nerve cells in the body. Their attempted disproof of the long-entrenched cell doctrine was one of the first of Dart's shocking and nonconformist notions. To these jejune adventures were added Dart's dabblings with the motor neuroblasts of the anterior horn of the spinal cord, the evolutionary complexity of the corpus striatum, the fate of the anterior end of the neural tube, and the trigeminal cranial nerve and its supposed relation to the cerebellum in the ancient Zeuglodont whales.

Figure 5.1. Raymond Dart. By courtesy
of the School of Anatomical Sciences,
University of the Witwatersrand,
Johannesburg.

It is no wonder that, before he was out of his twenties, Dart was
acquiring a reputation for spurning authority and for jumping to conclusions. Dart wrote much later of himself at that time: "Such a person,
I can see now in retrospect, was not only controversial, but upsetting
and potentially dangerous."[2] It was these tendencies toward heterodoxy that Dart took from his native Australia to England for a few
years, to the United States for a year, and to South Africa, far from
the mainsprings of the world's scholarly and scientific endeavours. It
raises the question, Is the rebelliousness of the rebel a flash in the pan,
a single opportunistic outbreak, or is there perhaps in some people
a lifelong tendency, virtually a personality trait, that expresses itself
intermittently, repeatedly, and serendipitously?

In this context, perhaps it was a quirky, ironic turn of the wheel
that Raymond Dart, who had been raised "in a devout Methodist and
Baptist family environment, sharing gladly also the fundamentalist

philosophy of Plymouth Brethren family friends," was to shake the world when he produced powerful evidence concerning the evolution of mankind.[3] Yet in one of his few published references to religion and evolution, he tells us that when, as an undergraduate student at the University of Queensland, he encountered the idea of evolution, it apparently occasioned him no mental distress. In his enthusiastic adoption of the evidence that humans had evolved, contrary to what one might have expected given his family's religious background, we discern another example of Dart's nonconformism.

It would convey a wrong impression if I were to paint a picture of my old mentor, predecessor, and friend as being dominated by a monomaniacal heretical streak. His life was painted in many different shades: he was a man of enormous compassion, helpfulness to his staff, students, and lame ducks, qualities of leadership, inspiration, vigor wedded to gentleness, histrionic excellence that he put to good effect in his lectures and addresses, resilience, and toughness, a man who could roar when the occasion demanded and who wept easily. He had the strength of character and the willpower to withstand the buffetings of a hostile world for twenty-five years after he first put forward his claims on behalf of the South African fossil child of Taung in 1925.

BRIEF SKETCH OF DART'S LIFE

Raymond Arthur Dart (1893–1988) was the fifth of nine children of Samuel Dart and Eliza Anne Brimblecombe of Indooroopilly. He was born in Toowong, Brisbane, Queensland, on 4 February 1893. He attended primary schools in Toowong and Blenheim and attended the Ipswich Grammar School from 1905 to 1909. In 1911 he entered the University of Queensland in its foundation year. There he completed a BSc(Hons) degree in 1913. This was followed by a medical degree from the University of Sydney in 1917. There, in the Anatomy Department, he fell under the spell of Professor James T. ("Jummy") Wilson (1861–1945). It is interesting to reflect, in the light of Dart's own later faculty for bringing out the best in his students, that Jummy Wilson was himself a distinguished "maker of men." Dart served as a resident at Royal Prince Alfred Hospital in Sydney, but the lure of anatomy was to outbid the competition of the wards and clinic. After serving in the Australian Army Medical Corps in France and England (1918–1919), Dart became a lecturer in anatomy at University College, London in 1919.

When the Rockefeller Foundation established fellowships for graduates from the British Empire, Dart became one of the first two foreign fellows, and he spent a year (1920–1921) in the United States. He returned to England and for eighteen months was a lecturer under Professor Grafton Elliot Smith (1871–1937) in the Anatomy Department at University College, London. Dart had by now definitely made up his mind to pursue anatomical research as his life's calling. He published his first paper at this time.[4]

Within a year of his return to England, Dart was on his way to succeeding E. P. Stibbe as professor of anatomy at the University of the Witwatersrand, Johannesburg. Dart assumed his chair in January 1923—a month before his thirtieth birthday—and held this position until the end of 1958. He made an immense impact in building up the infant medical school, serving concurrently as dean of the medical faculty from 1925 to 1943. A startling find almost at the beginning of his South African career changed the direction of his life's work and was ultimately to refashion thinking about human origins.

THE DISCOVERY OF 1924 AND DART'S CLAIMS OF 1925

Near the end of Dart's second year as head of the Department of Anatomy at the Witwatersrand University there fell into his hands—unwilling hands, as he avowed—two pieces of consolidated cave earth (*breccia*) that contained a fossilised child skull from the Buxton Limeworks at Taung between the towns of Kimberley and Mafikeng (formerly Mafeking). This skull came to be seen as the most important paleoanthropological discovery of the twentieth century. Although the record is painfully silent regarding the date of the discovery, it has been possible from contemporary press reports and other accounts to work out that the specimen in question was blasted out of the limestone deposits by a lime worker named M. de Bruyn early in November 1924. It was ensconced in the field office of the limeworks manager, A. E. Spiers, along with fossils of baboons and monkeys. Robert Burns Young, the professor of geology at the Witwatersrand University, brought the two tell-tale pieces to Johannesburg and handed them over to Dart on or about 28 November 1924. Together, Dart and Young found that the two fragments fitted perfectly.[5]

In four weeks Dart extracted the fossil from the calcified matrix in which it was incarcerated. It was a delicate and exacting task. The cleaning process finally revealed an almost complete face, extending from the lower jaw to the forehead, with a natural cast of the interior of the calvaria, or brain case. This superbly preserved "brain cast" was complete on one side from the frontal to the occipital poles and made a comfortable fit behind the forehead (frontal bone).

The development of the specimen out of the matrix revealed a beautiful, almost complete skull of a child. It was different from anything that had been found up to that time. Dart detected critical evidence that the head had been held on a nearly vertical spinal column: he inferred that it held its body nearly upright, as in the habitual stance of humans rather than in the obliquely quadrupedal posture that pertains to great apes. Its canine teeth were small, as in human subjects, not like the high, fang-like canines of apes. On the brain cast Dart believed he could detect the impression of a well-known groove (the lunate sulcus) on the hinder part of the cerebrum; he claimed that it lay in a posterior location, as in humans, rather than in an anterior position, as in apes. This supposed feature has remained doubtful to the present day. As against these human-like traits, the endocranial cast showed that the brain was much smaller than in modern human children of corresponding age and closer to the brain size of modern apes. From the state of eruption of the teeth Dart claimed that the child had died at about six years of age. Later studies with more advanced techniques, however, indicated that the child had been no more than three or four years old when it died.

So the Taung child was an unprecedented mixture of apish and human traits. Dart claimed that its blend of features might have characterized a putative "missing link" between human and nonhuman animals on the antiquated notion of a Chain of Being.

His article appeared in *Nature* on 7 February 1925.[6] Dart proposed to name the genus and species represented *Australopithecus africanus* (the southern ape of Africa). He claimed that although it was basically an ape, it showed significant departures, especially in the teeth, posture, and brain form, all of which seemed to have carried it to the threshold of the human family (as then recognized). He did not, however, claim that the new genus and species should be classified in the family of the hominids: that small brain size gave him pause, and he proposed to erect a new zoological family, intermediate between those of apes and of humans.

The discovery and especially Dart's claims evoked a storm of controversy from both the laity and the scientific world. In the scientific media his views were attacked as extravagant. There was little to distinguish "Dart's child" from a young anthropoid ape, especially a chimpanzee. Some critics asserted that the fossil child was too young for scholars to be sure what kind of adult it would have grown up to be. Some held, on very inadequate grounds, that it belonged to a geological epoch too recent for it to have been a human ancestor. Even the Graeco-Latin name *Australopithecus* was declared unacceptable. Arrayed against the young Raymond Dart (who turned thirty-two the day after the first public announcement of the discovery) was the authority of Arthur Keith, Arthur Smith-Woodward, Dart's former mentor Grafton Elliot Smith, and others.

If Dart's reception by the scientific world had been frigid, the attitude of some members of the lay public was frankly damning. Although the press was boundlessly enthusiastic and General J. C. Smuts—the author of *Holism and Evolution* and a lifelong amateur prehistorian—sent a personal letter of congratulations, there were others whose comments were not so complimentary. A Frenchman predicted that Dart would "roast in the quenchless fires of Hell" for claiming that the Taung skull represented man's ancestor. An Englishman wrote, "I hope you will be placed in an institution for the feeble-minded." A Dane warned that Professor Dart had signed his "perdition-warrant" by having the impertinence to account for the origin of man. Another critic accused Dart of being a priest of BAAL![7]

TWENTY-FIVE YEARS OF OPPOSITION

For many years few scholars accepted Dart's interpretation. Among his sparse early supporters was the paleontologist Robert Broom, who must be counted as one of the earliest scholars to be fully convinced by Dart's arguments. Broom's staunch, bulldog-like support over the years calls to mind Thomas Henry Huxley's espousal of Charles Darwin's cause. The avalanche that Dart's claims brought upon his head was such as to have deflected a lesser person from his standpoint. But Dart remained firmly by his original views, although, as he wrote, he set out on no campaign to convert a resistant world. Dart stuck to his guns for a quarter of a century before the tide began to turn. By the middle of the twentieth century, many more fossilized hominid remains had come to light. Many adult *Australopithecus africanus* fossils emerged from Sterkfontein, near Krugersdorp, in the Gauteng province, at the hands of Broom and John T. Robinson until about 1960, and of Phillip V. Tobias, Alun R. Hughes, and Ronald J. Clarke from 1966 to 2006. The new South African finds largely confirmed Dart's original claims. Broom had commented in 1925 that the Taung skull "was the most interesting fossil ever discovered."[8] Late in his life, Broom's enthusiasm had not dwindled: in *Finding the Missing Link,* Broom said of Dart, "Here was a man who had made one of the greatest discoveries in the world's history—a discovery that may yet rank in importance with Darwin's *On the Origin of Species.*"[9]

Another exceptional supporter of the Dartian heresy was William Johnson Sollas of Oxford University. After initially dismissing Dart's claims, largely on the basis of "the absurdly minute illustrations accompanying Dart's communication in *Nature,*" when he received from Broom a midline section of the skull showing the positions of four of the time-hallowed cranial landmarks, Sollas changed his view. He compared the Taung section with those of young apes and then wrote, "[T]he more I compare them [young apes] with Taungs [*sic*] the more difference I see, and the more human Taungs appears. I should have named it *Homunculus* (little man). The forehead is as thoroughly human as it is not anthropoid. It gives one a lot to think about."[10] A year later Sollas concluded, "*Australopithecus* is doubtless generically distinct from all known Apes, and in these important characters by which it differs from them it makes a nearer approach to the Hominidae."[11]

Elliot Smith recognized what he regarded as some characteristically human features in the brain cast. He added, "What above all we want Professor Dart to tell us is the geological evidence of age, the exact condition under which the fossil was found, and the exact form of the teeth."[12]

Yet although Broom, Sollas, and to a certain degree Elliot Smith came to agree with Dart, most other scientists did not. The correspondence columns of *Nature* virtually incandesced with the scorching comments that flew through the rest of that eventful year of 1925. Almost everyone, it seems, doubted Dart's outlandish claim that a creature with a brain size scarcely different from that of the apes and one-third as big as that of modern humans could have been a near relative of the human species. They also doubted that humanity had its origins in Africa.

As late as 1945, twenty years after the announcement of *Australopithecus,* George Gaylord Simpson, the distinguished American evolutionist and taxonomist, classified *Australopithecus* in the ape family then recognized, namely, the Pongidae. The foremost British authority, Sir Arthur Keith, was especially strenuous in his rejection of Dart's claim.

By the third quarter of the twentieth century the rejection of *Australopithecus* had lapsed. Although scientists still argue about the names to be assigned to various hominids (or hominins, as many investigators prefer to call them), there is nearly universal agreement that some populations of *Australopithecus* were most likely to have been ancestral to modern humans. With hindsight, we see that Dart was right when he claimed that, in a creature replete with other hominid traits, a small brain alone was not sufficient to disqualify these creatures from the exalted company of hominids.

What were the factors that made for the initial repudiation of Dart's claims, and what were the factors that led to their ultimate acceptance?

ANALYSIS OF THE NEGATIVE RESPONSE TO *AUSTRALOPITHECUS* AND DART'S CLAIMS FOR IT

When I first began examining these questions, I had been led to them by my own experience as one of the three "founding fathers" of the hominid species *Homo habilis.* That new species of the genus *Homo* was proposed by Louis S. B. Leakey, myself, and John R. Napier in 1964.[13] It was the same recipe as before: our claims that the new Olduvai fossils represented a new species were criticized and resisted. The period of rejection lasted for fifteen to twenty years before the tide began to turn. Pondering the delayed acceptance accorded to both *Australopithecus africanus* and *Homo habilis,* I chanced upon Gunther S. Stent's fascinating essay "Prematurity and uniqueness in scientific discovery," published in *Scientific American* in December 1972.[14]

Stent sought to understand the belated acceptance of the notion that DNA was the active principle in bacterial transformation and hence the basic hereditary substance. Oswald Avery's discovery that DNA fulfilled this role[15] was not appreciated in its day because, according to Stent, it was a "premature discovery." It would be easy to dismiss his statement as merely an empty tautology, a glib form of words. Yet Stent went on to address this problem, asking, "[I]s there a way of providing a criterion of the prematurity of a discovery other than its failure to make an impact?" This is Stent's criterion: "A discovery is premature if its implications cannot be connected by a series of simple logical steps to canonical, or generally accepted, knowledge."[16] When I applied this criterion to these two examples of seemingly premature fossil hominid discoveries, I found that both *Australopithecus africanus* and *Homo habilis* could readily be seen as premature discoveries.[17] The reception of both discoveries was in accord with Stent's criterion.

In order to judge the case of *Australopithecus africanus*, it was necessary to examine the tenets of the prevailing paradigm of 1925.

1. Asia was considered by most scholars to be the cradle of mankind. This was based on discoveries made in Java, Indonesia, by Eugene Dubois in late 1880s and 1890s. Those finds had consisted of an archaic form of hominid called *Pithecanthropus* and later renamed *Homo erectus*. At the time of their discovery, nothing so primitive had come out of Europe or Africa bearing on human evolution. In addition, a few fragments, mainly of fossil teeth assigned to *Sinanthropus*, later "lumped" into *Homo erectus*, had been found in China in the early years of the twentieth century. On the basis of this evidence from Java and China, it was expected that the earliest roots of humankind were to be found in Asia. Taung was unacceptable because it was in the wrong continent.

2. According to the historian Peter Bowler, many Europeans were prejudiced against Africa and all things African. This bias predisposed them to reject claims emanating from Africa. By contrast, when Peking Man came to light a few years later, it was avidly and widely accepted by European scientists. In brief, the Taung skull was an early victim of racial prejudice that militated against its acceptance.[18]

3. It was authoritatively believed by many scholars at the time that the brain had developed and enlarged early in hominid evolution, before the teeth and posture showed signs of appreciable change toward human morphology. In my childhood, cavemen were depicted in comic papers in accordance with this expectation: they were shown with large heads containing large, "clever" brains and high, fang-like canine teeth, and they walked in a stooping posture. The Taung child pointed to the opposite bodily structure: it was of the wrong kind of morphology to have been an ancestor.

4. The image of the expected hominid ancestor was a theoretical construct, and there were no fossil remains to substantiate it. For a few colleagues this lacuna was too great to be bearable, and a skull of the expected pattern was conveniently provided, doctored, stained, and seeded in gravel deposits at Piltdown in the south of England. This forgery, found a dozen years earlier than the Taung discovery, had a modern human-sized brain case (the salted cranium was that

of a big-brained modern human) and an ape-like jaw (we now know it had come from an orangutan). If Piltdown provided a key to the pattern of human evolution, as Arthur Keith and others believed, then Taung could have no bearing on hominid phylogeny. It would then be seen—as Keith did see it—as a somewhat aberrant ape. As long as Piltdown was considered to be an authentic and "ancient" fossil hominid, it was a serious deterrent to the acceptance of any claim on behalf of the Taung child. If Piltdown was valid, Taung would not have had any bearing on hominid evolution; if Taung showed the way, then Piltdown furnished a false image. The Piltdown forgery was uncovered only in 1953.[19]

5. It was assumed that features which were specific to any species could be unequivocally discerned only after puberty. At three to four or even six years of age, the Taung child died well before puberty. It was too young for its postpubertal and adult traits to be read from the fossil. Moreover, its supposed human-like features, it was averred, could simply have reflected its juvenile status—for it was well-known that younger members of a species more closely resembled those of other species. Wynfrid L. H. Duckworth of Cambridge University was one of four leading authorities to whom the editor of *Nature* referred Dart's original article with a request for comments to be published in the following week's issue (14 February 1925). Duckworth conceded that the Taung skull showed a number of advanced features but asked to what degree these features were the effects of the specimen's having belonged to a more advanced creature rather than a youthful individual. There can be no doubt that, had the first *Australopithecus* discovered been an adult, Dart's task of convincing an incredulous world would have been easier. Yet Dart had to read the signs of mankind in the child, while Duckworth and the rest of the world waited to be convinced by the discovery of an adult hominid of the same species.

6. One school of thought held that during hominin evolution all parts of the body would have become hominized more or less uniformly from an assumed ape-like ancestral form. Taung, on the other hand, showed *mosaic evolution,* some parts (such as the canine teeth and the posture) being more hominized and others (such as the expansion of the brain) less so. On the assumption that uniformity of development would be manifest, it was less likely that *Australopithecus* would have been an ancestral form.

7. The dating of a hominid ancestor, it was maintained, should be no later than the early Pleistocene. The dating of Taung, it was asserted

on insubstantial grounds, was too recent. From 1925 to 1947, Arthur Keith was strongly opposed to Dart's interpretation. One of the cardinal reasons cited by Keith was that the fossil was too young in geological time to have been an ancestor of humans. Keith's reasoning was based on stone tools that had been found in a river bed below the limestone tufa from which the Taung skull had been recovered. Keith assumed that the fossil skull in the tufa and the stone tools in the river bed were contemporaneous. From this assumption he inferred that *Australopithecus* was contemporary with tool-making humans. Because the human tool fabricators and their ancestors could not have been coeval, the species to which the Taung child belonged, said Keith, could not have been ancestral to the human species. Thus a conception of time that we now know to have been erroneous helped disqualify *A. africanus* from ancestral status. This is one of several instances in which the relative prominence accorded the time factor played an important role in the interpretation and classification of fossils.

When these tenets of the prevailing mindset are scrutinized, it is clear that the Taung skull was an example of a premature discovery in the sense of G. S. Stent, because its features and their implications could not be connected by simple, logical steps to the canonical or received wisdom prevailing at the time of the discovery.

On this basis, the measure of prematurity of Dart's interpretation of the Taung skull was between twenty-five and twenty-eight years, according to which date of ultimate acceptance one adopts. In contrast, the measure of prematurity of Avery and his colleagues was six to eight years. That of *Homo habilis,* as interpreted by Leakey, Tobias, and Napier, was almost twenty years.

OTHER DELAYING FACTORS IN THE ACCEPTANCE OF DISCOVERIES

Other factors apart from prematurity may account for delays in the acceptance of discoveries. Four kinds of delaying circumstances may be recognized: linguistic, political, personal, and theological or philosophical.

Linguistic delay: This instance depends on the failure of investigators working in one language to read scientific contributions in a foreign language. In 1950 one significant work that was to herald the approach to evolutionary biology in the second half of the twentieth century was published in German: Willi Hennig's *Grundzüge einer Theorie von Phylogenetischen Systematik.* For the next sixteen years it made hardly any impact on phylogeny or systematics, especially those of the higher primates, including Hominidae. Then it was translated into English by D. D. Davis and R. Zangerl and published in 1966 by the Illinois University Press with the title *Phylogenetic Systematics.* Quite suddenly, it produced a marked impact on paleontological, including paleoanthropological, studies. For many investigators, the application of Hennig's guidelines became virtually the only acceptable way to construct phylogenies. There seems little doubt that this was not a premature discovery in the sense of Stent: the delay was almost certainly occasioned by the language in which the "discovery" was first published.

Political delay: Aversion to the political system of a country may have a serious effect on the recognition and appreciation by scientists elsewhere of researches carried out there, irrespective of the personal political views of the researchers in that country. This is what happened during the apartheid regime in South Africa from 1948 to 1993, and especially in the 1970s and 1980s, when an academic boycott was in force. During these decades, much of the paleoanthropological work that was being carried out in South Africa was relegated to a secondary position in the scientific literature and the media. Many films and books on hominid evolution devoted only passing mention to the work of the South African investigators. The myth spread (or was propagated) that researches in that country had virtually come to a standstill. The situation was exacerbated by a boycott on publications emerging from South Africa that was observed by a number of periodicals, as well as prohibitions on visits to academic institutions and participation in international conferences in many parts of the world. It could be argued that the tendency to draw a veil of silence and exclusion over the not inconsiderable researches that were being carried on in South Africa acted to delay the advance of some branches of science, such as paleoanthropology. It is important to add that the observance of academic sanctions was not worldwide and that communication remained open with some countries and communities of scholars. This chapter in the history of infringements on the freedom of science and on the dissemination of knowledge is happily now past, with the replacement of South Africa's apartheid regime by a nonracial democracy. This is of course not the only such instance in the history of science in the twentieth century, and it is not inconceivable that the phenomenon could recur. So political attitudes could cause delays in the advance of science.

Personality factors: In his analysis of prematurity and uniqueness in scientific discovery, Stent points out that an unnamed microbiologist critic had offered a different explanation for the failure of the discovery of Avery et al. to be recognized.[20] This critic suggested that it was the "quiet, self-effacing, non-disputatious" personality of Avery that caused his contribution not to be recognized. Similarly, Erwin Chargaff attributed the seventy-five-year lag between Friedrich Miescher's 1869 discovery of DNA in the cell nucleus and the general appreciation of its importance to Miescher's being "one of the quiet in the land," who lived when "the giant publicity machines, which today accompany even the smallest move on the chessboard of nature with enormous fanfares, were not yet in place."[21] Stent himself adds the possibility that the delay in the appreciation of Mendel's discovery was due to Mendel's having been a modest monk living in an out-of-the-way Moravian monastery. Although Stent does not like the theory that lack of publicity could explain the retarded appreciation in these cases, a dispassionate look at the progress of hominid evolutionary studies of the past fifty years yields a strong contrast between the treatment of discoveries that were attended with great publicity and "hype"—in a field that lends itself to fanfares—and those whose discoverers did not seek or even eschewed publicity. The difference between the scientists' scientist and the popular scientist, between the extrovert and the modest discoverer, may be another aspect that should not be ignored in analyses of delaying principles in the history of science. I have already referred to the personality of Raymond Dart. So far from being one of the quiet in the land, Dart's tendency toward overstatement and bombast might have put colleagues off and militated against the ready acceptance of his writings and pronouncements.

Theological or philosophical agencies: Sometimes theological opposition retards the acceptance of a new discovery or hypothesis in clerical circles (though not necessarily in scientific milieux). A famous example is Galileo's work on the movement of the earth. In 1633 he had been forced by the church to recant because his views were at variance with the age-old cosmology, the biblical vision of the earth as the center of the universe. This condemnation of Galileo's teachings was reversed only in October 1992. Although it was specifically the Roman Catholic Church that denounced the "unacceptable" views of Galileo, the denunciation might well have deterred many scientists from pursuing such researches in the intervening centuries. Moreover, the papal statement of October 1992 pointed out that Galileo's condemnation led many to conclude that there was "an incompatibility between the spirit of science and its rules of research on the one hand, and the Christian faith on the other."[22]

In the nineteenth century, in the Judeo-Christian world, belief in Special Creation was widespread among the orthodox. The biblical six-day period of creation was accepted literally by very many people. The strength of this belief was preeminent in the early rejection of the evolutionary hypothesis. Evolution, the pious held, ran counter to the revealed word of God. The newfangled ideas about evolution that Lamarck, Haeckel, Wallace, Erasmus Darwin, Charles Darwin, and T. H. Huxley were promoting threatened to undermine organized religion and faith—such at least was the fear of many in Europe and North America. Similar ideas persisted into the twentieth century, and for some they persist to the present day.

Two aspects of this great antinomy are worth stressing. First, it was relatively easy to set Religion and Science up as two opposing forces. Evolution and theology (meaning the Judeo-Christian varieties) were, it seemed, clear-cut opposites. Few strove to seek reconciliation between the two forces, to show that religion did not stand or fall by the literal interpretation of Genesis 1, to seek the moral in the biblical tales rather than be dominated by the *ipsissima verba,* to open the heart and mind to the possibility that one could accept the evolutionary doctrine and still remain a religious person.

Second, in the Judeo-Christian world, few stopped to consider that it was unreasonable to accept the creation story of only one of the sacred writings, namely, the Bible, and to ignore those of the teachings of Hinduism, Islam, Confucianism, Zoroastrianism, Taoism, Buddhism, and the rest. This approach was by nature exclusivist, arbitrary, authoritarian; perhaps *arrogant* would not be too strong a word.

This was the theological underpinning of the Western world, steeped in the biblical Judeo-Christian tradition, when Charles Darwin published *On the Origin of Species* in 1859 and *The Descent of Man* in 1871, when Thomas Henry Huxley wrote *Man's Place in Nature* in 1863, when Eugene Dubois found in Java the first fossil of *Pithecanthropus* (now *Homo erectus*) in 1891, and when Raymond Dart presented the world with the first known specimen of *Australopithecus africanus* in 1925.

One reason why opinion was weighted against the acceptance of the Taung child was that, as a claimed potential missing link, it constituted a threat to the creationist viewpoint. Two cogent examples of the latter interlinkage follow.

THE TAUNG SKULL AND THE TENNESSEE MONKEY TRIAL

On 3 February 1925, news of the discovery and implications of the Taung skull was announced to the world by the Johannesburg daily newspaper, *The Star.* This was followed four days later by the appearance in *Nature* of the article in which Dart gave a preliminary account of the skull and set out his interpretation that it was an ape that had taken decided steps in the direction of humanity, thus conforming to the old idea of a "missing link." Even without the benefit of television, let alone the World Wide Web, the news of the discovery and of Dart's claim caused a stir throughout the world. Fast and furious debates followed.

Nowhere do they seem to have been faster or more furious than in Tennessee. Within weeks the state legislature enacted a ban on the teaching of evolution in Tennessee schools. In my mind the haste of that enactment was a direct response to the news of the skull. The Great Monkey Trial in Dayton, Tennessee, followed a short while later. John Thomas Scopes, a school teacher, had broken the new state law against the teaching of man's descent from a lower form. The presidents of Harvard and Yale Universities, leading clergymen, and a host of scientists offered to appear for the defense. The judge declined their help. The historical confrontation between Clarence Darrow and William Jennings Bryan was the highlight of the trial. Bryan placed his reputation, faith, and biblical knowledge at the service of the prosecution. Scopes was convicted and sentenced to pay a paltry fine ($100). Many years were to pass before radio broadcasting in the United States ventured to deal with evolution on the air. There can be little doubt that the discovery of the Taung skull and the claim that it was intermediate between apes and humans, followed as it was by worldwide press and wireless coverage, triggered the almost panicky reaction of the Tennessee legislators.

THE BANNING OF "EVOLUTION" IN SOUTH AFRICA

In South Africa, well into the second half of the twentieth century, museum exhibits dealing with evolution had to be counterpoised by exhibits about the biblical creation story. The word *evolution* was banned from the airwaves and from an exhibition at Cape Town in 1952 on the tercentenary of the first European settlement at the Cape of Good Hope. The teaching of evolution in schools was banned until recently—in the very country that until 1960 had yielded 90 percent of the world's evidence of early human evolution from nonhuman animals. For about ten years I was one of two regular anchormen on a weekly radio program called "The Voice of Science," presented by the South African Broadcasting Corporation. Although I was permitted to speak about *Australopithecus* and *Homo erectus,* I was adjured not to use the word *evolution* and was asked to resort to such bland euphemisms as *development.* Eyebrows were raised in certain quarters when, in 1971, during a commemorative lecture to the Royal Society of South Africa on the centenary of Charles Darwin's *Descent of Man,* I declared that one hundred years without evolution in our schools was enough! It is pleasing to report that, in the new democratic South Africa, school syllabi now permit the inclusion of evolution in teaching programs. It is most gratifying that I have lived long enough to hear the new South African government speak with pride of our country's and our continent's contributions to hominid evolution.

This study confirms that the concept of prematurity in scientific discovery is of value in the history of science. Could the concept be regarded as a limiting factor in the history of a branch of science? It might be held that the more premature discoveries characterize a particular field or the longer the index of prematurity, the more the progress of knowledge in that field is slowed down. The picture is, however, not as simple as that. The impact of premature discoveries on the advancement of a field of science must depend in part on the rigor with which the new (premature) discovery has been arrived at and enunciated and in part on the degree to which the prevailing paradigm is well based. If the paradigm itself is not rigorously founded and includes manifestly false or prejudiced views, like some aspects of the tenets limned above, the proposal of a new (premature) concept may be beneficial to the field in that it could help to shake off widely held views that are of limited validity or tenability, and so advance the field. On the other hand, if a paradigm has been strongly and rigorously established and is supported by a great deal of observational, experimental, and conceptual data, the role of new hypotheses that appear to be irreconcilable with the preexisting corpus of knowledge would be much more problematical, and they might in some less rigorous instances act as deterrents to the advancement of knowledge in the relevant field.

To return to Raymond Dart: above the personality traits there rises the image of a scientific visionary. For the revolution he wrought in the knowledge about hominid evolution, Dart's discovery of the Taung skull and what he made of it will be remembered as the most fundamental single breakthrough in the history of paleoanthropology. Next to it the labors of all those who have come after him have simply filled out the details of mankind's tortuous path of development over the course of the past five to seven million years.

ACKNOWLEDGMENTS

I am grateful to Oren Harman and Michael Dietrich for inviting me to contribute to this unusual book. My appreciation is extended to Heather White for preparing the manuscript with her customary dedication. Sadly, she died on July 10, 2007. My thanks are due also to Peter Faugust and Ruliang Pan, and Goran Strkalj for their help. I shall always remember with gratitude my years with Professor Raymond Dart (1944 to 1988).

NOTES

1. R. A. Dart, "Adventures with *Australopithecus*," *The Rationalist Annual* (1949) 16; R. A. Dart, "The status of *Australopithecus*," *American Journal of Physical Anthropology* 26 (1940): 169.

2. R. A. Dart, "Recollections of a reluctant anthropologist," *Journal of Human Evolution* 2 (1973): 417–427.

3. R. A. Dart, "Associations with and impressions of Sir Grafton Elliot Smith," *Mankind* 8 (1972): 171.

4. R. A. Dart, "A contribution to the morphology of the corpus striatum," *Journal of Anatomy* 55 (1920): 1–26.

5. P. V. Tobias, *Dart, Taung and the "missing link"* (Johannesburg: Witwatersrand University Press, 1984):, P. V. Tobias, "When and by whom was the Taung skull discovered?" in *Para Conocer Al Hombre: Homenaje a Santiago Genovés,* ed. L. L. Tapia (Mexico City: Universidad Nacional Autónoma de Mexico, 1990), 207–213.

6. R. A. Dart, "*Australopithecus africanus:* The man-ape of South Africa," *Nature* 115 (1925): 195–199.

7. R. A. Dart with Dennis Craig, *Adventures with the missing link* (New York: Harper and Brothers, 1959), 40; Tobias, *Dart, Taung, and the "Missing Link,"* 42.

8. R. Broom, "On the newly-discovered South African man-ape," *Natural History* 25 (1925): 410.

9. R. Broom, *Finding the Missing Link* (London: Watts, 1950), 27.

10. Letter cited by Broom in ibid., 26.

11. W. J. Sollas, "A sagittal section of the skull of *Australopithecus africanus*," *Quarterly Journal of the Geological Society of London* 82 (1926): 1–11.

12. G. E. Smith, "The fossil anthropoid ape from Taungs," *Nature* 15 (1925): 235.

13. L. S. B. Leakey, P. V. Tobias, and J. R. Napier, "A new species of the genus *Homo* from Olduvai Gorge," *Nature* 202 (1964): 7–9.

14. Gunther S. Stent, "Prematurity and uniqueness in scientific discovery," *Scientific American* 227 (1972): 84–93.

15. Oswald Avery, C. M. McLeod, and M. McCarty, "Studies on the chemical nature of the substance inducing transformation of pneumococcal types," *Journal of Experimental Medicine* 79 (1944): 137–157.

16. Stent, "Prematurity."

17. P. V. Tobias, "The species *Homo habilis:* Example of a premature discovery," *Annales Zoologici Fennici* 28 (1992): 371–380; P. V. Tobias, "Premature discoveries in science with especial reference to *Australopithecus* and *Homo habilis*," *Proceedings of the American Philosophical Society* 140 (1996): 49–64.

18. Peter Bowler, "Commentary on Piltdown: The case against Keith," *Current Anthropology* 33 (1992): 260–261; Peter Bowler, *Theories of human evolution: A centre of debate* (Baltimore: Johns Hopkins University Press/Blackwood, 1986).

19. J. S. Weiner, K. P. Oakley, and W. E. LeGros Clark, "The solution of the Piltdown problem," *Bulletin of the British Museum of Natural History (Geology)* 2 (1953): 141–146, J. S. Weiner, K. P. Oakley, and W. E. LeGros Clark, "Further contributions to the solution of the Piltdown problem," *Bulletin of the British Museum of Natural History (Geology)* 2 (1955): 225–287.

20. Stent, "Prematurity"; Avery, McLeod, and McCarty, "Studies on the chemical nature."

21. Stent, "Prematurity," 86.

22. Pope John Paul II, Address to Pontifical Biblical Commission, April 23, 1992.

FURTHER READING

Brain, C. K. *The hunters or the hunted?* (Chicago: Chicago University Press, 1981).

Broom, R. *Finding the missing link* (London: Watts, 1950).

R. A. Dart with Dennis Craig, *Adventures with the missing link* (New York: Harper and Brothers, 1959).

Findlay, G. *Dr. Robert Broom, palaeontologist and physician, 1866–1951* (Cape Town: Balkema, 1972).

Needham, J. *Time the refreshing river,* 2d ed. (Nottingham, U.K.: Spokesman, 1986) (first publication London: George Allen and Unwin, 1943).

Tobias, P. V. *Dart, Taung and the "missing link"* (Johannesburg: Witwatersrand University Press, 1984).

———, ed. *Hominid evolution: Past, present and future* (New York: Alan R. Liss, 1985).

CHAPTER SIX

In Weismann's Footsteps:
The Cyto-Rebellion of C. D. Darlington

OREN HARMAN

"The time in which men believed that science could be advanced by the mere collection of facts has long passed away."[1] Such was the judgment of the aging, increasingly cantankerous August Weismann, Germany's leading biologist, in 1886. A short four years after the death of Charles Darwin, Weismann was intent on describing a cellular theory to match the English naturalist's theory of evolution by natural selection. "The investigation of mere details," he wrote, "had led to a state of intellectual short-sightedness, interest being shown only for that which was immediately in view. Immense numbers of detailed facts were thus accumulated, but . . . the intellectual bond which should have bound them together was wanting."[2] What biology needed more than anything was a stroke of bold theory making, and that is what Weismann delivered: in order for sexual organisms to beget offspring, there must exist a special reduction division; otherwise the genetic material would double in each successive generation.[3] Not that Weismann, who was increasingly growing blind, or anyone else for that matter, had actually seen this division occur. This leap of faith was based entirely on unassailable genetic logic, and it gave meiosis—a term coined in 1905—to the world. It would also render Weismann a hero to a young English cytologist struggling to figure out what precisely happens during meiosis forty years later. I will return to Weismann below, but first, let us leap ahead in time to the real subject of our story.

Cyril Dean Darlington (1903–1981) was born by mistake. "My birth gave rise to a crisis in the family," he wrote in reminiscence. "My mother claimed that she didn't know how it happened. My father on the contrary was quite certain. He applied a rule without formally declaring it that never again would he cohabit with his wife. It was a rule that he never broke in his remaining forty years. Whether he regarded me as an accessory to this mishap or misdemeanor I am still not sure."[4] A retrospective biographer might claim that here, at the very beginning, was planted the seed of all that was to come. Darlington's very existence was from the outset shrouded in doubt; it was far from clear that he was wanted. There seems to have been a strong impetus for Darlington to prove to the world, and especially to his father, that the "mishap" or "misdemeanor" had not occurred in vain. It need not have resulted in the way that it did, of course; the infinite vicissitudes of life and of the psyche render such determinisms outrageous. But, in a certain, almost existential sense, Darlington was born a rebel.

Alfred, his lone and older brother, was always his father's favorite. Away from their home in Lancashire and then Ealing, fighting the war against the Germans and winning medals of honor, Alfred's very existence seemed to accentuate Cyril's own uselessness at home. Invariably, he caught the wrath of his dyspeptic father. His mother was overprotective and increasingly depressed. Childhood was a sullen, hollow affair. When he finally graduated from the South Eastern Agricultural College at Wye, Darlington put in an application for a scholarship to travel as far away as possible—to Trinidad, to become a farmer. A tough veneer had come to characterize his dealings with people. Already he was antiauthoritarian in the extreme, writing to his father at eighteen: "[N]othing galls me so much as to have other people's beliefs forced down my throat."[5] But because it showed no signs of its author's being either a promising student or a terribly talented agriculturalist, Darlington's application was rejected by the Empire Cotton Growing Corporation. Disaffected, arrogant, and at wit's end, Darlington was cajoled by a master at Wye to try the next best thing (and only thing): a volunteer position at the John Innes Horticultural Institution. After all, the only subject that had captured Darlington's imagination at Wye had been the new field people were now calling "genetics." The Innes was Britain's premier horticultural institute, but its Governing Council was able to secure the famous Mendelian William Bateson as its first director in 1910 only by promising to allow research in genetics as well. "The rewards from genetics are slight," the white-haired Bateson pronounced beneath his moustache as he approached a cocky but frightened Darlington in his carpet slippers in the institution's library in the fall of 1923; "only those who attained unheard-of heights of achievement can ever hope to make a living out

Figure 6.1. C. D. Darlington, ca. 1929. Courtesy of Clare Passingham.

of it."[6] Uncertain of his direction but determined nevertheless to succeed at all costs, Darlington shook the old man's hand. "Unheard-of heights of achievement" sounded strangely alluring.

Although he did not know it, by the time Darlington had arrived at the Innes the once-pioneering Bateson already had been surpassed by Thomas Hunt Morgan and his group in the "Fly Room" at Columbia University in New York. The chromosomal theory of heredity, which postulated that the genes responsible for heredity lay physically on the chromosomes in the nucleus of cells, was the lynchpin of the success of Morgan's genetics. But Bateson, a philosophical aesthete who had little time for materialism, denied the role of the chromosomes in the hereditary process.[7] Stubborn yet realistic, Bateson had journeyed in the winter of 1921–1922 to the United States to see for himself what the chromosome theory had to offer genetics. On his return, impressed and profoundly humbled, he appointed a cytologist, Frank Newton, less as an omen of a complete change of mind than as a fig leaf to cover what looked to be, in the new world of chromosomal genetics, his embarrassment.

Newton was skeptical of Bateson, and, when Darlington arrived in 1923, took him under his wing, teaching him how to examine hyacinth chromosomes under the old brass microscope his boss had

begrudgingly supplied to him. By 1926 Bateson was dead, and Newton, only thirty-three, succumbed to cancer less than a year later.
Left literally to his own devices, the young Darlington—his outward
overconfidence disguising a deep sense of awe at a mysterious microscopic world entirely unknown to him and frustratingly elusive—set
out to make sense of the chromosomes and their role in nature. He was
now a rebel with a field if not yet a cause, no matter how at sea or ill
prepared for the fight. "At the age of 18 most of the world seemed
stupid and annoyed me," he informed his diary. "At 24 I know it is
stupid and it ceases to worry me."[8] The outrageous confidence would
prove valuable. Within five short years, and working almost entirely
alone, Darlington authored *Recent Advances in Cytology* in 1932, a
book that was to have a profound impact not only on that field but
on genetics and evolutionary theory as well, catapulting Darlington
to world fame. It would also fashion him one of the greatest rebels
cytology had ever seen.

RECENT ADVANCES IN CYTOLOGY

What was the nature of Darlington's rebellion? And what was he rebelling against? When Darlington turned as an orphaned scientist to look
upon chromosomes, two problems continued to dominate the silent
world of cytology, which inhabited the space between the ocular lens
and the preparation slide. First, did the chromosomes pair side-by-
side during reduction division (meiosis), or did they pair end-to-end?
This question was important, for if the chromosomes paired end-to-
end, it would be unclear how genetic material passed between them.
The exchange of genetic material between homologous chromosomes,
called *crossing over,* was presumed to happen during meiosis, and
seemed to be the explanation for much of the variation begot by each
successive generation in nature and the true benefit of sexuality. In
1909 the Belgian cytologist and Jesuit priest Frans Alfons Janssens
had observed cross-like figures produced by the tangled homologous
chromosomes during meiosis and called them *chiasmata,* meaning
"cross-like" (see figure 6.2). Morgan assumed that Janssens's chiasmata evidenced the physical crossing over of genes between maternal
and paternal chromosomes and turned this assumption (it had yet to
be proved—see Chapter 8) into the very crux of the mapping procedure he and his group developed to suggest that genes lay physically
on the chromosomes. Its preoccupations small and technical, albeit

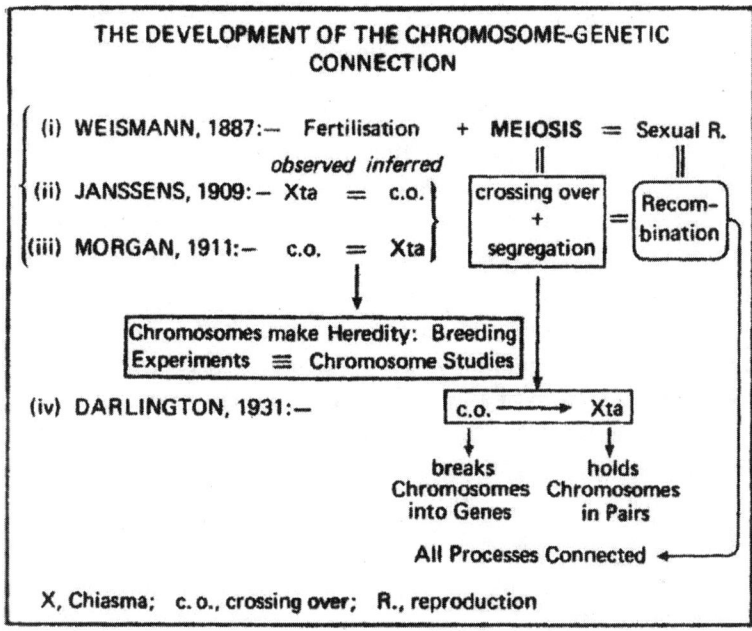

Figure 6.2. The scheme proposed by Weismann, Janssens, Morgan, and Darlington, Darlington's advance was to comprehend that crossing over was a necessary precondition for pairing, segregation, and reduction and therefore ultimately for sexual reproduction and evolution. From C. D. Darlington, "The Place of the Chromosomes in the Genetic System," *Chromosomes Today* 4 (1973): 1–13, at 6.

important, the second problem vexing cytology was whether the chiasmata were the result or rather the cause of such crossing-over.

Darlington was able to produce an axiomatic account of chromosome mechanics. In three simple laws he summed up the entire system: (1) All attraction between chromosomes is always in pairs, and those pairs are side-by-side, not end-to-end; (2) chiasmata are the conditions for orderly pairing and later segregation during meiosis; (3) chiasmata are always the consequences of genetic crossing-over between homologous chromosomes.

Why was any of this important?

Peering through the lens of a microscope at the tiny world of the chromosomes, Darlington was really after the grander questions of an evolutionary nature. Only after the laws of chromosome mechanics were described and understood could their role in evolution be theorized. In the preface to *Recent Advances in Cytology*, Darlington made clear where all this was leading: "The importance of the chromosomes as determining the hereditary functions has placed them outside the ordinary field of evolutionary enquiry. They have been considered as the very fount and origin of adaptive change and therefore not themselves capable of adaptation. Now they can be shown to be subject to the genetical variation which they themselves, by changes of their parts, determine."[9]

In the final chapter, "The Evolution of Genetic Systems," Darlington outlined these ideas in full. Not only did heredity *lead* to evolution, he argued; heredity itself was *subject* to evolution. Morgan and company had shown that genes exist on chromosomes, like beads on a string, and that they somehow carry the secrets of heredity. Following this lead, the mathematical population geneticists R. A. Fisher, J. B. S. Haldane, and Sewall Wright demonstrated with little more than pen and pencil that evolution works by nature's selecting the genes that are adaptive and selecting out those that are not. Darlington was now arguing that the chromosomes, far from being passive repositories of genes, are themselves under genetic control and determine by their movements the relative amount and kind of recombination of genes occurring in each successive generation. It was this quantity—the amount and kind of variation produced during meiosis on which selection can act—that more than any other variable determined the fate of organisms and ultimately the origin and propagation of species. Having begun five years earlier by looking at chromosomes for their own sake, Darlington could now begin to understand evolution itself largely as a function of their behavior.

RECEPTION

Making cytology relevant to evolution would seem to be a welcome development, but Darlington's ideas quickly met unbridled reactions of scorn and indignation. Hampton Carson, a young graduate student at the time, remembered the reaction to *Recent Advances of Cytology* in the Biology Department at the University of Pennsylvania: "The older members of this strongly cytological department received the Darlington book with stiff attitudes of outrage, anger, and ridicule. The book was considered to be dangerous, in fact poisonous, for the minds of graduate students. . . . Those of us who had copies kept them in a drawer rather than on the tops of our desks."[10] At the Sixth International Congress of Genetics at Ithaca in the winter of 1932, Darlington was given only five minutes to defend his views and was shouted down by a host of critics. On the West Coast, the leading American cytologist, John Belling, was working zealously on a scathing critical review. A Canadian colleague, Charles Leonard Huskins, offered to come across the Atlantic to punch Darlington's head off! In the relatively tranquil world of chromosomes, Darlington had done something to make a lot of people very mad.[11] What was all the ruckus about?

Clearly, Darlington made factual and experimental claims that fellow cytologists felt the need to challenge. In a science notorious for its intrinsic slipperiness—a "strange and difficult kind of visual chemistry with rules only dimly perceived," as one practitioner called it[12]—little could be legitimately claimed that was not immediately obvious to the eye. Darlington set down sweeping laws for much that was not visually obvious. Yet almost all the initial reservations about Darlington's supporting observations of the behavior of chromosomes (principally the number and frequency of chiasmata) were taken back one by one as chromosomal work done in the early 1930s tended to support his axiomatic claims. The initial shock that his book elicited within cytology ultimately gave way to acceptance, and by 1939 Sturtevant and Beadle, in their standard text *An Introduction to Genetics,* spoke of it as marking "the unification of chromosome cytology."[13]

If Darlington's microscopy was not the issue, then what was? Why was he being so vigorously resisted? Though not particularly interested in quantitative work himself, Darlington's chromosome-centered genetic system introduced too many dimensions for the mathematical population geneticists to cope with. The gene was a one-dimensional variable. The paper-and-pencil models of Wright, Haldane, and Fisher were major oversimplifications, whether or not they took them to be so. It was impossible for them to deal with all the effects of selection when the mechanism controlling selection (chromosome behavior during meiosis) and the unit that is being selected (the gene) are themselves both being changed by selection. Unlike the gene in the mathematical population geneticists' scheme, Darlington's chromosome-centered genetic system had no single dynamic focus. It integrated several interacting variables: the shape of the chromosomes, their number, the life cycles of organisms, their reproductive organizations, and their breeding behavior (outbreeding or inbreeding, sexually differentiated or hermaphrodite). All these were related in a system whose parts are mutually adapted and adaptively connected. Years later the Harvard University systematist and ornithologist Ernst Mayr wrote to Darlington: "Your thinking simply did not fit into the atmosphere of what I have dubbed 'beanbag genetics.'"[14]

Fisher had in fact realized that the hereditary mechanism could evolve under pressure of selection. In his 1932 address to the genetics congress at Ithaca, he stated: "Others have considered the bearing of the theory of heredity on evolution. I am going to consider the bearing of evolution on heredity."[15] His paper "Evolutionary Modification of Genetic Phenomena" reversed the accepted order manifested in Haldane's address, titled "Can Evolution Be Explained in Terms of Present Known Genetical Causes?"[16] But the level at which Fisher posed the question—the *gene* level—was different from Darlington's—that of the *behavior of whole chromosomes.* At Ithaca Fisher was unaware that Darlington had just published an account arriving at the same principal conclusion concerning the relevance of evolution to heredity, but when confronted by the young man with the notion of dynamic genetic systems he "never appeared to hear."[17]

Haldane was personally closer to Darlington, having acted as something of a mentor during the time he spent at the Innes as head of genetical research beginning in the spring of 1927. Haldane understood that the bell attached to Darlington's perspective must ring true in nature: the primary result of chromosome-centered genetic systems is the generation, preservation, and recombination of differences on which natural selection acts in furthering evolutionary change.[18] Though theoretical models could not yet be constructed to describe the full implications of the chromosome-centered genetic system for evolution in quantified terms, it was clear that such chromosomal phenomena as inversions, recombination, polyploidy, balanced lethals, and ring formation collectively play a more important role than simple point mutations, even if they constitute mere shuffling of existing genetic materials instead of creating entirely new ones. Evolutionary-minded geneticists had put all the focus on genes and mutations. Darlington now made it clear that chromosome mechanics play a more important role. Once again, Mayr summed it up: "[N]o one made a greater contribution to the understanding of recombination and its evolutionary importance than Darlington."[19]

The conceptual difficulties presented by Darlington's scheme, much like the initial attempts to challenge his microscopy, fail to explain the harsh reaction against his ideas. In science, such adjectives as dangerous and poisonous are reserved for more than mere unusual theories. They usually rear their ugly heads against what the established community of researchers dub scientific rebellions, full-on challenges to orthodox wisdom, daring confrontations that must, at all costs, be put down or else. What, then, can account for the unusual roughness with which Darlington, the young and brash scientific upstart, was treated? What was he rebelling against?

METHOD

In the winter of 1932, John Belling at the University of California at Berkeley was engaged in preparing a scathing attack on *Recent Advances in Cytology*. Belling died of heart disease at the end of February, and in May of that year, "Critical Notes on Darlington's *Recent Advances in Cytology*" appeared posthumously. "The method of the author," Belling wrote, "is to try to establish general propositions, and then to deduce from these. Unfortunately, to a general proposition there are sometimes enough exceptions to invalidate it. . . . Darlington's book contains too many conjectures."[20] Referring to Darlington's writing as "propaganda," America's leading cytologist suggested that Darlington cut out nine-tenths of his conjectures in the second edition.

Darlington was unshaken. It seemed apparent to him that the distinction between fact and theory (especially in cytology) was useless or was only useful so long as it was remembered that it was arbitrary. In a reply to Belling he offered a forceful justification of his method: "Belling maintains the morphological point of view, of which he is the ablest exponent. I maintain the analytical point of view, namely: every student should be aware of the conjectures that may reasonably be made in regard to the causes of the events he is studying. I maintain it partly, perhaps, because *to me the cause is more real than the event itself*."[21]

Darlington was now in America, a twenty-nine-year-old Rockefeller Fellow for the 1932–1933 academic year, hopping between labs on the East and West Coasts. On a visit to Woods Hole (where he met Morgan and E. B. Wilson, among others), a young lady approached him after a talk and suggested that they marry. Darlington knew that such impetuousness would gravely hurt his parents, and he seemed to revel in the thought. "Promiscuity only flourished under condemnation," he noted blithely in his diary. His socially conservative parents wrote frantic letters in an attempt to dissuade their rebellious son from tying the knot ("have you lost all your moral sense? My dear boy, think a good deal of what you are doing before it is too late! . . . Father has no respect left for you, and I am made to feel it. Your loving and affection-ate mother"). Darlington chose to drive the taunt further: "We bathed in the sun entirely in the nude and got a little sunburned," he wrote, no doubt relishing the moment. Finally, miffed, he barked: "I wonder whether you will ever learn not to expect your son to behave the same way as your next door neighbor's sons!"[22] The marriage lasted only a few months, but Darlington had already set himself on his course. He was self-consciously fashioning himself a rebel on all fronts. "Those who obey a general moral code," he wrote in his diary, "will always resent those who adopt an independent and changeable one. . . . *Vox populi* is always *vox diaboli.*"[23]

What was true for the bedroom was true for the preparation slide. When Belling attacked Darlington, he was representing an entire school of inductive, cautious cytology that was now under threat from a young, gun-slinging deductive upstart. Self-taught and arrogant, Darlington took pleasure in bucking the old-schoolers. He found their patience tiring, their focus on technical problems barren, and their caution sterile. "I always want to preach what I practice," Darlington now wrote, "a dangerous course for a rebel."[24]

But if the icon of careful induction had been Darlington's target, why, still, was the reaction against him so strong? What was it about the world of biology that rendered Darlington's methodological rebel-lion poisonous to young minds?

THE DISCIPLINARY LANDSCAPE

The history of cytology falls rather naturally into quarter-centuries, beginning with Oscar Hertwig's discovery of fertilization in 1875; life begins, it was finally shown, with the meeting of egg and sperm. From then until 1900, the mitotic apparatus, the chromosomes, gamete formation, and early embryology were all defined and recognized. It was an exciting period, the atmosphere charged with the conviction that science was on the verge of uncovering the secrets of the life pro-cesses and perhaps life itself. It was clear that cytology represented an independent field of research at the forefront of progress in biology.

When the laws of the Moravian monk Gregor Mendel were redis-covered at the turn of the century, cytology and genetics stood on al-most equal footing as far as the new study of heredity was concerned. The Sutton-Boveri hypothesis stated that chromosomes in the nucleus of cells and traits expressed in the bodies of organisms were two sides of the same coin, and cytology and genetics were the handmaidens leading biology toward the unlocking of the secrets of heredity. Before long, a close cooperation between cytologists and geneticists gave birth to the chromosomal theory of heredity.

And yet Janssens's 1909 theory of chiasma types marked the beginning of a trend that was to redefine the role of cytology for the generation ahead, for when Morgan took up the notion that the observed chiasmata were in fact the loci of genetic crossing-over, the statistically based breeding and mapping work of the geneticists suddenly became emancipated. As a divide between describers and experimenters increasingly bifurcated biology, willy-nilly cytology came to play the secondary role of supporter. Soon, cytologists' virtue came to lie in their utter reliability. Proceeding with extreme caution, they would no longer make claims unless they could be verified by increasingly stricter standards of empirical proof. This environment quickly engendered a theoretical sterility fostered by an acceptance of Bateson's exhortation to treasure one's exceptions. These soon amounted to hundreds of papers filling up the cytological literature.

A comparison of the first and third editions of E. B. Wilson's *The Cell* proves the point. The 1896 book was a highly speculative and bold account of cytological knowledge on the eve of Mendel's rediscovery. Twenty-nine years later, however, any sign of speculation had vanished without a trace, and the initial 371-page deductive treatise had been replaced by a 1,232-page inductive collection of disintegrated observation. The third edition, replete with hedging and prevarications about the laws of pairing and crossing over, was precisely what Weismann had been speaking of forty years earlier when he described the immense number of detailed facts that lacked an intellectual bond. The pioneer cytologist had become a doubting Thomas, and, in contrast to the budding geneticist, seemed to be doing little to help the cause.

Arriving on the scene shortly after the appearance of Wilson's book in 1925, Darlington immediately recognized his own situation reflected in that of his hero Weismann. Remembering Darwin's exhortation—"I can have no doubt that speculative men, with a curb on make far [*sic*] the best observers"[25]—Darlington, like Weismann before him, felt that in an age of empiricism it was important that someone have the courage to speculate. His disposition, like Weismann's, was theoretical, despite working in a highly detailed, descriptive field. Like his predecessor, Darlington was not afraid to advance conjectures unsupported by observation but rather deduced from genetic assumptions. Finally, like him, Darlington was really using cytology as a tool with which to probe the problems that were of greatest interest to him—those of evolution. The referentiality was intentional: Almost every Darlington paper written after 1931 quoted Weismann's great 1887 speculation about reduction division, implicitly drawing the connection to Darlington's bold speculations. When Darlington grasped that cytology needed to be emancipated and made relevant once more, he changed from a mere rebel with a field to a rebel with a cause; *Recent Advances in Cytology* was his manifesto. Haldane gave expression to this rebellious revolution, calling his young colleague's research "the beginning of a new epoch, the transition from an essentially descriptive to a largely deductive science."[26]

Old-school cytologists such as Belling were not nearly as pleased. To them, Darlington's new type of bold, deductive, and often conjectural hypothesis making represented a threat to their scientific autonomy and credentials. Cytology, after all, had come to be defined by its cautious, meticulous, "show-me" style. Anything else was not only not cytology, it wasn't even science! If they threw caution to the wind and faltered, what honor would remain in the profession? How would cytologists ever again be trusted?

By breaking the hold of the old inductive method, Darlington shifted cytology in the biological landscape of his day in relation to genetics and evolutionary theory. Cytologists could once again ask big questions about the origin of species and the evolution of taxonomical lines, rather than sinking deeper and deeper into the increasingly technical and unintelligible language of karyokinesis, heterotype, and the ins and outs of Acidic-Lacmoid Squash methods. And where cytology could once again dare, as in the days of Weismann, biology was sure to gain.

Nowhere was this more apparent than in what became known as the evolutionary synthesis. The catalyst of this enterprise was the 1937 book *Genetics and the Origin of Species* by the Russian-born American geneticist Theodosius Dobzhansky. This was the first coherent work to link the dry theories of the mathematical population geneticists to natural populations in the wild. In a series of experiments charting seasonal and yearly morphological and genetic changes in the chromosomes of natural populations of *Drosophila* (inversions and recombinations), Dobzhansky had shown how cytological changes could be selected and could influence the evolution of a population.

He relied heavily on Darlington's work, citing him more than any researcher other than Sturtevant and Sewall Wright. Never mind that Darlington would be afforded little credit for his role in the synthesis.[27] What counted was that Dobzhansky's book was immensely successful in placing the chromosome at the heart of evolution.[28] This had been Darlington's aim all along, but he was a cytologist. What business did he have tackling evolution? Only after Dobzhansky made it legitimate to consider chromosome behavior in relation to evolution could Darlington's original ideas be slowly understood and incorporated into the mainstream. Cytogenetics, the science Darlington in no small measure helped resurrect in the 1930s, had ultimately become, through Dobzhansky, an important catalyst of the evolutionary synthesis, arguably the greatest intellectual achievement in biology since Darwin's. By blazing a new and boldly deductive methodological path, Darlington succeeded in breaking down long-preserved boundaries between the different disciplines of the life sciences. Weismann, he no doubt believed, would have been proud.

EPILOGUE

Applying chromosome mechanics to evolutionary theory, Darlington had made bold and grand speculation the very totem of his science. Method, first and foremost, was the stuff of his rebellion. In the end, many of Darlington's daring speculations about chromosome mechanics would eventually have to be revisited or overturned.[29] And though rebellions of any kind invariably exact a price, truth, as Francis Bacon intoned, really does sometimes emerge more readily from error than from confusion. Thus the general perspective on the importance of chromosome mechanics in evolution would remain, if its complex details have yet to be worked out. Darlington's iconoclasm is important for the history of biology because it happened just as the legacies of the two greatest biologists of the nineteenth century—Darwin and Mendel—were finally being wed to each other, helping to construct a picture of life that has remained with us, notwithstanding a poke here and there, until this very day.

Studying Darlington's life, one gets the feeling that nothing but unhindered iconoclasm would satisfy this Lancashire lad's need to justify his own existence. When, after his revolution, cytology slowly returned to a more inductive, cautious style, in the early 1940s Darlington turned to cytoplasmic inheritance, once again exciting ire and controversy. In the 1950s and 1960s he turned to man, applying the determinism he had witnessed in the dance of chromosomes under the microscope to human history and culture, sparking greater controversy still. "The most important knowledge of any time is the knowledge whose truth is disputed," he wrote in his trusted diary. "No subject keeps my interest when I find my own view of it agrees with the accepted or majority view."[30] Spoken like a true rebel!

Darlington was a biologist of great intuition but often of less-than-responsible practice. He sometimes made daring claims based on little empirical evidence, even if rooted in solid theoretical ground. Like Weismann, he practiced and took pleasure in the art of the considered guess. What can we learn from his experience? Perhaps the most important lesson is this: science always needs both Promethean theoretical leaps and ant-like, blow-whistle, painstaking drudgery. At certain times, an emphasis on the one may come at the expense of the other, but this is not always a bad thing. In the end, the success of Darlington's rebellion in cytology meant that a stultified field could once again be made to thrive. That his empirics were reined in by the next generation of cytologists simply goes to show what he always believed to be true about his trade. "Science advances as though by the pulling out of a drawer which gives on one side only to jam on the other. . . . There is nothing to be gained except by pulling on the other side."[31] The blind August Weismann, one of the greatest cabinet-makers biology has ever known, would no doubt wink at that.

NOTES

1. August Weismann, *Die Bedeutung der sexuellen Fortpflanzung für die Selektionstheorie* (Jena: Gustav Fisher, 1886), 295.

2. August Weismann, *Studies in the Theory of Descent,* ed. and trans. R. Mendola (New York: AMS Press, 1975), xv.

3. August Weismann, "On the Significance of the Polar Globules," *Nature* 36 (1887): 607–609.

4. Autobiographical notes, 14 November 1977, the Darlington Papers (DP): C.1: A.3, Bodleian Library, Oxford University.

5. Darlington to his father, DP: C.5: A.169 (1922).

6. Taped interview with Darlington by Brian Harrison (25 July 1979), John Innes Horticultural Collection, John Innes Institute, Norwich, U.K.; DP: C.88: f.82.

7. William Coleman, "Bateson and the Chromosomes: Conservative Thought in Science," *Centaurus* 15 (1970): 228–314.

8. Darlington Diary, 1927, DP: g.32: A.65.

9. Cyril Darlington, *Recent Advances in Cytology* (London: Churchill, 1932), viii.

10. Hampton Carson, "Cytogenetics and the Neo-Darwinian Synthesis," in *The Evolutionary Synthesis: Perspectives in the Unification of Biology,* ed. Ernst Mayr and William B. Provine (Cambridge: Harvard University Press, 1998), 91.

11. "Personal Record" (November 21), DP:C.1: A.8.

12. Ibid., 89.

13. A. H. Sturtevant and G. W. Beadle, *An Introduction to Genetics* (Philadelphia: Saunders, 1939), 364.

14. Mayr to Darlington (13 May 1974), DP: C.104: H.160.

15. Apparently, Fisher's aphoristic style was an improvement on what appears in print. This quotation is from Cyril Darlington, "J. B. S. Haldane, R. A. Fisher, and William Bateson," in *The Evolutionary Synthesis: Perspectives in the Unification of Biology,* ed. Ernst Mayr and William B. Provine (Cambridge: Harvard University Press, 1998), 74.

16. See *Proceedings of the Sixth International Congress in Genetics* (Menasha, WI: Brooklyn Botanic Garden, 1932), 165–172, 185–189.

17. Darlington, "J. B. S. Haldane, R. A. Fisher, and William Bateson," 79.

18. See J. B. S. Haldane, preface to Cyril Darlington, *Recent Advances in Cytology* (London: Churchill, 1932), vi.

19. Ernst Mayr, "Prologue," in *The Evolutionary Synthesis: Perspectives in the Unification of Biology,* ed. Ernst Mayr and William B. Provine (Cambridge: Harvard University Press, 1998), 23.

20. John Belling, "Critical Notes on Darlington's *Recent Advances in Cytology,*" *University of California Publications in Botany* 17, no. 3 (1933): 76.

21. "Reply by C.D. Darlington," in ibid., 110 (emphasis added).

22. Letters between Darlington and his parents, August 1932, DP: C.6: A.184–190; Diary, DP: A.70: f.11.

23. Darlington, Diary entries from the 1930s, not all dated, DP: g.33: A: 66–70.

24. Ibid.

25. Charles Darwin to Charles Henry Lardner Wood, 24 February 1950, in *The Correspondence of Charles Darwin,* ed. F. Burkhardt and S. Smith (Cambridge: Cambridge University Press, 1985–91), 4:317.

26. Darlington, *Recent Advances,* v.

27. Darlington was cited fourteen times in Dobzhansky's 1937 book. Moreover, Chapter 4, "Chromosomal Changes," at fifty pages, was the centerpiece of the work. Nevertheless, Darlington is almost never mentioned as an important player in the evolutionary synthesis literature. The exception is Mayr and Provine (1998), yet Provine states in the epilogue: "The role of cytologists is especially difficult to assess. They seem to have played no major role in

the synthesis" (408). This is significant, for elsewhere he calls Darlington's *Evolution of Genetic Systems* "enormously influential" (70), meaning that his contribution is not understood as coming from cytology.

28. An immediate effect was the birth of a major branch of *Drosophila* work dealing with cytogenetics, evolution, and natural populations, pioneered at Texas.

29. The universal law of pairing two-by-two ceased to be thought of as universal. Chromosomes do often combine in threes and in fours, and evolution has found ways to safeguard orderly segregation nonetheless. Indeed, although not yet entirely understood or explained, it became apparent that other mechanisms besides chiasmata were playing important roles in keeping chromosomes paired in meiosis, and hence in their orderly segregation for gamete formation.

30. Darlington, Diary entries from the 1930s, not all dated, DP: g.33: A: 66–70; DP: e.11: A.82 (1962).

31. C. D. Darlington, *The Conflict of Science and Society* (London: Watts, 1949), 6.

FURTHER READING

Coleman, William. "Bateson and the Chromosomes: Conservative Thought in Science," *Centaurus* 15 (1970): 228–314.

Darlington, Cyril. *Recent Advances in Cytology* (London: Churchill, 1932).

———. *The Evolution of Genetic Systems* (Cambridge: Cambridge University Press, 1939).

Harman, Oren. *The Man Who Invented the Chromosome: A Life of Cyril Darlington* (Cambridge: Harvard University Press, 2004).

Hughes, Arthur. *A History of Cytology* (London: Abelard-Schuman, 1959).

Mayr, Ernst, and William B. Provine, eds., *The Evolutionary Synthesis: Perspectives on the Unification of Biology* (Cambridge: Harvard University Press, 1998).

CHAPTER SEVEN

Striking the Hornet's Nest:
Richard Goldschmidt's Rejection
of the Particulate Gene

MICHAEL R. DIETRICH

In 1938, L. C. Dunn, acting as the managing editor of *Genetics,* rejected a manuscript by Richard Goldschmidt for fear that it would "lead to unprofitable controversy." The manuscript was a preliminary report of some experiments on spontaneous mutation that Goldschmidt thought decisively weighed against the existence of the particulate gene. In a letter asking the opinion of Curt Stern, who had been Goldschmidt's assistant a few years earlier, before both had been forced to leave Nazi Germany, Dunn expressed his concerns. "If a paper like this had come from a less well known worker," Dunn wrote, "my first impulse would be to ask for better evidence that these events were not simple contaminations, or at least that the data should be reported in such a form that the reader would be free to test his own hypothesis by the data." Given the uproar that the attack on the gene would cause, Dunn feared that Goldschmidt's reputation would be severely damaged. Moreover, Dunn continued, "while I can see that there is a kernel of great interest and truth in it, no single point is proved by itself, and this will lead to unprofitable controversy and possibly to condemnation without a fair hearing, because he has presented his material in so difficult a form, has dismissed so airily any of the details, and has given so great an air of certainty to his interpretation."[1] The manuscript rejected by *Genetics* was accepted by *The American Naturalist* and appeared in 1939.[2] Goldschmidt was quickly enveloped in controversy.[3] But was Dunn right? Was this controversy "unprofitable"?

When Richard Goldschmidt sparked this debate he was sixty years old. Only a few years earlier he had been forced to resign his position at the Kaiser Wilhelm Institute of Biology in Berlin because of his Jewish ancestry.[4] From middle-class beginnings in Frankfurt, Goldschmidt had become a German Mandarin, a member of a social and cultural elite whose status owed more to education than to heredity.[5] This status shaped Goldschmidt into a formidable figure who could be both generous and authoritarian. Losing his position was one of the bitterest episodes in Goldschmidt's life. Although his reputation won him a position at the University of California at Berkeley, returning to teaching at a public university after twenty-three years as a research scientist was a shock and convinced Goldschmidt that he had been spoiled in Germany.[6]

Goldschmidt's expulsion from Germany and his feelings of loss may have led him to seek controversy as he tried to regain some of his past prestige.[7] He certainly knew that he was seeking controversy and explained his behavior to L. C. Dunn in terms of his "genetic makeup," which led him to "run ahead of the facts with conclusions" and assign new facts "their place within the whole." To Dunn, who was already worried about Goldschmidt's reputation, he wrote, "I prefer being wrong to a terribly cautious agnostic. I work after all because I love to do it and because I thoroughly enjoy galloping around the field of ideas. If others enjoy teaching my horse to draw a bus, let them have their fun. Time will tell who goes farther."[8] While in Germany Goldschmidt had been outspoken, but in the United States his vehement denial of the particulate gene seriously threatened his good reputation.[9] Indeed, Goldschmidt's rejection of the particulate gene and his related rejection of gradual evolution earned him a new reputation as one of the leading heretics of twentieth-century biology.

GOLDSCHMIDT'S GERMAN GENETICS

As a student, Goldschmidt had been trained in the German morphological tradition. His early work on the nervous system of the nematode *Ascaris* gave way to research on heredity around 1909, when genetics itself was becoming a field of research. As Richard Hertwig's assistant at the University of Munich, Goldschmidt decided to apply genetic thinking to the classic problem of sex determination.[10] His experimental system for this research was the gypsy moth, *Lymantria dispar*. Typically, gypsy moths are sexually dimorphic: females have white wings with dark bands, whereas males are smaller and have brown wings. When different geographic varieties of gypsy moth are mated, however, they produce a continuum of sexual types that range from female to male. Goldschmidt coined the term *intersex* to describe

Figure 7.1. Richard Goldschmidt in his laboratory at the Kaiser
Wilhelm Institute for Biology in Berlin-Dahlem, ca. 1932.
By courtesy of Curt Stern Papers, American Philosophical Society,
Philadelphia.

these intermediate forms and proceeded to develop a genetic theory
that could account for this continuum.

When American geneticists such as Thomas Hunt Morgan were
embracing the problems of transmission genetics, Goldschmidt approached the more complex phenotypic expression in gypsy moths
in terms of the developmental and physiological effects of genes and
gene products.[11] Goldschmidt proposed that all gypsy moths had both
male and female genes. Whether a moth appeared male or female
depended on the balance between its male and female genes. The

relative strengths of the female and male genes determined how long an organism would develop as one sex before switching to the other. This Time Law of Intersexuality allowed Goldschmidt to explain how a continuous range of sexual phenotypes could be produced by moving this critical turning point in development.[12]

In order to support his theory of intersexuality, Goldschmidt developed an experimental system for breeding and crossing different geographic varieties of gypsy moths. The premise of this system was that different geographic varieties had evolved factors of different strengths. Within a region, the male and female factors tended to be in balance, but if a European male moth was crossed with a female Japanese moth, for instance, an intersex would be produced. Between 1911 and 1931, Goldschmidt traveled around the world to collect gypsy moths and to study their geographic distribution, especially in the Japanese archipelago. The moths he collected would be systematically bred and crossed to determine the strength of their sex determining-genes. The result was one of the first studies of the genetics of geographic variation, and it became widely cited in the evolutionary biology literature.[13]

In the course of his research on gypsy moths, Goldschmidt articulated a theory of physiological genetics that gave tremendous agency to developmental and physiological processes. Borrowing ideas from enzymatic activity and the impact of hormones in development, Goldschmidt proposed that gene expression could be understood in terms of timing. Each gene produced a substance (an enzyme or a hormone) at a certain rate. The phenotype associated with that gene would be expressed when enough substance was produced—when a reaction threshold had been crossed. The same gene producing the same amount of substance could produce very different phenotypic effects if the rate of production was fast or slow. More specifically, Goldschmidt postulated that there were critical periods during development. If a gene product crossed its expression threshold during that period, a "normal" phenotype would be produced. If the threshold was crossed early or late, a different but usually related phenotype would be produced. A single gene, then, could produce a wide range of phenotypes depending on its reaction system—the developmental and physiological processes that determined expression thresholds and critical periods.[14]

Note that genes played a major role in Goldschmidt's physiological genetics before 1934. Even after he began to have serious doubts about the particulate gene concept, he would fall back into discussion of genes in order to defend the power of developmental and physiological processes influencing phenotypic expression.

"THE THEORY OF THE GENE IS—DEAD!"

Goldschmidt began to question the existence of the particulate gene in 1932 when Theodosius Dobzhansky convinced him of the seriousness of position effects.[15] According to the particulate theory of the gene, genes were indivisible and independent units of structure, function, mutation, and recombination.[16] Position effects eroded this concept by demonstrating that a gene's neighbors had an influence on its expression.[17] According to L. C. Dunn, by the 1930s the particulate gene was showing "some signs of disappearing in a cloud of position effects."[18]

In 1934, H. J. Muller and his collaborators in the Soviet Union also began to question the nature of the classical gene through their research on mutation. Muller and his colleagues claimed that different kinds of chromosomal rearrangements could produce what appeared to be a bristle mutation in *Drosophila* called scute.[19] Unlike other instances of position effect, however, the locations of the breaks associated with Muller's scute rearrangements were distributed over a much larger section of the chromosome than had been thought possible.[20] For Goldschmidt, Muller's research and similar research in other labs suggested that most mutations were in fact rearrangements of different sizes, each having different phenotypic effects. When Morgan and his co-workers articulated their theory of the particulate gene, they often invoked the metaphor of a string of beads to explain the arrangement of genes on a chromosome. The genes were indivisible, individual beads strung together in a linear fashion to form a chromosome. A mutation, in this theory, was a chemical change in a single bead. Muller and his co-workers were now suggesting that breaks, not simply chemical changes, were producing mutation-like effects. More important, Muller's results suggested that whereas the breaks involved in a rearrangement could be localized, their distribution over a relatively large area of chromosome suggested that the function of producing scute bristles was also distributed over an area much larger than a single "bead."

As Goldschmidt toured the United States in 1935, searching for a new position and raising money from lecturing fees, he began to articulate his doubts about the particulate gene. At the time his views were almost exclusively based on the experiments of others on position effects and on the ability of x-rays to produce minute rearrangements in chromosomes. Goldschmidt reported that the response to these ideas about the gene was that he had gone crazy.[21]

In the summer of 1936, however, new experimental evidence about the mutability of the *Drosophila* genome was introduced and brought to bear on the question of the gene. Milislav Demerec, Harold Plough, and C. Holthausen announced at the meetings of the Genetics Society of America that they had independently observed a high frequency of spontaneous mutation in an inbred Florida stock of *Drosophila*. Goldschmidt and his family were en route from Germany to Berkeley, where Goldschmidt would become a professor at the University of California, and so he did not attend the meeting. Before he left Berlin, however, he had also observed a higher-than-normal number of spontaneous mutations in the same Florida stock. In 1937, the three labs involved all published the results of their experiments.[22] Goldschmidt's article, however, stood out from his peers'.

His short article in the *Proceedings of the National Academy of Sciences* claimed that increased spontaneous mutability was evidence against the existence of the classical gene. The sudden appearance of many different kinds of mutations, Goldschmidt argued, was the result of multiple rearrangements at a number of different loci. In Goldschmidt's experiments, parents with the blistered mutation would produce offspring without blisters but with plexus, dumpy vortex, thoraxate, and purple in every individual. Goldschmidt interpreted these events as chromosomal rearrangements with an insertion or break in the blistered locus.[23] He admitted that his results were preliminary, but he was willing to conclude that all gene mutations were in fact rearrangements and that all mutations were in fact position effects. This had important consequences for the gene concept, because although a break could be correlated with the appearance of a phenotypic effect, that did not imply that the functional ability of a wild-type allele could be limited to just the site of the break or mutation corresponding to the produced phenotypic effect. The unit of normal genetic expression, for Goldschmidt, was the "whole, wild type chromosome."[24] He promised to clarify his ideas in a future publication; though he did claim in 1937 that the chromosome was the "unit," he also recognized that the chromosome had to have some internal structure or "texture" in order to ensure normal development.

Writing to L. C. Dunn, Goldschmidt almost bragged that the 1937 article would make him "an outcast in genetics" because he "now stated in writing, as before only orally, that there is no such thing as a gene. Horror!"[25] The following year, when Goldschmidt attempted to publish a more detailed account of his experiments in *Genetics,* he argued that although his experiments were not yet finished, he wanted to start with a general publication describing his theory of gene action and rearrangement followed by detailed and supporting analysis of particular mutants. As editor of *Genetics,* L. C. Dunn could not go along with this plan and refused to publish Goldschmidt's generalizations. Instead, Dunn urged Goldschmidt to write up his analysis of results regarding specific mutants and their representation on the salivary gland chromosomes.

Goldschmidt heeded Dunn's advice, and his experimental research program from 1939 to 1945 was devoted to completely analyzing the spontaneous mutations that he had observed in *Drosophila.* At the same time, Harold Plough and Milislav Demerec attacked Goldschmidt's claims for the importance of rearrangements. At the 1941 Cold Spring Harbor Symposium, for instance, Plough and Demerec argued that spontaneous mutations were not associated with any rearrangements and were only occasionally associated with small deficiencies.[26] Four years later Goldschmidt published his complete analysis, not in *Genetics,* but as a 549-page monograph in the *University of California Publications in Zoology* series.[27] Goldschmidt's vast analysis did not quell his critics, however, for while his laboratory group had been experimentally analyzing mutants, Goldschmidt had been following what he saw as the evolutionary consequences of his new perspective on the gene.

THE CASE FOR SYSTEMIC MUTATIONS

In 1939 Goldschmidt was invited to give the prestigious Silliman Lectures at Yale University. He decided to articulate the "phylogenetic consequences" of his rejection of the gene by proposing a mechanism of saltational evolution based on genetic rearrangements. In 1935 Muller and his co-workers had hesitated to accept that all mutations could be rearrangements because they believed that "the range of possibilities of phenotypic change through intergenic rearrangements alone must be far from adequate for any indefinitely continued evolution."[28] This was the gauntlet that Goldschmidt took up. A lifetime of work on gene action convinced him that rearrangements provided a wide spectrum of phenotypic possibilities for evolution to do with as it may. But he also had ideas about evolution's actions that significantly broadened his controversial standing within American biology.

The principal claim of Goldschmidt's Silliman lectures, published in 1940 as *The Material Basis of Evolution,* is that microevolution and macroevolution are distinct and require different mechanisms. Evolution below the species level could proceed by the gradual accumulation of small mutations, but macroevolution required a different means to bridge the gap between species. Goldschmidt proposed two new mechanisms for macroevolution: systemic mutations and developmental macromutations.[29]

For Goldschmidt, systemic mutations were massive rearrangements of the entire reaction system associated with a chromosome. The result of such mutations would be a new and significantly different chromosomal pattern with a well-integrated set of associated physiological reactions. This systemic repatterning would become apparent phenotypically as a sudden shift to a new form. Calling on his extensive research in physiological genetics, Goldschmidt argued that the genotypic effects of repatterning would appear when a stable chromosome arrangement created a new system of reactions such that the expression thresholds for the new phenotype could be crossed. So the chromosomal repatterning and rearrangement of reaction pathways associated with a systemic mutation could take place over an extended period of time, but the resulting new phenotypes would appear suddenly when the expression threshold was crossed.

Goldschmidt's argument for the role of systemic mutations was intended to make the best case possible for an alternative to the emerging neo-Darwinian synthesis. In particular, Goldschmidt was directly responding to the claims of the influential evolutionary geneticist Theodosius Dobzhansky. Goldschmidt believed that Dobzhansky had failed to decide whether species were formed by the accumulation of genetic mutations or by the accumulation of chromosomal rearrangements.[30] In Goldschmidt's eyes, Dobzhansky knew a decision had to be made when he wrote in *Genetics and the Origin of Species*:

> To what extent the differences between such species as *Drosophila pseudoobscura* and *D. miranda* are due to position effects is also a matter of speculation; the greatly different gene arrangements in these species may be responsible for many alterations in the morphological and

physiological properties of their carriers. In any event, position effects show that gene mutations and chromosomal changes are not necessarily as fundamentally distinct phenomena as they at first appear.[31]

Dobzhansky's failure to take the next step and admit that chromosomal repatternings could be the decisive changes needed for speciation is, according to Goldschmidt, the result of "a dogmatic belief in the inflexibility of the classical theory of the gene."[32] Unfettered by the dogma of the gene, Goldschmidt could promote a theory of evolution that was compatible with the "facts," meaning Dobzhansky's and Muller's work on position effects and chromosomal differences. Framing his position in this way issued a direct challenge to Dobzhansky and the neo-Darwinians: they had to come to terms with the possible evolutionary impact of chromosomal rearrangements.[33]

Goldschmidt summarizes his case for macroevolution by systemic mutations as follows: "Whether this model is good or bad, possible or impossible, the fact remains that an unbiased analysis of a large body of pertinent facts shows that macroevolution is linked to chromosomal repatterning and that the latter is a method of producing new organic reaction systems, a method which overcomes the great difficulties which the actual facts raise for the neo-Darwinian conception as applied to macroevolution."[34] At this point in *The Material Basis of Evolution* he seems to hedge his bets by introducing a second mechanism for macroevolution that did not depend on his rejection of the particulate gene.

Drawing on his work in physiological genetics, Goldschmidt proposed that mutations in developmentally important genes could produce large phenotypic effects and that these large changes could result in rapid speciation. He called the results of these developmental macromutations "hopeful monsters." These were not necessarily the result of systemic mutations. Instead, Goldschmidt used hopeful monsters to argue, by analogy, for evolution by systemic mutations. The possibility of mutations in developmentally important genes was intended to make the genetic mechanism of systemic mutation more plausible.[35]

Unlike his theory of systemic mutations, Goldschmidt's idea of developmentally significant mutations with large effects was widely recognized and accepted. Contemporaries such as Curt Stern, W. Dwight Davis, George Gaylord Simpson, and Sewall Wright each recognized the importance of developmental macromutations.[36] Whereas Goldschmidt's theory of speciation via systemic mutation had no support

among geneticists or evolutionary biologists, biologists could engage
with the idea of hopeful monsters without debating the significance
of chromosome rearrangement and the nature of the gene. The idea
of hopeful monsters was not, however, without its critics. Simpson
pointedly argued that a single example did not constitute evolution.[37]
In the emerging evolutionary synthesis, the dynamics of evolving
populations was centrally important, and Goldschmidt's emphasis
on chromosomes and the potentialities of development did not incor-
porate this turn to thinking in terms of populations instead of individuals or types. Goldschmidt's immediate response to these charges
was to appeal to what he knew—the physiological dynamics of gene
expression. Only later did he persuade his old friend Sewall Wright
to demonstrate how mutations with a large phenotypic effect could
spread through a population. Indeed, Wright incorporated macromutations into his shifting balance theory as one possible mode of
evolutionary change.[38]

In retrospect, Goldschmidt described the reaction of the architects
of the emerging evolutionary synthesis as savage. It seemed to him
that he "had struck a hornet's nest."[39] Almost every review of *The
Material Basis of Evolution* was negative, and yet important figures
such as Dobzhansky realized that Goldschmidt's challenge had to be
taken seriously.

A GENETIC ALTERNATIVE

From 1940 until his death in 1958, Goldschmidt pursued two lines
of research in support of his rejection of the gene and his theory of
evolutionary change. On one hand, he developed a research program
concerning homeotic mutants, creating an interpretation of their effect
on developmental processes and their role as important mechanisms
of evolutionary innovation.[40] The homeotic mutants of *Drosophila*
were the embodiment of the hopeful monster. On the other, Goldschmidt worked to articulate an alternative to the particulate gene.
Emphasizing that hereditary units were those necessary to produce
normal development, he rejected genetic definition based on the localized effects of mutation in favor of a hierarchy of genetic units.[41]

In the place of the classical gene, Goldschmidt proposed a genetic
hierarchy with five levels. At the bottom were subgenes—structures
that were only then being visualized in salivary gland chromosomes
of *Drosophila* after treatment at high pH.[42] Next came genes, which

he associated with structures visualized as bands on salivary gland chromosomes. Above the gene level were extended regions of gene action such as the yellow-scute region that Muller and others characterized as spreading over many bands on the X chromosome in *Drosophila.* Above this were even larger sections of the chromosome that might have identical action. At the top of the hierarchy were the blocks of euchromatin and heterochromatin discovered by Emil Heitz.[43]

This proposed hierarchy of structures also corresponded to different types of genetic action. At the three lowest levels, rearrangements of different sizes produced position effects. The two higher levels were associated with the action of homeotic mutants that were observed by Goldschmidt to coexist on the third chromosome of *Drosophila* and with effects produced by rearrangements involving heterochromatin. Muller had shown that rearrangements with one break in heterochromatin and the other in euchromatin behaved differently from those with both breaks in euchromatin. In particular, rearrangements involving the placement of the terochromatin next to euchromatin seemed to have position effects that stretched over long distances.[44]

The culmination of Goldschmidt's efforts to articulate a genetic hierarchy came in his 1955 book *Theoretical Genetics.* Based on his own study of yellow mutants, Muller's study of scute mutants, and Demerec's study of other mutants, such as white and Notch in *Drosophila,* in 1955 Goldschmidt argued that the ability to produce these phenotypes was spread over segments of chromosome and that these segments could overlap. This overlap led Goldschmidt to believe that "it cannot be the morphological segment which counts [as the hereditary unit], but a field-like function of the segment which under certain conditions . . . reaches from the center of the segment to different distances."[45] This interpretation in terms of fields still associated chromosomal structures and functions but gave primacy to function or gene action.

The same interpretation in terms of fields was offered for position effects produced by breaks in heterochromatin. Rearrangements with one break in heterochromatin (and the other in the euchromatin) were known to have a greater effect than rearrangements with both breaks in euchromatin. In 1941 Demerec had explained this difference in terms of what he called sensitive regions. According to Demerec, when a rearrangement occurred new parts of the chromosome were brought into contact with each other, resulting in a position effect. The sensitive region was the region surrounding the place where this new contact occurred and in which position effects could be detected. Using his research on Notch and white mutants, Demerec concluded that rearrangements involving placement of euchromatin next to heterochromatin resulted in a sensitive region that was five to ten times bigger than a similar rearrangement involving only euchromatin.[46] Demerec explained this effect as a change in either the gene or in gene activity. Goldschmidt thought that explanations that appealed to heterochromatin's action on distant genes provided no insight and were "devoid of meaning."[47] Instead Goldschmidt preferred to think of heterochromatin as stretching the fields of adjacent segments to produce what he thought of as a case of "an extreme type of overlapping." This stretching effect convinced Goldschmidt that chromosomes had to be understood in terms of segments with associated fields of action.

As he had done in 1944 for genetic structures, Goldschmidt ordered these segments and their fields into a hierarchy. These genetic fields would range from those associated with submicroscopic segments of the chromosome to larger fields covering possibly the entire chromosome.[48] The action of genic material at any time could be the result of a field at any one of these levels.

Whereas Goldschmidt's ideas of a hierarchy of units and fields was favorably received by some,[49] his retheorizing of the units of heredity did not have a significant impact because most biologists embraced a neoclassical concept of the gene that preserved the structure-function definition of the particulate gene without a definitional role for mutation or recombination.[50] Strictly speaking, Goldschmidt was correct —the particulate gene concept was dead as a concept that unified structure, function, mutation, and recombination. The way in which he made his case for rejecting the gene, however, seems to have detracted from his ability to make a case for his alternative. Typically Goldschmidt's alternative to the particulate gene has been characterized as the chromosome-as-a-whole hypothesis. Such characterizations reveal that most commentators have failed to come to terms with the views that Goldschmidt adopted after 1940.[51] Perhaps, as Stephen Jay Gould has noted, the result of Goldschmidt's controversial conjectures was that he was fated to be both "ridiculed and unread," at least with respect to his later work.[52]

CONCLUSION

In his rejection of the particulate gene and his extension of his idea of gene action from chromosomal to systemic mutations and macro-evolution, Richard Goldschmidt set himself against central ideas in both genetics and evolutionary biology. Goldschmidt was fully aware that these views would be controversial and, indeed, seemed to relish these disputes. He was taken aback by the overwhelmingly negative reaction with which his views were met, but he did not stop defending, researching, and articulating his views.

Dunn feared that his friend would create "unprofitable" controversy and damage his reputation in the process. Although he recognized the legitimacy of position effects, Dunn worried that Goldschmidt's forceful assertions would create useless opposition instead of suggesting further research. This kind of dissent for dissent's sake would lead nowhere. Despite some difficulty replicating Goldschmidt's spontaneous mutation experiments, Goldschmidt and others did produce a series of further experiments evaluating his claims about chromosomal rearrangement and their implications for the particulate gene. In this sense, Goldschmidt's dissent was profitable in that it did not lead to an irresolvable dispute.

A stronger claim can be made, however, concerning the value of dissent. By challenging a dominant, indeed foundational concept in genetics, Goldschmidt's dissent created an opportunity for innovation—for the articulation of alternatives to the accepted foundations of genetics. Whereas other geneticists such as H. J. Muller were willing to question some features of the particulate gene, such as its appropriateness as a unit of mutation, Goldschmidt was willing to rethink the gene more completely. Drawing on his knowledge of gene action and his credibility as an expert in physiological genetics, Goldschmidt's rejection of the gene was at once radical and not without scientific foundation. That his colleagues were not willing to take the same intellectual leap as he was does not mean that they did not change as a result of Goldschmidt's dissent and the resulting controversy. After World War II, the particulate gene was abandoned in favor of a gene concept that emphasized structures and functions that were being increasingly redefined as genetic fine structure and genetic regulation became important new areas of research.

Richard Goldschmidt stands out as an iconoclast because he deliberately opposed foundational aspects of genetics and evolution. His tendency for dramatic overstatement and forceful advocacy certainly helped reinforce his image as a rebel or a heretic. Many strenuously opposed him during his lifetime, but Goldschmidt's reputation as a heretic endures today because many are still invested in maintaining this reputation. Scientific creationists seeking biologists who were accomplished and well-recognized frequently invoked Goldschmidt as a scientist who disagreed with the evolutionary "orthodoxy."[53] Goldschmidt's dissent from neo-Darwinian orthodoxy was taken to legitimize their own dissent. Goldschmidt's reputation as a heretic was most firmly cast by Stephen Jay Gould.[54] Like the creationists, although he would shudder at the comparison, Gould used Goldschmidt's reputation as an opponent of neo-Darwinian gradualism to promote his own dissention. As Gould promoted his theory of punctuated equilibrium, he allied himself with Goldschmidt. He even arranged for *The Material Basis of Evolution* to be reprinted by Yale University Press and to be reviewed in *Paleobiology* by a number of leading biologists. Gould's identification with Goldschmidt was cut short, however, when biologists began to ask seriously if he was advocating Goldschmidt's mechanism of systemic mutations for macroevolution. In the hands of Gould and many others, the value of Goldschmidt's dissent was not that he opened new dialogues about the nature of the gene or the mechanisms of evolution, but that, by association, his iconoclasm can become a marker of innovation.

NOTES

1. L. C. Dunn to Curt Stern, June 28, 1938. Curt Stern Papers, American Philosophical Society Library, Philadelphia.

2. Richard Goldschmidt, "Mass Mutation in the Florida Stock of *Drosophila melanogaster:* Details of an Old Experiment Reinterpreted," *American Naturalist* 73 (1939): 547–559.

3. Portions of this article are derived from earlier publications, including Michael R. Dietrich, "On the Mutability of Genes and Geneticists: The 'Americanization' of Richard Goldschmidt and Victor Jollos," *Perspectives on Science* 4 (1996) 321–345; Dietrich, "Richard Goldschmidt's 'Heresies' and the Evolutionary Synthesis," *Journal of the History of Biology* 28 (1995): 431–461; Dietrich, "From Hopeful Monsters to Homeotic Effects: Richard Goldschmidt's Integration of Development, Evolution, and Genetics," *American Zoologist* 40 (2000): 28–37; Dietrich, "Richard Goldschmidt: Hopeful Monsters and Other 'Heresies,'" *Nature Reviews Genetics* 4 (2003): 68–74; and Dietrich, "From Gene to Genetic Hierarchy: Richard Goldschmidt and the

Problem of the Gene," in *The Concept of the Gene in Development and Evolution: Historical and Epistemological Perspectives,* ed. Peter Burton, Raphael Falk, and Hans-Jörg Rheinberger, 91–114 (Cambridge: Cambridge University Press, 2000).

4. Richard Goldschmidt, *In and Out of the Ivory Tower* (Seattle: University of Washington Press, 1960); Curt Stern, "Richard Benedict Goldschmidt (1878–1958): A Biographical Memoir," *National Academy of Sciences, Biographical Memoirs* 39 (1967): 141–192, reprinted in *Richard Goldschmidt: Controversial Geneticist and Creative Biologist, Experientia Supplementum* 35 (1980): 68–99.

5. Fritz Ringer, *Decline of the German Mandarins: The German Academic Community, 1890–1933* (Cambridge: Harvard University Press, 1969), 5. Also see Jonathan Harwood, *Styles of Scientific Thought* (Chicago: University of Chicago Press, 1993).

6. Goldschmidt, *Ivory Tower,* 305–306; Richard Eakin, "Contributions to the Department of Zoology, University of California, Berkeley," in *Richard Goldschmidt: Controversial Geneticist and Creative Biologist, Experientia Supplementum* 35 (1980), 64.

7. Stern, "Richard Benedict Goldschmidt," 183.

8. Richard Goldschmidt to L. C. Dunn, May 27, 1940, L. C. Dunn Papers, American Philosophical Society Library, Philadelphia.

9. See L. C. Dunn to Curt Stern, May 28, 1938 and Stern to Dunn, June 6, 1938, Curt Stern Papers, American Philosophical Society Library, Philadelphia.

10. Marsha Richmond, *Richard Goldschmidt and Sex Determination: The Growth of German Genetics, 1900–1935* (Ph.D. diss., Indiana University, 1986); Jane Maienschein, "What Determines Sex? A Study of Converging Approaches, 1880–1916," *Isis* 75 (1984): 457–480.

11. Garland Allen, "Opposition to the Mendelian-Chromosome Theory: The Physiological and Developmental Genetics of Richard Goldschmidt," *Journal of the History of Biology* 7 (1974): 49–92; Richmond, *Richard Goldschmidt and Sex Determination.*

12. Richard Goldschmidt, "Intersexuality and the Endocrine Aspect of Sex," *Endrocrinology* 1 (1917); 433–456; Richard Goldschmidt, *The Mechanism and Physiology of Sex Determination,* trans. William Dakin (London: Methuen, 1923).

13. Richard Goldschmidt, "*Lymantria,*" *Bibliographia Genetica* 11 (1934): 1–185. Ernst Mayr, *Animal Species and Evolution* (Cambridge: Harvard University Press, 1963).

14. Richard Goldschmidt, *Physiological Genetics* (New York: McGraw-Hill, 1938).

15. Richard Goldschmidt, "Spontaneous Chromatin Rearrangements and the Theory of the Gene," *Proceedings of the National Academy of Sciences*

23 (1937): 621–623. Also see Theodosius Dobzhansky, "Position Effects on Genes," *Biological Review* 11 (1936): 364–384; Herman J. Muller and A. Prokofyeva, "The Individual Gene in Relation to the Chromomere and the Chromosome," *Proceedings of the National Academy of Sciences* 21 (1935): 16–26; Goldschmidt, *Physiological Genetics,* 308–309.

16. E. A. Carlson, *The Gene: A Critical History* (Ames: Iowa State University Press, 1966).

17. Position effects were discovered in 1927 by A. H. Sturtevant as he sought to explain bar-eyed mutants in *Drosophila.*

18. L. C. Dunn to Richard Goldschmidt, November 15, 1937, L. C. Dunn Papers, American Philosophical Society Library, Philadelphia.

19. Dietrich, "From Gene to Genetic Hierarchy."

20. H. J. Muller, A. Prokofyeva, and D. Raffel, "Minute Intergenic Rearrangement as a Cause of Apparent 'Gene Mutation,'" *Nature* 135 (1935): 253–255.

21. Goldschmidt, *Ivory Tower,* 323.

22. M. Demerec, "Frequency of Spontaneous Mutations in Certain Stocks of *Drosophila melanogaster,*" *Genetics* 22 (1937): 469–478; H. H. Plough and C. Holthausen, "A Case of High Mutational Frequency Without Environmental Change," *American Naturalist* 71 (1937): 185–187; Richard Goldschmidt, "Spontaneous Chromatin Rearrangements and the Theory of the Gene," *Proceedings of the National Academy of Sciences* 23 (1937): 621–623.

23. Goldschmidt, "Spontaneous Chromatin Rearrangements," 622.

24. Ibid.

25. Richard Goldschmidt to L. C. Dunn, November 3, 1937, L. C. Dunn Papers, American Philosophical Society Library, Philadelphia.

26. H. H. Plough, "Spontaneous Mutability in *Drosophila,*" *Cold Spring Harbor Symposia on Quantitative Biology* 9 (1941): 127–137; M. Demerec, "Unstable Genes in *Drosophila,*" *Cold Spring Harbor Symposia on Quantitative Biology* 9 (1941): 145–150.

27. Richard Goldschmidt, R. Blanc, W. Braun, M. Eakin, R. Fields, A. Hannah, L. Kellen, M. Kodani, and C. Villee, "A Study of Spontaneous Mutation," *University of California Publications (Zoology)* 49 (1945): 291–549.

28. Muller, Prokofyeva, and Raffel, "Minute Intergenic Rearrangement," p. 255.

29. Richard Goldschmidt, *The Material Basis of Evolution* (New Haven: Yale University Press, 1940).

30. Ibid., 204.

31. Theodosius Dobzhansky, *Genetics and the Origin of Species* (New York: Columbia University Press, 1937), 117, quoted by Goldschmidt, *Material Basis,* 204.

32. Goldschmidt, *Material Basis,* 242.

33. Dietrich, "From Gene to Genetic Hierarchy," 97.

34. Goldschmidt, *Material Basis,* 249. This passage is italicized in the original.

35. Goldschmidt, *Material Basis;* Dietrich, "Richard Goldschmidt's 'Heresies'"; Dietrich, "From Hopeful Monsters to Homeotic Effects."

36. Dietrich, "Richard Goldschmidt's 'Heresies.'"

37. Sewall Wright, "The Material Basis of Evolution by R. Goldschmidt (review)," *Scientific Monthly* 53 (1941): 165–170. See Dietrich, "From Hopeful Monsters to Homeotic Effects."

38. Richard Goldschmidt, *Theoretical Genetics* (Seattle: University of Washington Press, 1958).

39. Goldschmidt, *Ivory Tower,* 324.

40. Dietrich, "From Hopeful Monsters to Homeotic Effects"; Gregory K. Davis, Michael R. Dietrich, and David Jacobs, "Homeotic Mutants and the Assimilation of Developmental Genetics into the Evolutionary Synthesis," in *Descended from Darwin: Insights into American Evolutionary Studies, 1925–1950,* ed. Joe Cain and Michael Ruse (Philadelphia: American Philosophical Society, in press).

41. Richard Goldschmidt, "On Some Facts Pertinent to the Theory of the Gene," in *Science in the University* (Berkeley: University of California Press, 1944): 197.

42. M. Calvin, M. Kodani, and R. Goldschmidt, "Effects of Certain Chemical Treatments on the Morphology of the Salivary Gland Chromosomes and Their Interpretation," *Proceedings of the National Academy of Science* 26 (1940): 299–301.

43. Emil Heitz, "Heterochromatin, Chromocentren, Chromomeren," *Deutsche botanische Gesellschaft, Berlin* 47 (1929): 274–284. Heterochromatin is condensed chromosomal material that was considered inactive, whereas euchromatin was considered to be active.

44. Richard Goldschmidt, "Position Effect and the Theory of the Corpuscular Gene," *Experientia* 2 (1946): 197–230, 250–256.

45. Goldschmidt, *Theoretical Genetics,* 162.

46. M. Demerec, "The Nature of the Gene," in *Cytology, Genetics, and Evolution* (Philadelphia: University of Pennsylvania Press, 1941): 6.

47. Goldschmidt, *Theoretical Genetics,* 162.

48. Ibid., 180.

49. Kenneth Mather, "Genes," *Scientific Journal of the Royal College of the Sciences* 16 (1946): 64–71; Kenneth Mather, "Nucleus and Cytoplasm in Differentiation," *Symposia of the Society for Experimental Biology* 2 (1948): 196–216.

50. Petter Portin, "The Concept of the Gene: Short History and the Present Status," *Quarterly Review of Biology* 68 (1993): 173–223.

51. Carlson, *The Gene,* 125–128.

52. Stephen Jay Gould, "The Uses of Heresy: An Introduction to Richard

Goldschmidt's *The Material Basis of Evolution*," in *The Material Basis of Evolution*. (New Haven: Yale University Press, 1982), xiv.

53. Henry Morris, *Scientific Creationism* (El Cajon, Calif.: Institute for Creation Research Press, 1982).

54. Gould, "The Uses of Heresy," xiii–xlii; Stephen Jay Gould, "The Hopeful Monster Revisited," in *The Panda's Thumb* (New York: Norton, 1980), 186–193.

FURTHER READING

Dietrich, Michael R. "Richard Goldschmidt: Hopeful Monsters and Other 'Heresies,'" *Nature Reviews Genetics* 4 (2003): 68–74.

———. "Richard Goldschmidt's 'Heresies' and the Evolutionary Synthesis," *Journal of the History of Biology* 28 (1995): 431–461.

Goldschmidt, Richard. *In and Out of the Ivory Tower* (Seattle: University of Washington Press, 1960).

Stern, Curt. "Richard Benedict Goldschmidt (1878–1958): A Biographical Memoir," *National Academy of Sciences, Biographical Memoirs* 39 (1967): 141–192.

CHAPTER EIGHT

Rebellion and Iconoclasm in the Life and Science of Barbara McClintock

NATHANIEL COMFORT

As the essays in this volume suggest, rebelliousness and iconoclasm often go together. Yet a rebel is not necessarily an iconoclast, and an iconoclast need not be a rebel. Rebelliousness is a personal trait, a habitual challenging of authority or flouting of convention. Iconoclasm, in contrast, is a sort of professional activity, a demolishing of cherished dogmas or institutions. Some scientific icons have been shattered by utterly conventional researchers; most rebels never demolish any significant beliefs or institutions. But where rebelliousness and iconoclasm meet, the historian can get a purchase on such elusive yet compelling dimensions of scientific practice as the making of scientific reputations and the mutual influence of personality and professional activity. In short, the study of scientific rebels helps us address our field's greatest challenge: integrating our actors' life, work, and context.

The geneticist Barbara McClintock (1902–1992) is an emblem of iconoclasm. As a rising star of maize genetics in the late 1920s and 1930s, she worked out, almost single-handedly, the basic cytology of the corn plant, distinguishing among its ten chromosomes and associating them with the known clusters of genes, or linkage groups. She followed with a string of classic papers in the 1930s that ensured her place as one of the great figures of classical genetics. Among other early distinctions, she became, in 1944, only the third woman and one of the youngest scientists to be elected to the National Academy of Sciences. Thus, in the first two decades of her career, she was not an iconoclast but a pioneer. But then, in the late 1940s, she discovered a startling new type of genetic element that is now known as a transposable genetic element, or transposon. Imagine the genes arrayed along the chromosomes like pearls on a necklace. McClintock found

Figure 8.1. Barbara McClintock in the 1960s. Courtesy
of Marjorie Bhavnani and the American Philosophical
Society Library, Philadelphia.

special pearls that broke the string, hopped off, rejoined the string
where they had been, and then reinserted themselves at another site.
In 1983, thirty-five years after the initial discovery, she won a Nobel
Prize in the category of Physiology or Medicine for this work. Only
Peyton Rous, the virologist who discovered the first tumor virus in
1912 and won a Nobel in 1966, waited longer before receiving his
prize. According to the story of McClintock's iconoclasm, the radical
nature of her suggestion that genes could move, her gender, and her
supposedly intuitive, feminine experimental style prevented the truth
of transposition from being accepted by the scientific community.[1]

Add McClintock's personal charm, her wit, and the humility with which she accepted the world's highest scientific honor, and you have a satisfying story of iconoclasm. McClintock's late Nobel has been seen simultaneously as the legitimation of a scientific iconoclast and as proof of her marginalization.[2]

The central irony of her career is that although her colleagues agreed with her that she was an iconoclast, they disagreed with her about which icons she shattered. Further, when she became famous, public opinion coincided with McClintock's image of herself as a rebel, but again for different reasons than those she gave. Teasing apart McClintock's rebellion and iconoclasm in their private and public versions affords an opportunity to examine the building of complex scientific reputations and some aspects of the mutual influence of the scientist's temperament, sense of identity, and experimental style. A slightly technical use of the term *myth* makes it easier to talk about these four types of narrative. I use *myth* to refer to an origin story that may be fact, fiction, or a mixture. A private myth is a story one constructs for oneself; it is a part of one's deep identity. A public myth is the story that becomes public knowledge. The private and public myths of rebellion and iconoclasm in Barbara McClintock's life story are interconnected. All causal arrows among them have two heads. The ur-myth, though—the myth from which all the others spring—is what McClintock believed she had actually discovered—her private myth of iconoclasm.[3]

PRIVATE ICONOCLASM

McClintock's scientific iconoclasm began mid-career, when she was a staff member at the Carnegie Institution of Washington's Department of Genetics at Cold Spring Harbor, on Long Island, New York. In the summer of 1944, in what was for her a straightforward gene-mapping experiment, she discovered a large number of so-called mutable genes: genes that oscillate spontaneously between alternate forms during the growth of the plant. Mutable genes were known only in corn. The first had been discovered in 1914 by McClintock's postdoctoral advisor, Rollins Emerson. Anyone who has admired the multicolored spots and stripes on the kernels of an ear of Thanksgiving "Indian corn" has seen its effects. McClintock's close friend Marcus Rhoades had discovered a second, called *Dotted,* in 1936. In her 1944 experiment, McClintock more than tripled the number of known mutable alleles.

Examining them in the following years, she quickly zeroed in on one genetic locus that seemed to be responsible for much of the disruption. She called this new gene *Ds,* "*Dissociator*," for its tendency to break chromosomes, although later she regretted the name.

Dissociator soon proved to do far more than break chromosomes. In early 1948, McClintock realized that the mutable genes were not mutating in the usual sense of a mutation as a permanent chemical alteration of a gene. Rather, the *Ds* element was reversibly altering the other genes near it. In trying to map *Ds,* she discovered that it was in different places in different plants. It could move, or transpose. So could its companion element, *Ac,* or *Activator,* which was required for *Ds* to act. Now she understood: the mutable genes were created by *Ds* or *Ac* jumping into them, which disrupted them; when the mobile element jumped back out again, the gene returned to normal. Apparently, such transpositions could occur many times during the development of even a single kernel. This explained the stripes and spots in the leaves and kernels.

To McClintock it seemed that the mutable genes must be aberrant examples of a normal process. Over about two years' time, she developed a sweeping theory of mobile elements that encompassed genetics, development, and evolution. She imagined that she had massively disrupted a system that is usually invisible because it is the very mechanism of normal development. The mobile elements, she reasoned, must be the means by which different genes are activated at different times and in different tissues. She hypothesized that the mobile elements she had discovered were renegades, hopping around the chromosomes creating spurious mutable alleles. In the normal plant, she speculated, such elements must be under a tight system of control. She called her elements "controlling elements" to indicate her theory of their function in development.[4]

She thought her theory would make obsolete the style of classical genetics she had grown up with, that of mapping and identifying genes without knowledge of their physiological function. She portrayed the coming revolution in a letter to Marcus Rhoades in April, 1950: "The 'good old days of mapping genes' without a main clarifying objective are over. From now on . . . there must be a phase of integration where the various isolated phenomenon [*sic*] are drawn together and where the biochemical, histochemical, chromosomal, cytological, developmental, etc. phases are more clearly integrated. Some of the geneticists see this very clearly but a great many still cling to the good old chemical change in the gene to explain the change in phenotypic expression. The dilemma is becoming more obvious each year."[5]

McClintock was challenging the very notion of mutation—the "good old chemical change in the gene"—as the primary mechanism for altering a hereditary trait. She had come to believe that changes in traits resulted from changes in the regulation of genes, not changes in the structure of the gene itself. This regulation, she believed, was achieved through the action of her controlling elements. The idea had profound implications for genetics, development, and evolution.

The icon she felt she was smashing was that of gene autonomy: the notion, which went back to the early days of *Drosophila* genetics, that genes functioned independently and incessantly. According to this model, if the gene product is altered, the gene itself must be mutated, and any regulation of gene products must occur outside the nucleus, in the cytoplasm. In contrast, McClintock was among the first geneticists to think concretely about the set of chromosomes—what we call the genome—as an integrated, self-regulating unit responsive to outside stimuli. For the rest of her career, she continued to insist that the significance of her elements lay in their role in regulating gene expression. When, in 1960, the French bacterial geneticists François Jacob and Jacques Monod proposed their "operon" model of gene regulation—a model that involved no transposition—McClintock responded immediately with an article in *American Naturalist* comparing the gene-control systems in bacteria and maize. Transposition, then, was a novel mechanism, but McClintock's main interest was in the biological process of gene regulation or control. As late as 1980, she continued to insist, "the real point is control."[6]

PUBLIC ICONOCLASM

In the public myth, the icon McClintock shattered was not gene autonomy but chromosome stability. Although her colleagues accepted transposition immediately, they never accepted her model of the genetic control of development. Through the 1950s and 1960s her work remained a curiosity, restricted to maize. In the late 1960s the discovery of transposition in bacteria began a slow shift in transposition's received meaning. A new generation of young geneticists publicly credited McClintock with having discovered the phenomenon years earlier, in maize. This set transposition in a new context. Bacteria have several mechanisms for shuffling genetic information around.

Transposition came to be understood as a way of creating genetic novelty, of mixing up genes and altering their function. Controlling elements became known as "transposons." They were seen as genetic parasites that hopped around genomes and between bacteria, sometimes taking host genes—such as antibiotic resistance genes—with them. In the 1970s transposition was discovered in other organisms as well. It was soon recognized as a fundamental biological mechanism.

During that decade this new interpretation of McClintock's discovery superseded the old. McClintock and her cohort, as well as a few other geneticists who worked on corn and fruit flies, continued to argue that McClintock had discovered gene regulation and that the significance of her work was as a predecessor to the operon. But the younger generation, consisting mostly of bacterial and viral geneticists, tended to say that McClintock had discovered transposition. Their voices increasingly dominated the textbooks, lecture halls, and other venues where scientific interpretations are canonized. At the same time, with transposition coming to be seen as so important, McClintock received an accelerating number of honors and prizes. She was nominated at least four times for a Nobel Prize. Three times, the nominations stressed gene regulation and the predecessor-of-the-operon argument. The last time, the argument shifted: McClintock was nominated for her discovery of transposition. And her Nobel citation, awarded in 1983, read, "for her discovery of transposable genetic elements."

PRIVATE REBELLION

As McClintock became famous for transposition, she was asked about her childhood, her views about science, and the story of her discovery. In interviews conducted shortly before and after the Nobel Prize, she began to reveal a private myth as the unwilling rebel, the exception to the rule, the outsider. She often portrayed herself as the stranger. As a child, she said, she used to play for hours by herself without any toys; she used to run on the beach in a style that, she later discovered, Buddhist monks also used. She said she felt alien even in her own family: "I didn't belong to that family, but I'm glad I was in it. I was an odd member." Her mother, she said, did not want her to go to college for fear she would become "a strange person, a person that didn't belong to society." She never could conform to established rules: "I have never been able to function well in an organized group where the organization is handed down to you. Under any circumstances I can't function in a group like that. Not that I'm against that kind of organization at all; I'm not. But I can't function in a personally successful way. Sooner or later I upset the apple cart, somewhere along the line, and somebody else's apple cart, not always mine."[7]

She saw herself as literally exceptional—as someone for whom exceptions were made or for whom the usual rules did not apply. Telling interviewer Franklin Portugal about her experiences in college, she related a story about a girl who was trying to get into Cornell University's Agriculture College but was being denied admission because she was female. "So I went over and saw the dean and said, 'Why aren't you letting her in. I came through on Ag [the Agriculture School]. Why aren't you letting her in?' He said to me—now this is strange, too. He said, 'You were an exception; we made an exception for you.' Why did they make an exception when I first entered? . . . These things have happened to me right straight along."[8] She rebelled because she knew no other way to behave. She portrayed herself as incapable of belonging to groups, of following rules, of being conventional.

Of course, McClintock was well aware that she was being interviewed. In some sense, this myth of exceptionalism is the myth McClintock wanted the public to hear. Without a candid and reflective private journal, untangling how she really viewed herself from how she wanted others to think she viewed herself is probably impossible. But the line is blurry in any case. On one hand, it is unlikely that she— not a self-conscious or self-promoting person—concocted a public persona out of whole cloth, unconnected to her self-identity. On the other, a story comes to seem true through repeated telling. McClintock had good reason to believe her own stories of exceptionalism.

This private myth may have given her a way to explain the rejection of her developmental theory. She framed her entire career as a series of insights and discoveries that people thought "crazy" but that later proved true. Sometimes her story is not credible. For example, her first major scientific contribution, in 1930–31, was to correlate the known genetic linkage groups—groups of genes that are inherited together—with specific chromosomes. It was an obvious and long-standing problem in maize genetics. Once she solved it, other maize geneticists joined in to finish the job. And its completion has long been understood as the breakthrough that ushered in the so-called golden age of maize genetics in the 1930s. She acknowledged as much, though modestly, in 1947—just before she discovered transposition. In her acceptance speech for the achievement award from the American Association of University Women that year, she said, "It seems that in 1930 I had the good fortune to develop the technique for working out some particular phases of the life cycle of maize that we really needed very much." Although her idiosyncratic speech patterns introduce some ambiguity into the anecdote—she never worked on the maize life cycle per se—it can only refer to her exploitation of a particular stage of the cell cycle, the pachytene stage, for cytology and her subsequent correlation of the linkage groups to the chromosomes.[9] But looking back on the event in 1980, McClintock said, "It was so new at the time that I was ostracized. . . . It was decided by the people in genetics that I was doing something very crazy; they couldn't understand it. Not only that, they thought me a little mad for doing this." This is inconceivable; it was a project that for years had been of intense interest to her graduate supervisor and was an obvious hurdle to further progress in maize genetics. In 1947, then, she viewed her pathbreaking work in cytogenetics as something "we really needed very much," but by 1980, she thought she had been ostracized for it. Perhaps the rejection of her theory of development—the transformation of her private iconoclasm into her public iconoclasm—triggered or at least enhanced her private myth of rebellion.[10]

PUBLIC REBELLION

As McClintock began to tell these stories publicly, a new myth emerged: her public rebellion. Interviewers wanted to talk to the discoverer of transposition, not the inventor of gene regulation. McClintock obligingly related her private myth of rebellion. But oxidized by the public myth of iconoclasm, her stories of exceptionalism changed into a myth of marginalization.

McClintock helped this transformation by acquiescing in the shift from genetic control to transposition. When she was asked how her colleagues had reacted when she first told them about transposition, she did not confuse the interviewer by saying that it was not transposition that really mattered. She allowed transposition to stand for her discovery. Eventually, she adopted the terminology herself. For example, she began the introduction to the 1987 volume of her collected papers—titled *The Discovery and Characterization of Transposable Elements*—with the words, "Transposable elements were discovered."[11] She went on to describe the reaction to her presentation at Cold Spring

Harbor in 1951 as "puzzlement and, in some cases, hostility" (x). In the same essay she concluded, "In retrospect, it appears that the difficulties in presenting the evidence and arguments for transposable elements in eukaryotic organisms were attributable to conflicts with accepted genetic concepts. That genetic elements could move to new locations in the genome had no precedent and no place in these concepts" (x). It seems unequivocal that she thought transposition was not accepted because her colleagues could not then accept that genes could move. She did acknowledge, a few sentences later, that "a further difficulty in communication stemmed from my emphasis on the regulatory aspects of these elements" (xi). This makes it appear as though the emphasis she put on gene regulation distracted people from the importance of transposition. In fact, regulation was emphasized much more literally: in the fifties, she had said only enough about transposition to convince her audience that she had a plausible mechanism. The rest of her 1987 essay discusses genetic control. She concludes by spanning the gap between the public and the private iconoclasm: "Only now, more than forty years after the discovery of transposable elements, are we beginning to understand enough about the ways that they can affect genes to decipher some intriguing new aspects of gene control from their study" (xi).

Such subtleties were easily, perhaps naturally, lost on the wider public. When McClintock claimed to have been ridiculed for her discovery, there was by then only one discovery: transposition, for which she won the Nobel Prize. Because transposition was now understood as true, fundamental, and nearly universal, she appeared to be vindicated. Her colleagues, those who had been skeptical, were vilified. The public myth of rebellion snowballed. McClintock became known as a maverick whose insight was so profound that none of her colleagues could see it; only with the perspective of three decades' work in molecular biology could the truth of her discovery be recognized. She had challenged the notion of the fixity of the genes and stuck with it through the years, though her peers marginalized her. The failure of McClintock's peers to accept transposition became an example of the fallibility of science.[12] The Nobel simultaneously legitimated her decades of struggle against the opponents of transposition and, by its apparent long delay, served as an acknowledgment of her marginalization by the scientific community.

The iconoclast became an icon. Discovered by a generation of feminist writers in the 1980s, McClintock became a symbol of intuitive, nonreductionist, feminine—even feminist—science. This scientist who fought vainly for decades to persuade her colleagues that gene regulation via transposition was just as rigorous and well supported as her earlier discoveries was transformed into a case study of intuitive antireductionism and emotional attachment. As her story went public, she became the scientist with a "loving identification" with her plants who became "intimately involved" with them, "a mystic by nature" whose style of science was "utterly different" from the logical thinking that characterized most scientists' work. Such portraits appalled McClintock—although of course she played a role in their creation.[13]

The connections among these four dimensions—the private iconoclasm of gene regulation, the public iconoclasm of transposition, the private rebellion of exceptionalism, and the public rebellion of marginalization—are intricate (see figure 8.2). I can only suggest a few points of contact. But a brief look at their interplay reveals some important relations among the individual's identity, her experimental style, her scientific networks, and her public role and reputation.

The process by which McClintock's central contribution shifted from gene regulation to transposition occurred over a span of about fifteen years and had essentially nothing to do with her own science or field of research. Through the 1950s and 1960s, a number of maize geneticists worked with McClintock's elements, but they treated them as tools for doing maize genetics. None expanded on transposition's broader implications for genetic evolution. Rather, it was the independent discovery of transposition in bacteria, in a context far removed from gene regulation and by a group of scientists unconnected to the traditions of maize genetics, that forced recognition of transposition as a general phenomenon. The example highlights the role of social factors, such as the increasing insularity of scientific networks clustered around model organisms and standards of etiquette and attribution of credit, in the making of canonical scientific history narratives.

This is not to deny that reputations are connected to real intellectual achievements. The accepted meaning of a given finding almost always evolves in the context of later discoveries and theories. At some point, opinion coalesces around a particular interpretation of an experimental result or discovery. The knowledge becomes canonical; the scientist's reputation crystallizes. Yet the meaning of the data or theory may continue to evolve. Thereafter, preserving the canonical interpretation requires adjusting the internal historical narrative. The

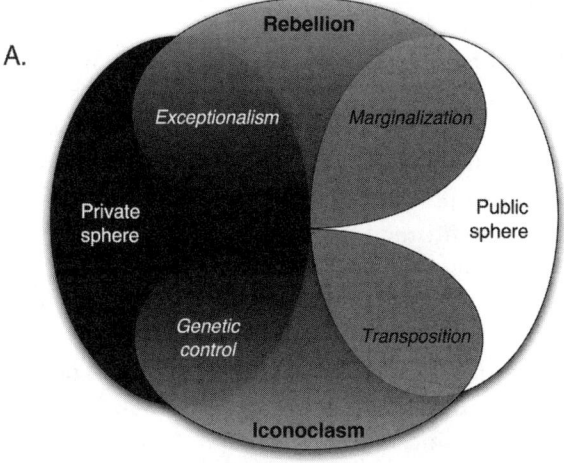

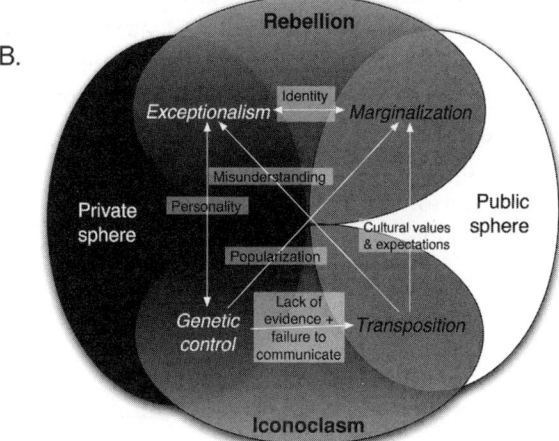

Figure 8.2. A. Myths of rebellion and iconoclasm in the private sphere and the public sphere. McClintock's own sense of her iconoclasm was that she discovered genetic control. As her achievement moved into the wider scientific arena and the public sphere, her iconoclasm morphed into the discovery of transposition. McClintock repeatedly expressed an image of herself as one for whom exceptions were made— her private rebellion. In the public sphere, her rebelliousness tends to be portrayed as marginalization by the scientific community.

B. The connections among these four domains can generate hypotheses about elusive but important forces in science. This figure illustrates several of them, such as different forms of communication and the interaction of scientific findings with scientists' identities and with cultural perceptions and expectations of science and scientists.

accepted model of gene regulation, the operon, was canonized almost immediately: Jacob and Monod shared in a Nobel Prize in 1965,
only four years after their major English-language presentation of the
model. Jacob and Monad's reputation crystallized around the discovery
of gene regulation as presented in their 1960 and 1961 publications. The
operon, however, continued to evolve. New regulatory elements were
discovered, both positive and negative feedback mechanisms were found
for controlling gene expression, and regulatory elements were found
sometimes to occur far from the structural gene they control. Though
potted histories still point to the operon as the basic unit of gene regulation, the current understanding of gene regulation is almost beyond
recognition as Jacob and Monod's model.

Compared to the operon, the one gene–one enzyme hypothesis
took longer to be canonized, but its canonical meaning remains close
to its original formulation. In 1941, George Beadle and Edward Tatum
proposed that a gene specifies a single enzyme, responsible for a single
chemical reaction in a metabolic pathway. The idea remained controversial for a decade; some participants remember it as the major issue
at the 1951 Cold Spring Harbor Symposium, where McClintock gave
her famous presentation of controlling elements.[14] Beadle and Tatum
shared a Nobel Prize with Joshua Lederberg in 1958. In 1962, Vernon
Ingram, younger than Beadle and Tatum but working on a comparable
set of problems, refined the model, saying that a gene could specify
a *part* of a complex enzyme. The more accurate form of the model
was one gene–one polypeptide. This remains the canonical form of
Beadle and Tatum's discovery, and their reputation has crystallized
around it.[15] Yet the accepted relationship between genes and proteins
has continued to evolve. Multiple genes encode single enzymes and
even single polypeptides. Single genes encode multiple polypeptides.
In an alternate historical universe, one gene–one enzyme could be
considered simply wrong. Yet it remained a canonical truth of genetics
long after further research had eroded its literal truth to a few special cases.[16] One gene–one enzyme is a pedagogically useful concept
—students can learn the principle and then be taught the exceptions—
and one that retains a position in the great creation myth of molecular
biology.

In contrast, McClintock's controlling elements did not become canonical knowledge. Her theory of their biological function was not
persuasively supported by published evidence, it did not connect to
existing theories, and scientists interested in such questions did not

work on maize.[17] The received meaning of McClintock's elements evolved significantly *before* they became canonical knowledge. Also, their meaning was canonized by researchers a generation younger than McClintock, working on a vastly different organism, with techniques and concepts that did not exist when McClintock discovered them. The distance between 1950s maize genetics and 1970s bacterial genetics made it easy to dissociate McClintock's interpretation from her result and therefore to attach a new meaning to that result.

Ironically, had McClintock's reputation crystallized twenty years later, in the 1990s, she might be better known for her broader vision than for her empirical discovery. Her vision of fluid genomes and multiple levels of gene regulation is now fashionable in the fast-growing hybrid field called evolutionary developmental biology. Her notion that changes in gene expression during development, rather than physical changes in structural genes, underlie much of the variation among species is now an accepted principle, articulated in widely used textbooks such as Scott Gilbert's *Developmental Biology*.[18] It comes as no surprise that McClintock is not indexed in such books. Today, changes in gene expression are understood in terms of conventional, stable genes, not transposons. Thus, both McClintock's theory of gene regulation and her discovery of transposition have become canonical knowledge in biology, but they have been uncoupled, with her reputation crystallizing around the empirical discovery incorrectly interpreted, rather than the theoretical insight unsupported by evidence. The reasons she is known for the one and not the other are historical more than logical.

Similarly, the public myth of McClintock's rebellion is historically contingent. Her private myth of exceptionalism went public at a moment when feminist essentialism was on the rise. Yet McClintock bridled at any suggestion that she was less rigorous, less mechanistic, less reductionist, or more emotional than any other scientist. In an interview with Evelyn Fox Keller, she rejected the notion that she had any sentimental attachment to her plants. Elaborating on her remark that she had a feeling for her research organism, she said, "I think you'd have the same feeling towards some piece of apparatus if you had used it enough that it was giving you something and you knew that it had to be taken care of well." Referring to a microscope in her lab, she continued, "It looks like I have an affection for it, but it isn't that."[19] McClintock's plants were instruments, not friends. Yet by the filtering of her words and ideas through the minds of scholars and readers, the meaning of her life and science was transmuted. To those seeking such examples, McClintock's eccentricity became marginalization, her insight became intuition, and her sensitivity became sentiment.

Uncoupled from McClintock's private myth of rebellion, the public myth of her rebellion was reinforced by the public myth of her iconoclasm as the discoverer of transposition. Once she became misunderstood as having been talking about transposition rather than gene regulation, then she in fact appeared to have been marginalized. Reading backward, from the Nobel Prize for transposition to McClintock's narrative to the experiments that revealed transposition, neglect and marginalization are all but unavoidable. Such dynamics among public and private myths illustrate a hazard of retrospection, one that may be particularly great for historians who can interview their subjects. Those of us who aim to write histories our actors will recognize as true must interrogate even seemingly obvious connections between rebellion and iconoclasm.

Finally, attention to the links between private myths of rebellion and iconoclasm can reveal connections between temperament and styles of experimentation and reasoning. I have suggested, for example, that the public myth of McClintock's rebellion is exactly backward.[20] It was precisely an *incapacity* for intimacy, a kind of emotionless elimination of the ego, that enabled McClintock to have such remarkable insights into gene action.[21] Further, her sense of identity as exceptional and her unusual attention to anomaly in maize and other organisms suggest an extension from the introspective world of personal identity outward to the broader view of nature, from the microcosm of the self to the macrocosm of nature. In a tape-recorded laboratory discussion with several cytogeneticists in 1970, McClintock referred simultaneously to her dyslexia and to her private myth of rebellion: "My orientation in space is always backwards. . . . This happens time and again. I've found repeatedly I wonder, 'Well now, why does a person do it this way?' and I go around and ask a number of people, 'Now how would you orient this?' and I find I'm all by myself." McClintock punned on her literal perception and her scientific, analytical perception. The laughter indicated in the transcript suggests that her audience got the joke.[22]

Experimental style, of course, depends heavily on experience with mentors and peers and the random events that can influence choice of organism, techniques, and research problem. But a given set of experiences would translate differently in the minds of different people.

It was not inevitable that a scientist faced with the results of Mc-Clintock's 1944 experiment would interpret them as the disruption of a hidden system of developmental regulation. Indeed, that was the source of McClintock's frustration: when she showed her colleagues the spots and streaks in her corn kernels, they saw randomness where she saw disrupted order. What prompted what she called her "dive into the deep," her leap into speculation? Her work with controlling elements marks a dramatic shift in her style of reasoning, which I see as reflected in her emerging private myth of rebellion.

The role of identity in experimental style or style of thought is difficult to get a purchase on; it varies in both tempo and mode from scientist to scientist. For some, identity may play only a minor role, yet for others it may be quite significant. It may be important at some points of the career and not at others. Yet though subjective, such qualities are not indeterminate and the historian can, occasionally, access them. Attempting to do so is a risky but necessary task for the biographer who seeks to get a feeling for his own human organism.

NOTES

1. This myth has been told in countless popular, scientific, and scholarly accounts. For a survey of this literature, see Nathaniel Comfort, *The Tangled Field: Barbara McClintock's Search for the Patterns of Genetic Control* (Cambridge: Harvard University Press, 2001), chap. 1, "Myth," and notes therein.

2. For meticulous accounts of some of McClintock's early work, see E. Coe and L. B. Kass, "Proof of physical exchange of genes on the chromosomes," *Proceedings of the National Academy of Science USA* 102, no. 19 (2005): 6641–46; L. B. Kass and W. B. Provine, "Genetics in the roaring 20's: The influence of Cornell's professors and curriculum on Barbara McClintock's development as a cytogeneticist," *American Journal of Botany* 84, no. 6, supplement (1997): 123; L. B. Kass and Christophe Bonneuil, "Mapping and seeing: Barbara McClintock and the linking of genetics and cytology in maize genetics, 1928–1935," in *Classical Genetic Research and Its Legacy: The Mapping Cultures of Twentieth-Century Genetics,* ed. Hans-Jörg Rheinberger and Jean-Paul Gaudilliere (New York: Routledge, 2004), 91–118. For her later science and for broader accounts of her life and its meaning for her science, see Comfort, *Tangled Field,* and Evelyn Fox Keller, *A Feeling for the Organism: The Life and Work of Barbara McClintock,* 10th anniv. ed. (New York: Freeman, 1993).

3. For the introduction of the term "private myth," see Leon Edel, "The

figure under the carpet," in *Telling Lives: The Biographer's Art,* ed. Marc Pachter (Washington: New Republic Books, 1979), 16–34.

4. Rollins A. Emerson, "The inheritance of a recurring somatic variation in variegated ears of maize," *American Naturalist* 48 (1914): 87–115; Marcus Rhoades, "The effect of varying gene dosage on aleurone color in maize," *Journal of Genetics* 33 (1936): 347–54; Barbara McClintock, "Mutable loci in maize," *Carnegie Institution of Washington Yearbook* 47 (1948): 155–69.

5. McClintock to Rhoades, 3 April 1950, folder 20, box 12, Marcus Rhoades Collection, Lilly Library, Indiana University, Bloomington.

6. Barbara McClintock, "Some parallels between gene control systems in maize and in bacteria," *American Naturalist* 95 (1961): 265–77; Barbara McClintock, "Interview with William B. Provine and Paul Sisco," Cold Spring Harbor Laboratory, Cold Spring Harbor, N.Y., 1980. In Division of Rare and Manuscript Collections, Carl A. Kroch Library, Cornell University Library, Ithaca, N.Y.

7. Sharon Bertsch McGrayne, "Barbara McClintock." In *Nobel Women in Science* (New York: Birch Lane, 1993), 144–75, 147; Comfort, *Tangled Field,* 22; Evelyn Fox Keller, Interview with Barbara McClintock, 1 December 1978, American Philosophical Society, Philadelphia, 46.

8. Franklin H. Portugal, Interview with Barbara McClintock, 21 August 1980, Cold Spring Harbor, N.Y., American Philosophical Society Library, Philadelphia.

9. Barbara McClintock, American Association of University Women Achievement Award acceptance speech, American Association of University Women Archives, Washington, D.C., Convention Papers collection, file "1947 Convention Proceedings," p. 742. The speech later refers to a visit in the fall of 1930 from Lewis Stadler, a maize geneticist from the University of Missouri. Stadler did in fact visit her then, so her chronology seems reliable.

10. Portugal interview.

11. Barbara McClintock, *The Discovery and Characterization of Transposable Elements: The Collected Papers of Barbara McClintock,* Genes, Cells, and Organisms: Great Books in Experimental Biology (New York: Garland, 1987), vii.

12. "Fallibility of science": Keller, *Feeling for the Organism,* 197.

13. "Loving identification": Helen E. Longino, "Subjects, power, and knowledge: Description and prescription in feminist philosophies of science," in *Feminism and Science,* ed. Evelyn Fox Keller and Helen E. Longino (Oxford: Oxford University Press, 1996), 264–79, 264; "Intimately involved": Linda Jean Shepherd, *Lifting the Veil: The Feminine Face of Science* (Boston: Shambhala, 1993), 70; "Mystic by nature," "utterly different": Joan Dash, *The Triumph of Discovery: Women Scientists Who Won the Nobel Prize* (Englewood Cliffs: Julian Messner, 1991), 92.

14. Norton Zinder, "Forty years ago: The discovery of bacterial transduction," *Genetics* 132 (1992): 291–94.

15. For example, James D. Watson, Nancy Hopkins, Jeffrey Roberts, Joan Steitz, and Alan Weiner, *Molecular Biology of the Gene,* 4th ed. (Menlo Park: Benjamin Cummings, 1987), 220; Chris Evers, "The one gene—one enzyme hypothesis," *Access Excellence: The National Health Museum; The Site for Health and Bioscience Teachers and Learners* (http://www.accessexcellence .org/RC/AB/BC/One_Gene_One_Enzyme.html), accessed April 28, 2006.

16. Jan Sapp makes a similar point in his *Genesis: The Evolution of Biology* (Oxford: Oxford University Press, 2003), 163.

17. Comfort, *Tangled Field,* chaps. 5–8.

18. Scott Gilbert, *Developmental Biology,* 7th ed. (Sunderland, Mass.: Sinauer, 2003).

19. Barbara McClintock, interview with Evelyn Fox Keller, Jan. 13, 1979, American Philosophical Society Library, Philadelphia.

20. Comfort, *Tangled Field,* chap. 2.

21. Comfort, *Tangled Field,* pp. 30–31.

22. Barbara McClintock, interview by Charles R. Burnham and Ron L. Phillips, University of Minnesota, circa 1970, courtesy Ron L. Phillips, private collection.

FURTHER READING

Comfort, Nathaniel. *The Tangled Field: Barbara McClintock's Search for the Patterns of Genetic Control* (Cambridge: Harvard University Press, 2001).

Federoff, Nina, and David Botstein, eds. *The Dynamic Genome: Barbara McClintock's Ideas in the Century of Genetics* (Cold Spring Harbor: Cold Spring Harbor Laboratory Press, 1992).

Kass, L. B., and Christophe Bonneuil. "Mapping and seeing: Barbara Mc-Clintock and the linking of genetics and cytology in maize genetics, 1928–1935," in *Classical Genetic Research and Its Legacy: The Mapping Cultures of Twentieth-Century Genetics,* ed. Hans-Jörg Rheinberger and Jean-Paul Gaudilliere (New York: Routledge, 2004), 91–118.

Keller, Evelyn Fox. *A Feeling for the Organism: The Life and Work of Barbara McClintock,* 10th anniv. ed. (New York: Freeman, 1993).

McClintock, Barbara. *The Discovery and Characterization of Transposable Elements: The Collected Papers of Barbara McClintock,* ed. John A. Moore (New York: Garland, 1987).

Chapter Nine

Challenging the Protein Dogma of the Gene: Oswald T. Avery, a Revolutionary Conservative

UTE DEICHMANN

Oswald Theodore Avery distinguished himself by outstanding research in twentieth-century biomedicine whose results had a major impact on immunochemistry and, in particular, early molecular biology. The demonstration by Avery and his younger associates Colin M. MacLeod and Maclyn McCarty in 1944—that the substance capable of bringing about a lasting transformation of pneumococcal types that apparently consists of heritable changes in bacteria is DNA—for the first time clearly associated a genetic phenomenon to a nucleic acid. It challenged the generally accepted view that proteins form the material of genes. Avery and his associates' discovery thereby became the basis of all further studies on the structure and genetic functions of DNA.

In spite of the revolutionary nature of Avery's discovery, his research practices as well as his general attitudes were conservative. He was a microbiologist who used already well-established empirical methods. Unlike, for example, Max Delbrück, who, by initiating quantitative genetic phage research in order to explain fundamental properties of the gene such as identical replication and mutation, developed a new experimental and theoretical approach in genetics, Avery did not create completely new research practices. He was already in his mid-sixties and nearly retired when he made his most important discovery.

Before dealing with Avery's work and his research practices in detail, I provide some background about the early history of the two biological fields that Avery brought together, affecting them drastically: genetics and microbiology.

CHANGING CONCEPTS OF THE NATURE OF GENES

Early Assumptions about Genes as DNA

A phosphorus-containing substance with a high molecular weight, DNA was discovered in 1869 by the Swiss biochemist Friedrich Miescher at the University of Tübingen in Germany. The "nuclein," which he isolated from nuclei of lymphocytes, consisted—as chemists showed shortly afterwards—predominantly of DNA and some percentage of protein. The late nineteenth century saw a significant improvement in microscopes, which, together with newly available industrial dyes, provided powerful tools for the study of processes in the cell nucleus such as the behavior of chromosomes in mitosis. Several biologists, most notably the American cytologist Edmund Beecher Wilson, assumed that nuclein was identical with chromatin, the stainable part of the cell nucleus, and they considered it possible that nuclein played a central role in hereditary transmission.

Within a few years, however, biologists' interest in the material nature of the gene and the biological role of nuclein or DNA declined. Very few scientists, the most outspoken being the German-American biochemist Jacques Loeb, continued to be interested in physical-chemical explanations for phenomena of inheritance. Nuclear chemistry did not provide experimental support for speculations about the material nature of genes, and the apparent disappearance in microscopic studies of chromatin during cell division in germ line cells (meiosis) gave rise to the opinion that chromatin was not an essential part of the chromosome structure.[1] Mainly for this reason even Wilson abandoned his belief in DNA as the hereditary substance, assuming instead that it was proteins.[2] In the 1920s the assumption that genes are proteins became the almost unanimous opinion of researchers who were still interested in the material nature of genes.

Chemical analysis of DNA for many years did not provide any evidence that it possessed the diversity required for the carrier of hereditary information. Early on, researchers mistakenly believed that DNA is a small, uniform molecule of four nucleotides (the tetranucleotide hypothesis). When the macromolecular nature of DNA was demonstrated in the late 1930s, DNA was thought to be a polymer comprised of repeating units of tetranucleotides. This hypothesis was critically examined only after Avery's 1944 discovery.

Genes as Abstract Factors

The "rediscovery" in 1900 of Gregor Mendel's 1865 work on plant hybridization marks the beginning of classical, or formal genetics, which focused on the transmission of traits from parents to offspring. Major concepts and a large part of the early terminology (including the term "genetics") were developed by William Bateson in England until 1906. The term *gene* as the "genotypic" basis of a distinct "phenotypic" trait was introduced by the Danish researcher Wilhelm Johannsen in 1909 (see Chapter 4). His notion of the gene had a far-reaching impact: while realizing that the behavior of genes had something in common with "chemical bodies," he concluded nevertheless that this did not mean that genes themselves were chemical entities. He suggested that the term *gene* be used merely as an abstraction, "for the time being only something like a unit of calculation."[3] This gene concept became one of the most powerful abstractions in biology.

Around 1910 a second phase of Mendelian genetics began in the United States with the work of Thomas Hunt Morgan and his collaborators Alfred H. Sturtevant, Calvin B. Bridges, and Hermann J. Muller. They developed the chromosome theory of inheritance, according to which genes are located on chromosomes and transmitted in linkage groups unless crossover occurs, and they established gene maps of the chromosomes.[4] Thus they endowed genes with a location and some physical existence. But the search for the material nature of the gene was not relevant for this highly successful approach, as the following quotation from Morgan's 1934 Nobel Prize lecture shows: "There is no consensus of opinion amongst geneticists as to what the genes are—whether they are real or purely fictitious—because at the level at which the genetic experiments lie, it does not make the slightest difference whether the gene is a hypothetical unit, or whether the gene is a material particle."[5]

Max Delbrück later made it clear that in the mid-thirties genes were still "algebraic units of the combinatorial science of genetics and it was anything but clear that these units were molecules analyzable in terms of structural chemistry."[6] As Raphael Falk shows in Chapter 4, the concept of an abstract gene turned out to be extremely fruitful for solving genetic questions.

Figure 9.1. Oswald Avery, ca. 1944. Courtesy of
the National Library of Medicine.

Genes as Proteins

Proteins were chemically analyzed beginning in the early nineteenth
century. Owing to the large variety of proteins in the cell and the
increasing evidence that enzymes and antibodies consist entirely or
largely of proteins, the assumption that only proteins are the carriers
of biological specificity became almost universally accepted in the
early twentieth century. The fact that proteins consist of twenty different component amino acids, whereas DNA has only four differing
building blocks, might have contributed to the notion of a greater
variability of proteins. But it did probably not play an important role
because until the 1950s biological specificity was conceived of only in
terms of spatial structure, and the relationship between the sequence

of different building blocks and the spatial structure of a molecule was unknown before that time.

When scientists began to experimentally examine the question of the physical and chemical nature of genes in the 1930s, first by radiation studies, then by virus research, their unanimous opinion was that genes must be proteins. This phase was marked by the crystallization of tobacco mosaic virus (TMV) in 1935 by Wendell Stanley, who wrongly identified it as a protein. Tobacco mosaic virus, capable of identical replication and mutation, became the model of a gene.[7] When, two years later, Frederick Bawden and Norman Pirie showed that TMV contained also RNA, genes were thought to be nucleoproteins, the specificity lying in the protein part of the molecule.

In the late 1930s a number of different experiments showed the crucial importance of DNA for cell replication and mutation. The Swedish biologist Torbjörn Caspersson in 1936 demonstrated by UV absorption that DNA replication took place at the onset of cell division,[8] and between 1939 and 1941 several research groups showed that the UV-mutation spectrum was identical with the DNA absorption spectrum.[9] Researchers concluded, however, that DNA had only an auxiliary function. The hypothesis that DNA is the material of genes, obvious as it may sound with hindsight, fell victim to the dogma that genes must be proteins. When Erwin Schrödinger in his famous book *What Is Life?* (1944) speculated on how biological specificity might be stored in a linear fashion in genes, he, too, took it for granted that genes were proteins. Gunther Stent later reviewed the widespread acceptance that genes are proteins in the 1940s.[10]

CHANGING NOTIONS OF BACTERIAL INDIVIDUALITY AND VARIABILITY

Until the middle of the nineteenth century most biologists and medical scientists believed that bacteria are extremely primitive organisms.[11] The various bacterial forms were considered to be merely different manifestations of one type or very few types (the theory of bacterial polymorphism). Ferdinand Cohn and Louis Pasteur were among the few scientists who believed in the individuality and biological stability of different bacterial types, a concept that was forcefully propagated by Robert Koch. In 1876 the establishment of what became called the Cohn-Koch dogma of bacterial monomorphism replaced the doctrine of polymorphism and marked the beginning of professional medical microbiology.

The new dogma also implied, however, that bacteria reproduce only asexually and do not have genes, a notion that was retained well into the 1940s. Bacterial chromosomes were unknown. When in the 1940s mutation studies showed that hereditary properties of bacteria might be best explained by postulating the existence of genes, geneticists at first doubted whether these genes should be homologized with the Mendelian factors of higher organisms.[12] These doubts were dispelled only after Joshua Lederberg and Edward Tatum discovered genetic recombination (exchange of genes) in the bacterium *E. coli* in 1946.

OSWALD THEODORE AVERY AND HIS MAIN RESEARCH BEFORE 1944

Born in Halifax, Nova Scotia, in 1877, a son of a clergyman, Avery received his M.D. from the College of Physicians and Surgeons of Columbia University in 1904. After practicing clinical medicine for several years, he went in 1907 to the Hoagland Institute in Brooklyn, one of the first privately funded medical research laboratories in the United States, where, as associate director of the Division of Bacteriology, he began research on questions of immunity. In 1913 he accepted an invitation from the director of the Hospital of the Rockefeller Institute for Medical Research in New York, Rufus Cole, to participate in studies on pneumococcus and its relation to lobar pneumonia. He soon became a member of this institute, where he continued his research until his retirement in 1948.

Avery did not belong to the "club" of geneticists, biochemists, or biophysicists (including, for example, George Beadle, Max Delbrück, Torbjörn Caspersson, and Alfred Mirsky) who had set out to tackle questions of genetics on the molecular level. His main work was in bacteriology and immunology, and he always considered himself a microbiologist. While devoting large parts of his research to the study of a single microorganism, pneumococcus, the breadth of his work and its fruitfulness for further research were nevertheless of profound significance for many fields. Among his research topics were bacterial nutrition and growth, intracellular enzymes, and immunological classification of bacterial strains. Apart from transformation, the hereditary change of bacteria by isolated DNA (see below), he will be remembered mostly for his work in immunochemistry, then a new field, which he helped create.

Avery's work on questions concerning the chemical basis of

immunology fitted well the focus of research at the Rockefeller Institute on the search for the chemical basis of life phenomena. Chemical analyses that Avery conducted with Michael Heidelberger on substances related to specific serological types of pneumococcus showed that, surprisingly, they were not proteins but various carbohydrates (polysaccharides). It was the first time that carbohydrates were shown to be involved in immune reactions (in addition to proteins, the material of antibodies). In the fundamental papers they produced between 1923 and 1929, Avery and Heidelberger showed that they were interested in the general problem of the relation between bacterial specificity and chemical constitution or, as Avery's later collaborator McCarty phrased it, the molecular basis of immunological specificity.[13]

Avery's and Heidelberger's announcement, in 1923, that immunological specificity is at least in one case related to a carbohydrate and not a protein met with strong skepticism and criticism. Avery contradicted for the first time the generally accepted view that only proteins are complex and variable enough to account for biological specificity. Avery and Heidelberger refused to become involved in controversies but finally convinced their colleagues by the accumulation of new facts.[14] In a similar way Avery later responded to criticism concerning his experiments with transformation.

The physician Fred Griffith in London first carried out transformation experiments in bacteria in the 1920s.[15] He showed that pneumococcal types could exist in either encapsulated S forms or in R forms, which had lost their capsules. Nonvirulent colonies (R) appeared rough under the microscope, whereas the colonies of virulent strains (S) were smooth (see figure 9.2). In 1928 he published the results of experiments in which he transformed an R form to an S form pneumococcus by adding heat-killed S form to R pneumococcus. Griffith, who was primarily interested in the implications of these experiments for the epidemiology and disease patterns of pneumonia, considered a stimulus provided by the dead S strains as a possible mechanism. Although he himself did not allude to the fact that the change was permanent, the results nevertheless raised doubts about the widespread notion of the stability of pneumococcal types.

Avery at first met Griffith's report with disbelief and skepticism, a reaction that was understandable in one, as Avery's collaborator René Dubos put it, "who had devoted so much effort and skill to the doctrine of immunological specificity."[16] Nevertheless Avery encouraged his young co-worker Martin H. Dawson to repeat Griffith's ex-

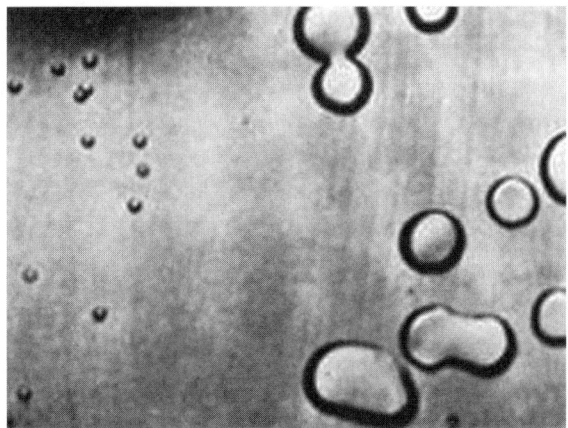

Figure 9.2. Rough (left) and smooth (right) colonies
of pneumococcus. Reproduced from the *Journal of
Experimental Medicine* 79 (1944): 137–158. Courtesy
of the Rockefeller University Press.

periments. After Griffith's results had been duplicated by Dawson and
also by Richard Sia and J. Lionel Alloway at Avery's laboratory and by
Fred Neufeld at the Robert Koch Institute in Berlin, Avery accepted
the validity of the claim for transmissible changes of immunological
specificity and recognized the phenomenon's major importance. Five
papers about transformation were published from Avery's laboratory
by 1933. For various reasons—one being a serious disease—Avery
temporarily stopped work on transformation.

THE CHALLENGE AND ITS IMPLICATIONS FOR MICROBIOLOGY AND GENETICS

In 1940 Avery resumed work on transformation together with Colin
MacLeod. This time the framework was genetics. As was true with
his work in immunochemistry, Avery not only aimed at finding the
chemical nature of a particular substance, the transforming factor, but
he also did it as part of a general problem. From 1930 to 1948 he read
and commented on the assumptions of leading geneticists and biologists about transformation.[17] His notes record the reactions to Griffith's
and his own paper. The genetic framework becomes obvious in a letter Avery wrote to his brother Roy in 1943 after his experiment was
completed: "By means of a known chemical substance it is possible to
induce predictable and hereditary changes in cells. This is something

that has long been the dream of geneticists. . . . Nucleic acids are not only merely structurally important but functionally active substances in determining the biochemical activities and specific characteristics of cells . . . sounds like a virus—may be a gene."[18]

In 1944 Avery, MacLeod, and Maclyn McCarty, who had joined Avery in 1942 in order to replace MacLeod (who had accepted a position elsewhere), published their results concerning transformation in the *Journal of Experimental Medicine.* They showed that the factors underlying the property of serospecific polysaccharide capsules can be transferred from one colony of bacteria to another by cell free extracts and that the transformed cells transmit their new phenotype to succeeding generations. They demonstrated clearly that what would later be called the genes of this phenotype consist of DNA and DNA alone, to the exclusion of proteins and other macromolecules. In order not to be speculative, they did not use the term *gene* but simply wrote that DNA alone is the fundamental unit of the transforming principle of Pneumococcus Type III.

The sketches given above of the status of research on the material nature of genes and bacterial variability make it understandable that Avery's observations, claims, and tacit extrapolations implied major challenges to generally accepted views in microbiology and the emerging chemical genetics. In microbiology, Avery's experiment contributed to further questioning of the dogma of the stability of bacterial types and opened the door to the concept of *sexuality of bacteria.* In the new biochemical and biophysical genetics the demonstration of the biological action of DNA was a "challenging breakthrough" and an "altogether unexpected discovery."[19] The unexpected changes brought about by the experiment were the following:

- By providing clear evidence for a genetic role of DNA it challenged the dogma that genes consist of proteins.
- By demonstrating the biological specificity of DNA it challenged the assumption that only proteins were capable of such specificity.
- It showed that the physical nature of genes can be analyzed directly in contrast to the highly favored indirect methods of radiation and virus (including phage) research, and it called for chemistry, particularly the chemistry of DNA, to be added as a tool to analyze the gene. Whereas x-ray radiation experiments finally failed to be useful for elucidating the material nature of genes and their action, virus (in particular, phage) research provided important insights. But chemical

and biochemical methods became increasingly important, too, for the elucidation of the problems of gene structure, replication, coding, and protein biosynthesis.

- By replacing the "protein dogma" of the gene by DNA it opened the door to a new conceptual and experimental approach in the newly developing molecular genetics, such as the interaction of nucleic acids and proteins in gene replication and protein biosynthesis.[20]

Note that Avery's experiment, when its genetic implications finally became fully accepted, did not demand an immediate change in classical or formal genetics, where the question of the *chemical* basis of genes, from Morgan onward, remained unimportant.[21]

RECEPTION: APPRECIATION, SKEPTICISM, AND ACCEPTANCE

As in other cases of unexpected discoveries, it took scientists some time to fully understand the meaning and far reaching implications of Avery's experiment.[22] A period in which appreciation prevailed was followed by one of growing skepticism and neglect, which a few years later gave way to the general acceptance and understanding of Avery's results.

Avery's paper was immediately widely discussed. A citation analysis shows that it was frequently cited in the first ten years after its publication and that most citations were appreciative.[23] Their authors included scientists who perceived the paper's implications for genetics. Among the appreciative statements were those of Sir Henry Dale and Hermann J. Muller. In 1945 the Royal Society awarded Avery its highest award, the Copley Medal. Though this was done in recognition of his work in immunochemistry, Dale, the society's president, also pointed to Avery's 1944 discovery and its general significance for genetics: "Here surely is a change to which, if we were dealing with higher organisms, we should afford the status of a genetic variation; and the substance inducing it—the gene in solution, one is tempted to call it—appears to be a nucleic acid of the deoxyribose type. Whatever it be, it is something which should be capable of complete description in terms of structural chemistry."[24]

Two years later, in 1947, Muller, then the most renowned geneticist, clearly recognized the far-reaching importance of the discovery for genetics: "Avery, MacLeod & McCarty (1944) have gone further [than Griffith], and have given evidence which they believe points to the conclusion that the effective substance in this treatment [which leads to the transformation of pneumococci] is the nucleic acid itself, of the variety to be imitated, in practically protein-free condition, and in fact that nucleic acid in its naturally polymerized form. If this conclusion is accepted, their finding is revolutionary."[25] Generally, the paper by Avery, MacLeod, and McCarty was immediately appreciated by members of various disciplines.

The paper motivated new research concerning transformation, the chemistry of DNA, and bacteria genetics. The chemical basis of DNA's biological specificity, which Avery could not yet provide, was successfully analyzed, first by Rollin Hotchkiss and then by Erwin Chargaff. By the end of the 1940s Chargaff demonstrated the species-specificity of DNA's composition and in 1950 formulated his well-known rules concerning the base ratios in DNA (the molar amount of guanine is equal to that of cytosine, and that of adenine is equal to that of thymine). Avery's experiment also played an important role in motivating Maurice Wilkins and John Randall, the head of the Physics Department at King's College, to continue x-ray studies on DNA despite the opposition of the Medical Research Council, which wanted them to study the structure of cells rather than that of molecules.[26] Wilkins and Randall were convinced that Avery had shown that genes were made of DNA and not proteins. Wilkins and Rosalind Franklin, who joined the Physics Department in 1951, were the only researchers of renown in England to conduct x-ray studies on DNA. Wilkins succeeded in making James Watson excited about this work, and Watson, in return, motivated Francis Crick, who was carrying out x-ray research on proteins for his dissertation, to shift his interest from proteins to DNA.

Because Avery's paper implied gene exchange between bacteria, it also initiated a genetic approach to variation in bacteria. Most important, it made Joshua Lederberg become interested in bacterial reproduction; he started work on *E. coli* in order to verify whether bacteria indeed reproduce sexually and therefore could become objects of genetic research. Lederberg and the biochemist Edward Tatum demonstrated in 1946 that contrary to prevailing dogma, these microorganisms are capable of reproducing sexually and possess (Mendelian) genes. The papers about bacterial transformation by Avery et al. and subsequent researchers also stimulated a new hypothesis, according to which the occurrence of recombination in bacteria (that is, gene exchange) does not require cell fusion as in higher organisms. It can be merely the result of an exchange of DNA between bacteria, which are genetically distinct.

Despite widespread appreciation of Avery's paper, skepticism and neglect grew when problems of transferring transformation to other systems occurred, in particular, when early successes with *E. coli* turned out not to be reproducible. Prominent criticism by the biochemist Alfred Mirsky fueled this change in attitude. Mirsky did not accept Avery's evidence for the proposition that DNA was the sole carrier of transformation. Even purified DNA, according to Mirsky, would still be contaminated by a tiny percentage of protein, and this protein would bring about transformation. Although this problem could not be solved to the end—it is impossible to exclude very small amounts of protein contamination in the DNA preparation—Mirsky's argument ignored major implications of Avery's experimental work.[27]

Mirsky's powerful questioning of the validity of Avery's conclusions led to indifference toward them by some of the scientists who, like Muller, had responded enthusiastically in the beginning. The neglect was particularly strong on the part of members of the genetic phage group working with Max Delbrück (with the exception of Salvador Luria). As the microbiologist René Dubos, a collaborator of Avery, put it, "certain members of the 'phage group' regarded the orthodox chemical approach to the understanding of biological phenomena as pedestrian, too slow, and not revolutionary enough for their intellectual ambition . . . they did not seem able to do much with or build on [Avery's experiment]."[28]

The disciplinary gap between chemically oriented microbiologists, on one hand, and geneticists including the emerging group of "molecular geneticists," on the other, seems to have been the main reason for the neglect of Avery's work by the latter. They were largely committed to new genetic and physical methods (in particular, phage research and x-ray studies) and clung dogmatically to the assumption that proteins were the sole carriers of biological specificity. Only when members of this group demonstrated the importance of DNA for phage replication did Avery's conclusion become acceptable to them.[29]

In less than ten years after their publication Avery's results were generally accepted and understood. The research that they initiated, in particular that by Chargaff and Wilkins, was toward Watson and Crick's elucidation of the double-helix structure of DNA, which in turn opened up another phase in molecular genetics.

AVERY'S RESEARCH PRACTICES AND GENERAL ATTITUDES

Avery's far-reaching results and the revolutionary nature of his discovery stood in stark contrast to his conventional research practices,
his cautious way of formulating his results, and his personal modesty.
Generally, he refrained from producing grand speculative theories. He
also shunned philosophical discussions about science and scientists
in spite of having received extensive philosophical training.[30] He used
scientific concepts only in the domains that could be converted into
experimental practice. Thus he never discussed general questions
such as the nature of the universe, of life, or of free will unless they
could be converted into questions that were amenable to experimental
tests. To him, scientists cannot sensibly discuss the *nature* of life, but
they can examine the mechanisms, for example, of growth and self-
reproduction. Notwithstanding the fact that he avoided theoretical
discussions about the scientific method—an attitude that he shared
with many empirical scientists—a careful choice of problems, appropriate methods, and their optimization played a decisive role in his
work.

Though remaining close to clinical applications, Avery focused
his work on problems of basic science, where he tried throughout
his life to experimentally find simple causes for apparently complex
phenomena. According to René Dubos, Avery agreed with most of
his colleagues at the Hoagland Laboratory and Rockefeller Institute
that biological phenomena are only complex expressions of physico-
chemical processes and that physics and chemistry offer the only
pathways leading to a real understanding of the phenomena of life.[31]
He therefore made it his scientific ideal to formulate pathological and
biological problems in physicochemical terms and to define chemically the substances and reactions involved.

The focus on basic research and on biochemical and physicochemical methods was typical of the research at the Rockefeller Institute. Of
special importance was the influence of Jacques Loeb. This German
Jewish biochemist and biophysicist, who became a member of the
Rockefeller Institute in 1910, had an enormous influence on an entire
generation of young scientists.[32] Loeb propagated forcefully his view
that phenomena of life can be explained by chemical and physical
methods, and he was convinced that biologists had to use and develop
further these methods if they wanted to work scientifically. As early
as the second decade of the century Loeb had stimulated the earliest

attempts to associate (but not equate) genes and enzymes. We do not know the effect he had on Avery, but the latter's focus on examining biological problems by chemical means and on the close relation between chemistry and genetics renders a direct or indirect influence very probable.

With all problems, Avery recognized and chose the aspects that had large biological significance and could be solved in principle. For them he designed experiments taken from a variety of areas of research. His emphasis on exact experimentation made his work very reliable. When he had found a problem that interested him and that appeared relevant, he continued this line of experimentation, even when major obstacles arose. This persistence explains in part his success: at a time when most microbiologists considered the transformation of pneumococcal types an "oddity of little interest, he had the persistence and the vision to convert type transformation into a precise and elegant laboratory model of a phenomenon with great significance for theoretical biology."[33]

Avery's predilection for pursuing a line of research beyond initial success is expressed in his distinction between two scientific styles: on one hand are scientific investigators "who go around picking up the surface nuggets, and wherever they can spot a surface nugget of gold they . . . grab it and put in into their collection." On the other hand is the more unusual investigator "who is not really interested in the surface nugget. He is much more interested in digging a deep hole in one place, hoping to hit a vein. And of course if he strikes a vein of gold he makes a tremendous advance." Some of the admonitions he constantly gave to his young co-workers illustrate his concepts of how to conduct and not to conduct research: "Be fearless when it comes to hypotheses, but humble in the presence of facts" or "It is great fun to blow bubbles, but you must be the first to prick them."[34]

That Avery did follow his own concepts is demonstrated best by the low-key wording with which he announced his findings in 1944:

> If the results of the present study on the chemical nature of the transforming principle are confirmed, then nucleic acids must be regarded as possessing biological specificity the chemical basis of which is as yet undetermined.
>
> . . . The data obtained by chemical, enzymatic, and serological analyses together with the results of preliminary studies by electrophoresis, ultracentrifugation, and ultraviolet spectroscopy indicate that, within

the limits of the methods, the active fraction contains no demonstrable
protein, unbound lipid, or serologically reactive polysaccharide and consists principally, if not solely, of a highly polymerized, viscous form of
desoxyribonucleic acid.

. . . The evidence presented supports the belief that a nucleic acid
of the desoxyribose type is the fundamental unit of the transforming
principle of Pneumococcus Type III.

By contrast, James Watson and Francis Crick made it very clear that
their discovery of the DNA structure might be of fundamental importance for solving the problem of gene replication: "It has not escaped
our notice that the specific pairing we have postulated immediately
suggests a possible copying mechanism for the genetic material."[35]

As was pointed out by Avery's biographers, Maclyn McCarty and
René Dubos, and by others who knew him, Avery's modesty, his insistence that scientific results should speak for themselves, and his reluc-
tance to engage in far-reaching theories and speculation—his "scientific Puritanism," as a co-worker called it—contributed to the fact that
the reception his paper received was, on the whole, more reserved than
it deserved. Another reason is the fact that he retired shortly afterward (in 1947) and only rarely presented the findings at conferences.

Avery's demonstration in 1944 of the transformation of pneumococcal types by DNA was iconoclastic for two reasons. It challenged
the generally accepted views that the carriers of biological, including
genetic, specificity must be proteins and that bacteria are unable to
undergo stable and predictable hereditary changes. As the microbiologist Bernard Davis perceived it, "the Avery discovery was truly revolutionary" because of its intrinsic significance and unexpectedness.[36]
After meeting with appreciation followed by neglect and rejection,
the views of Avery et al. became generally accepted. Their 1944 paper marked the beginning of a new phase in the emerging molecular
genetics in which an emphasis on biochemical methods and DNA
replaced the emphasis on physical methods (in particular, radiation
studies) and proteins.

The means by which Avery challenged orthodoxy were, interest-
ingly enough, entirely conservative. His approach was empirical, and
he devised a special set of experiments that were taken from different
fields of research. But he did not create a completely new methodology,
nor did he formulate an iconoclastic theory. His revolutionary work
was not accompanied by a rebellious attitude but by modesty and "sci-

entific Puritanism." His attitudes were completely opposite to the attitudes of many other iconoclastic or would-be iconoclastic scientists.[37] Richard Goldschmidt, for example, did not only formulate heretical new theories, but he also propagated them forcefully, even when major doubts were raised concerning the validity of his experiments (see Chapter 7). In contrast to Goldschmidt's work, however, Avery's had a decisive influence on scientific development and the emergence of a flourishing new field of research: molecular genetics.[38]

Avery's success can be explained by reliable experimentation in which he used chemical methods to solve biological questions, the choice of problems that, though they addressed specific questions, were highly relevant for biology in general, and persistence in sticking to a certain problem once he was convinced of its relevance and solubility in principle. Reliable and exact experimentation, a competent bridging of various scientific fields, the focus on chemistry as a basis for the solution of biological questions of high relevance, and modesty also characterized many of the biomedical researchers who developed influential new concepts or methods, among them Avery's Rockefeller Institute colleagues Jacques Loeb, Michael Heidelberger, Karl Landsteiner, and Leonor Michaelis.[39] This is true despite their stronger emphasis on theory as well. The scientific success of Avery and his colleagues makes it clear that major scientific advance and revolutionary findings are often dependent on these characteristics, among other things. They need not be accompanied by iconoclastic attitudes.

NOTES

1. Only in the 1930s did Jean Brachet succeed in demonstrating the persistence of DNA by Feulgen-staining throughout meiotic prophase.

2. E. B. Wilson, *The Cell in Development and Heredity,* 3d ed. (New York: Macmillan, 1925), 351, 653.

3. Wilhelm Johannsen, *Elemente der exakten Erblichkeitslehre* (Jena: Gustav Fischer, 1909), 124–125.

4. Already in 1903 and 1904, Walter Sutton and Theodor Boveri, looking for a cytological basis of genetics, had formulated early versions of the chromosome theory.

5. Morgan was awarded the Nobel Prize for Physiology or Medicine. http//nobelprize.org/nobel_prizes/medicine/laureates/1933/morgan-lecture.pdf, p. 3.

6. Max Delbrück, "A Physicist's Renewed Look at Biology: Twenty Years Later," *Science* 168 (1970): 1312–1315.

7. Details about the history of TMV research are in Angela N. H. Creager, *The Life of a Virus: Tobacco Mosaic Virus as an Experimental Model, 1930–1965* (Chicago: University of Chicago Press, 2002).

8. Robert Olby, *The Path to the Double Helix: The Discovery of DNA* (Seattle: Dover, 1974), 105–107.

9. Ute Deichmann, *Biologists Under Hitler* (Cambridge: Harvard University Press, 1996), 219.

10. Gunther Stent, "Prematurity and Uniqueness in Scientific Discovery," *Scientific American* 227 (1972): 84–93. Stent's claim that Avery's discovery was premature because the then-current view of the molecular nature of DNA made it inconceivable that DNA could be the carrier of hereditary information appears, in hindsight, as a justification of the neglect of Avery's work by Delbrück and the phage group and cannot be upheld in view of an examination of the discussions at the time. Ute Deichmann, "Early Responses to Avery et al.'s 1944 Paper on DNA as Hereditary Material," *Historical Studies in the Physical and Biological Sciences* 34, no. 2 (2004): 207–233.

11. A summary of the changing concepts of bacterial variability is in René J. Dubos, *The Professor, the Institute, and DNA: Oswald T. Avery, His Life and Scientific Achievements* (New York: Rockefeller University Press, 1976), chap. 10.

12. Joshua Lederberg, "The Transformation of Genetics by DNA: An Anniversary Celebration of Avery, MacLeod and McCarty (1944)," *Genetics* 136 (1994): 423–426.

13. Maclyn McCarty, *The Transforming Principle: Discovering That Genes Are Made of DNA* (New York: Norton, 1985), 105.

14. Ibid., 106.

15. Ibid., chap. 4.

16. René J. Dubos, "Oswald Theodore Avery, 1977–1955," *Biographical Memoirs of Fellows of the Royal Society* 2 (1956): 36–48.

17. Dubos, *The Professor,* 152. In 1954, Avery turned his notes over to his last collaborator, Rollin Hotchkiss.

18. 17 May 1943, The Oswald Avery collection, http://profiles.nlm.nih.gov/CC/.

19. Rollin Hotchkiss, "The Decade Before the Double Helix," in *DNA: The Double Helix,* ed. Donald Chambers (New York: Annals of the New York Academy of Sciences, 1995), 55.

20. For details, see Robert Olby, "The Molecular Revolution in Biology," in *Companion to the History of Modern Science,* ed. Robert Olby (London: Routledge, 1990), 503–520.

21. Many concepts of classical genetics were later defined in molecular terms. In 1955 Seymour Benzer was the first to give a molecular definition of the functional concepts of transmission, recombination, and mutation. See Philip Kitcher, "Genes," *British Journal for the Philosophy of Science* 33

(1982): 337–359; Raphael Falk, "The Gene—a Concept in Tension," in Peter Beurton, Raphael Falk, and Hans-Jörg Rheinberger, *The Concept of the Gene in Development and Evolution* (Cambridge: Cambridge University Press, 2000), 317–348.

22. The reception of this paper has been extensively analyzed in Ute Deichmann, "Early Responses to Avery et al.'s 1944 paper on DNA as Hereditary Material," *Historical Studies in the Physical and Biological Sciences* 34, no. 2 (2004): 207–233. References for this section are taken from this paper unless stated otherwise.

23. Between 1945 and 1954, the paper by Avery et al. was cited 239 times, including mentions by ten existing or future Nobel laureates in physiology and chemistry. For details see ibid.

24. Sir Henry Dale, "Anniversary Address to the Royal Society," 1945, quoted after István Hargittai, *The Road to Stockholm. Nobel Prizes: Science, and Scientists* (New York: Oxford University Press, 2002), 225.

25. H. Muller, "The Gene," *Proceedings of the Royal Society of London* (1947): 22.

26. Maurice Wilkins, "DNA at King's College, London," in *DNA: The Double Helix,* ed. Donald Chambers (New York: Annals of the New York Academy of Sciences, 1995), 200–204.

27. For example, the removal of protein during purification led to an increase, not a decrease, in specific transformation activity, and the insensitivity of the transforming substance to deproteinization procedures and protein splitting enzymes stood in sharp contrast to its extreme sensitivity to the DNA splitting enzyme (see McCarty, *Transforming Principle,* 216–217). For Mirsky's motives to reject so strongly Avery's conclusions, see ibid. and Deichmann, "Early responses."

28. Dubos, *The Professor,* 158.

29. Alfred Hershey and Martha Chase demonstrated this in 1952. The fact that the argument of protein contamination was not raised against their conclusion although a large amount (around 20 percent) of protein contaminated the infectious DNA shows that scientists—by and large—had meanwhile accepted Avery's findings.

30. Dubos, *The Professor,* 168–173.

31. Ibid., 174–175.

32. Arnold W. Ravin, "Genetics in America: A Historical Overview," in Edward Garber, *Genetic Perspectives in Biology and Medicine* (Chicago: University of Chicago Press, 1985), 17–34. In order to illustrate Loeb's approach, Ravin mentions Loeb's special fondness of Mendelian genetics, whose exactness, quantitative precision, and experimental testability he liked. Loeb called Mendel's treatise "one of the most prominent papers ever published in biology" (18).

33. Dubos, *The Professor,* 175.

34. Ibid., 172–173.

35. James D. Watson and Francis H. C. Crick, "A Structure for Deoxyribose Nucleic Acid," *Nature* 171 (1953): 737–738.

36. Bernard Davis, untitled article in *BioEssays* 9 (1988): 130–131.

37. For a critical examination of the related popular concept of scientific genius see, e.g., Ulrich Charpa and Ute Deichmann, "Jewish Scientists as Geniuses and Epigones—Scientific Practices and Attitudes Towards Them: Albert Einstein, Ferdinand Cohn, Richard Goldschmidt," *Studia Rosenthaliana* 38, forthcoming.

38. On Goldschmidt's questionable genetic research see ibid. and Deichmann, "Richard Goldschmidt's Physiological Theory of Heredity—a Critical Examination," paper presented at a workshop about Jewish scientists in German and Israeli academia titled "Integrity of Research," Hebrew University, 29 January 2006.

39. Except for Heidelberger, all were German Jewish émigrés. Concerning the phenomenon of special scientific success of German Jewish biochemists see Ute Deichmann, "For Me His Type of Working Is Disgusting: Leonor Michaelis (1875–1949), Emil Abderhalden (1877–1950), and Jewish and Non-Jewish Biochemists in Germany," in Ulrich Charpa and Ute Deichmann, *Jews and Sciences in German Contexts* (Tübingen: Schriftenreihe of the Leo Baeck Institute London, 2007), 72:101–126.

FURTHER READING

Avery, Oswald T., Colin M. MacLeod, and Maclyn McCarty. "Studies on the Nature of the Substance Inducing Transformation of Pneumococcal Types: Induction of Transformation by a Deoxyribonucleic Acid Fraction Isolated from Pneumococcus Type III," *Journal of Experimental Medicine* 79 (1944): 137–158.

Chambers, Donald, ed. *DNA: The Double Helix* (New York: Annals of the New York Academy of Sciences, 1995).

Charpa, Ulrich, and Ute Deichmann. "Jewish Scientists as Geniuses and Epigones—Scientific Practices and Attitudes Towards Them: Albert Einstein, Ferdinand Cohn, Richard Goldschmidt," forthcoming in *Studia Rosenthaliana*, vol. 38.

Deichmann, Ute. "Early Responses to Avery et al.'s 1944 Paper on DNA as Hereditary Material," *Historical Studies in the Physical and Biological Sciences* 34, no. 2 (2004): 207–233.

Dubos, René J. *The Professor, the Institute, and DNA: Oswald T. Avery, His Life and Scientific Achievements* (New York: Rockefeller University Press, 1976).

Lederberg, Joshua. "The Transformation of Genetics by DNA: An Anniversary Celebration of Avery, MacLeod and McCarty (1944)," *Genetics* 136 (1994): 423–426.

McCarty, Maclyn. *The Transforming Principle: Discovering That Genes Are Made of DNA* (New York: Norton, 1985).

Robert Olby. *The Path to the Double Helix: The Discovery of DNA* (Seattle: Dover, 1974).

CHAPTER TEN

Roger Sperry and Integrative Action in the Nervous System

TIM HORDER

Roger Sperry was "something of a maverick."[1] But what does such a judgment mean? Being in the position of a rebel in the scientific community of one's own time is, of course, one of the potential routes to recognition. To adopt a heretical role in that community is a risky business: it may not ensure fame, and one is very likely to be wrong. Major steps in scientific progress require original thinkers, however, and therefore all great scientists need, in a sense, to be rebels in relation to the scientific orthodoxy of their time. Moreover, the really great scientist may have been so far ahead of his time that he is overlooked in the usually hoped-for entry into the accepted listings of the "great names" in the subject; full recognition may only come posthumously. Roger Sperry, I argue, may well fall into this latter category.

Little has been written about how scientists acquire their reputations. In this chapter I seek to address this general theme via a case study of Sperry as maverick. Sperry is of additional special interest and importance because, although barely remembered now, he was a pioneer in what has since become the extremely fashionable subject of cognitive neuroscience. His work addressed subjects of fundamental significance: perception, memory, learning, control of behavior, and consciousness. His research comes as close as anyone's to tackling the classic "mind-body" problem in a scientific manner. Can mind be reduced to and explained by brain function? Sperry arrived at what he presented as a scientifically rigorous answer, while in the process confronting us with some remarkably stark alternatives and fresh insights.

LIFE

Roger Wolcott Sperry, born in 1913 in Hartford, Connecticut, graduated with a degree in English from Oberlin College in Ohio in 1935. Influenced by the polymath psychologist R. H. Stetson, whose influence on Sperry is made clear in Sperry's 1952 paper, he switched to psychology, working first on motor phonetics.[2] From 1938 to 1941 he completed a Ph.D. at the University of Chicago under Paul Weiss on the effects of transposition of motor nerves in rats. This work was part of a broad program of investigations concerning the extent to which behavior is "hard-wired" (based on how parts of the nervous system become connected during embryonic development) as opposed to being open to modification through learning.

Between 1941 and 1946 Sperry joined the leading psychologist Karl Lashley at Harvard University and at the Yerkes Laboratory of Primate Biology in Florida, a period interrupted by wartime service in the army (1942–45) working (alongside Weiss) on peripheral nerve injury cases. He returned to Chicago in 1946. In 1952 he was appointed section chief for Neurological Diseases and Blindness at the National Institutes of Health, but in 1954 moved to the Biology Division at Caltech, where he remained. From the 1950s onward he gradually began working on humans with "split brains," that is, patients in whom the two cerebral hemispheres have been surgically separated, in order to relieve epilepsy, by division of the corpus callosum that links them. Issues raised in this work involved "localization," that is, the extent to which particular brain functions are performed in particular brain regions, a primary example of which is specialization of function on either side of the brain ("lateralization").[3] This was another long-standing and controversial theme in brain research, and one in which Lashley had been a key figure. Sperry, along with David Hubel and Torsten Wiesel, received the Nobel Prize for Physiology or Medicine for this work in 1982. He retired in 1987 and died, after a long neurological illness, in 1994 at age eighty-one.[4]

THE CONTEXT OF SPERRY'S EARLY SCIENTIFIC THINKING

As I have documented, the decade of the 1930s was the high point of "organicist" thinking, a perspective that countered the mounting physico-chemical reductionism in biology and one that emphasized the integrative, organized, holistic aspects characteristic of living systems.[5]

Figure 10.1. Roger Sperry. Courtesy of the Archives, California Institute of Technology.

Lashley's work marked the extreme expression of this form of thinking as applied to neuroscience and psychology.[6] On the basis of studies which showed that a rat's learning correlated with the mass of brain tissue irrespective of the regions of brain tissue removed, he argued that intelligence, memory storage (the "engram"), and learning were mediated diffusely in the brain in terms of "equipotentiality"; this was his "law of mass action." Similar holistic theories were fashionable elsewhere in psychology; examples are systems theory, field theory in social psychology (Kurt Lewin), and in particular Gestalt theory. Kurt Goldstein (in a 1939 book introduced by Lashley), on the basis of medical evidence found in neurological case studies, argued influentially against localization of functions in the brain.[7] Neuropsychology moved from a structuralist emphasis on instincts and (Sherringtonian) reflexes toward functionalist conditioned reflexes and learning theory.

Organicism was particularly well exemplified in embryology, and Paul Weiss, originally trained in Vienna as an engineer, was a typical example among embryologists.[8] To interpret his studies of the coordinated behavior of transplanted limbs ("homologous response"), Weiss argued repeatedly for his well-known but almost mystical notion of a "resonance" between the limb and the controlling spinal cord that innervates it.[9] The metaphor and the mechanistic implications were remarkably similar in kind to the "radio-broadcast model" of the Gestaltists, who thought of the brain as performing its functions by setting up patterns of cortical activity ("electric fields") that were "isomorphic" with the pattern of perceived sensory input.[10] By the 1940s such top-down organicist thinking was applied right across biology, medicine, and psychology at a time when the bottom-up approaches represented by genetics, molecular biology, and behaviorism were already threatening to take over.[11]

Sperry's work embroiled him in some of the most basic and enduring quandaries in biology, in particular, the nature-nurture problem and its parallel in psychology, hard wiring and innate fixed behavior versus plasticity and learning. Whether he realized it or not, Sperry was taking on a problem that still haunts modern biology: how to reconcile and strike a balance between reduction and holism. What methods and conceptual strategies do we have to deal with the different levels at which biological systems can be considered? In the context of the brain, these issues (the "Humpty Dumpty dilemma") boiled down to whether what was known about the minute constituents of brain tissue, laboriously resolved by the efforts of generations of neuroscientists, could be reassembled such that integrated function and behavior could in the end be explained.[12]

SPERRY'S HUMPTY DUMPTY DILEMMA

As the embryologist Viktor Hamburger quipped, "I know of nobody else who has disposed of cherished ideas of both his doctoral and his postdoctoral sponsor, both at that time the acknowledged leaders in their fields."[13] The two main lines of Sperry's research would indeed position him as a rebel in relation to the orthodoxies of the established and influential figures of Weiss and Lashley.[14] Moreover, they would confront him with two seemingly contradictory possible avenues of advance, one reductionist, one holistic.

Neural Specificity

Sperry's first sustained body of research was a continuation of questions raised by Paul Weiss's work concerning the way in which amphibian limb movements are related to their nerve supply. Sperry's early work concerning movement control in rats (first published in 1935) and monkeys—and on the effects of human wartime peripheral nerve injuries—was brilliantly reviewed in 1945.[15] He concluded that even when behavior had been made dramatically maladaptive by transplanting muscles or the nerves innervating them, learning could not compensate for the effects of incorrect connections. He moved on to study the control of the regeneration of nerve connections centrally, that is, between brain parts and especially, from 1942, between the eye and the brain. (Figure 10.2 shows samples of his diagrams of such concepts.) He had been struck by Matthey and Stone's findings that the optic nerve could regenerate and restore visual function in adult amphibians and fish—it was usually assumed that only peripheral nerves could regenerate. Picking up on maladaptive behavior in some of Stone's results, he showed that, after rotation of an eye by 180°, a frog failed to relearn and persisted in responding to the visual world (for example, food lures) in precisely the opposite direction to normal.[16] The fact that this occurred prior to and after optic nerve regeneration demonstrated that regenerating optic nerve fibers had restored their normal pattern of connections in the brain insofar as these mediated the locating of visual stimuli.[17] This classic experiment is still recalled in many current neuroscience texts.

Sperry also continued to work on the neural connections mediating oculomotor control and on sensory nerves to the skin. He was struggling with increasing difficulty to establish an interpretation that was compatible with Weiss's theories.[18] By the early 1960s, however, the work on the visual system had inexorably led to a position diametrically opposed to Weiss's. Around 1965 something like a final denouement occurred.[19] Now he definitively asserted his general conclusion that individual nerve fibers achieve appropriate connections with their targets by a "lock and key" mechanism that requires all neurons to be individually different ("chemospecific"); this was the chemoaffinity theory: "[L]argely as a result of Sperry's work, there had been a transformation of the theory of neuronal plasticity by imperceptible stages into its opposite, specificity."[20] Sperry's theory now dominated the field; his theory of nerve connection control has remained the point of departure for understanding ever since.

Other laboratories were rapidly joining the fertile investigative territory thus opened up, and almost immediately they began to challenge his hypothesis in its details if not in its foundations. Sperry had

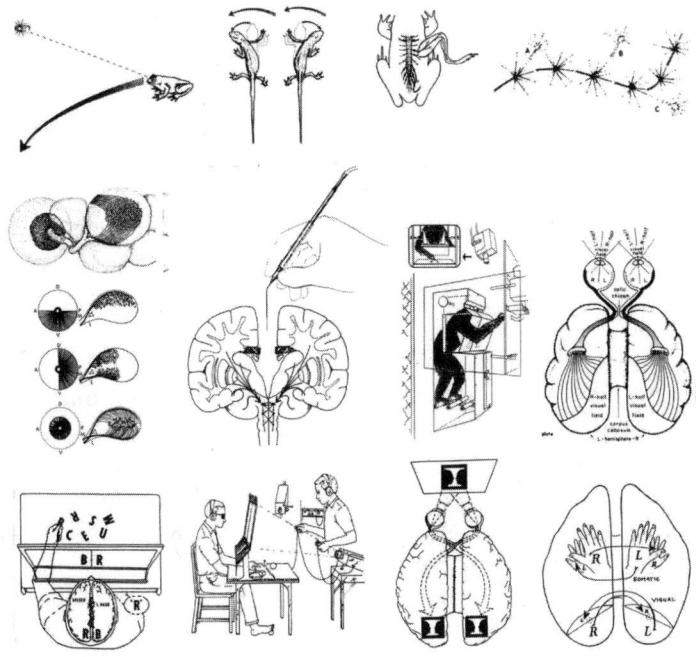

Figure 10.2. Samples from Sperry's diagrams. Reprinted with the permission of Cambridge University Press.

arrived at his theory largely on the basis of behavioral testing. Only in 1959–60 did he begin to verify his inferences by direct anatomical study of optic projections. Other laboratories exploited the more modern methods of electrophysiological mapping and neuroanatomical tracing and began to undermine the extreme terms in which Sperry had originally stated his model. Increasingly the chemospecificity model gave way to more complex, multifactorial explanations.[21] By the 1980s this once-vibrant area of study had become unpopular; an air of fractious competition and conflict came to surround the field. The problem was made worse by rumors of fabricated results elsewhere and by the apparently greater credibility and higher status of parallel studies of visual system development in cats and monkeys associated with the technically sophisticated work of Hubel and Wiesel, which emphasized environmental modifiability rather than formation of fixed visual maps in the brain.[22] Sperry seems to have abandoned his interest in the subject; the last experiments in which he attempted to resolve

Weissian issues in motor and cutaneous systems were published in 1975. A new student named R. L. Meyer and a postdoctoral fellow named M.-Y. Yoon introduced electrophysiological methods for mapping visual projections beginning in 1973. Sperry reviewed the subject on a number of occasions, doing so for the last time in 1979.

But the end result of Sperry's path-breaking work had been the most resounding assertion of a hard-wired view of the nervous system, founded on a principle of extreme localization: topographic maps.

Split Brains and Consciousness

Running in parallel with his work on neural specificity was Sperry's concern to test the concepts of Gestalt psychology and mass action. Beginning in 1947 he conducted a continuing series of direct tests of "electric field theory." Important papers published circa 1955 pointed clearly against isomorphism and field theory, leaving the importance of the *connections* between parts of the nervous system increasingly likely as the critical alternative.

In early 1952 Sperry suggested experiments involving transection of the corpus callosum (CC) to a Ph.D. student, R. E. Myers; the first experiments (on so-called split-brain cats) were reported in 1953. Failure to detect any function for this massive fiber tract had long puzzled neurologists and psychologists.[23] Lashley is said to have remarked that its principal function is to spread epilepsy—Warren McCulloch disagreed; it must be to keep the two hemispheres from physically falling apart![24] Initially the aim was the Lashlian one of searching for the localization in the brain of the engram; the focus was the transferability of learned motor skills from one side of the brain to the other and hence the identification of the functional role of the CC in memory. From 1958 onward the experiments were repeated on monkeys. By ingenious design of experiments it was possible to present different stimuli and learning tasks independently to one or the other of the separated hemispheres and so to test the different functional capacities of each on its own. The hemispheres could be caused to undertake entirely incompatible tasks or store contradictory engrams.

In 1961 a neurosurgeon named Joseph Bogen, conducting research in a room near Sperry's, suggested repeating Akelaitis's observations on patients in whom the CC had been cut to alleviate their epilepsy. Bogen's chief, Philip Vogel, first performed this operation in 1962. At first Sperry's group was sensitive to the possibility of differential capacities in the two hemispheres and the laterality of functions; as expected from long-standing neurological doctrine, speech was impossible for the right ("nondominant") side of the brain. It was not long before some extravagant generalizations were being made concerning the specialized and distinct nature of each half-brain; the "laterality explosion of the late 1960s and 1970s" began.[25] During the 1970s Sperry's group had shown that "the so-called subordinate or minor hemisphere, which we had formally supposed to be illiterate and mentally retarded and thought by some authorities to not even be conscious, was found to be in fact the superior cerebral member when it came to performing certain kinds of mental tasks."[26] But from the mid-1970s onward Sperry "undertook to test the right hemisphere more specifically for the presence of self-recognition and related forms of self and social awareness . . . and emotion";[27] the extent to which the whole range of mental capacities (including emotions) could be shown (often indirectly and nonverbally) to be intact on both sides led Sperry to talk increasingly about separate "consciousness" in the two hemispheres. This conclusion—toward which he would have been inclined on the basis of his earlier findings concerning the equivalence of the two hemispheres in animals—was increasingly reinforced as it was ever more dramatically revealed how a patient would—unknowingly—accept and act on entirely incompatible information in the two sides of the brain.[28]

The split-brain studies could be read in two seemingly contradictory ways: as evidence for lateralization (and therefore localization) or as evidence for the way in which the brain or its separate parts operate as a diffuse, integrated whole. Sperry seems to have convinced himself that he could satisfactorily combine the two.

Free Will and Values

To the rhetorical question "Is it possible, in theory or in principle, to construct a complete, objective explanatory model of brain function without including consciousness in the causal sequence?"[29] Sperry's answer was clearly no. The verbal (and nonverbal) testimony of humans (as opposed to mute animals) further justified and legitimized his reference to consciousness as a key overall accompaniment of—and contributor to—brain function. Largely managing to avoid the many ambiguities of the word, Sperry stated that his motivation was that it solved his long-term aim: resolving the nature of the chain of causality in behavioral acts, the completing of the intermediary link between inputs and outputs in the brain that lies at the heart of neuropsychology.[30] Ultimately deciding against determinism (influenced by the Gestaltist J. J. Gibson and by John Dewey), he held to the "Chicago law of behavior," namely, that an animal behaves "as it damn pleases."[31]

This objective perhaps explains why, after 1972, he talked increasingly about values. If, as he argued, mental processes formed part of the neuropsychological causal chain, then decision making (especially as revealed where free will was exercised on the basis of values) was direct evidence for this proposition. His initial hesitancy in presenting these views disappeared; indeed, he forcefully expressed his personal values in a leading neuroscience review in 1981 at a time when such perspectives would have been distinctly novel to most scientists.[32] His 1983 book expanded further: "Instead of renouncing or ignoring consciousness, the new interpretation gives full recognition to the primacy of inner conscious awareness as a causal reality."[33] Moreover, he wrote, "Recent conceptual developments in the mind-brain sciences are seen to bring changes in world view perspectives that revise the ultimate criteria and frame of reference for determining human value priorities."[34] This "leads to a stand in which science becomes the best source, method, and authority for determining the criteria and reference frame for ultimate values."[35]

It is clear that Sperry saw his conceptions of consciousness and values as arising explicitly from his split-brain studies. He derived direct consequences, including implications for public attitudes toward science and the ability of science to address world problems. In his later years he led international and UN initiatives whose theme was duties; this seemed to emerge from his ecological concerns in particular.

THE FINAL MESSAGE OF SPERRY'S SCIENTIFIC WORK: DILEMMAS RESOLVED OR SHARPENED?

Already, circa 1965, the steady progression of Sperry's thinking reached a crucial choice-point (which had been foreshadowed explicitly in 1952) when he finally and spectacularly broke from the Weissian yoke by definitively presenting his interpretation of neural specificity.[36] This break coincided with the period of increasing confidence about the implications of the split-brain work. These two themes were pointing in different directions: the first pointed to hard wiring and localization, the second to a form of holistic thinking. Sperry seems to have sensed the potential conflict.[37]

A number of factors may have inclined him toward his final concentration on the themes of consciousness and values. Perhaps in reaction to Weiss's overweening influence, his neural specificity model had been expressed in unrealistically bold terms; this provoked a number of contentious counterpositions. Being primarily a psychologist and biologist, Sperry did not rush to keep up with rapid developments in physiological and neuroanatomical techniques. For some reason, he appears to have moved on. Moreover, the wide public interest in the split-brain work encouraged concentration on this subject, and this took Sperry back to his early, broad psychological concerns. A host of popularizers and commentators of varying degrees of expertise entered the fray.[38] Sperry felt particularly challenged by a long-running debate with the Nobel Prize–winning neurophysiologist and heir to the Sherringtonian tradition Sir John Eccles, a religiously motivated dualist.[39]

Although his arguments have the merit of a foundation in superb experimental work and a remarkable consistency of purpose, Sperry's final position in bridging the divide of brain and mind, or reduction and holism, remains conspicuously incomplete and implicitly contradictory. His earlier work emphasized the central importance of neural mapping. But "'[t]he search for psyche' in our own case, at least, has been directed mainly at higher-level configurations of the brain, such as specialized circuit systems."[40] Elsewhere he wrote, "The fiber systems of the brain mediate the stuff of conscious awareness as well as the switching mechanisms, synaptic interfaces, or other interaction sites of the gray matter."[41] Statements such as these are as close as he ever got to defining his final model in any concrete detail. Although always suspicious of Gestalt theories, he ended up presenting a perspective that was little different from Lashley's. Given a choice among the levels of analysis of brain function that were available, Sperry's instinct was to stay with the highest levels.

ON LASTING REPUTATIONS

The creation of reputations in science is—surprisingly—rarely discussed.[42] Being a Nobel Prize winner is no guarantee of lasting fame within the annals of a scientific discipline. A survey of the many biographical encyclopedias and dictionaries of scientists will reveal that Sperry is only sporadically included—falling somewhere between Sherrington or Pavlov (invariably included) and Lashley or Weiss (almost never included).[43]

Opportunities to investigate the inner workings of the human brain are rare and always controversial and complex in their interpretation: compared to the similarly important evidence from blindsight, the work of Penfield, or modern methods of brain scanning, Sperry's approach is possibly unique in the extent to which it actually succeeds in addressing the mind within.[44] But there remains a genuine paradox in his work overall—the potential incompatibility of his hard-wiring theory of brain development as against the holistic elements in his thinking—that may have contributed to puzzlement in evaluating his scientific contribution. It is an advantage to have discovered one single, clear-cut, easily remembered scientific truth, preferably a law, principle, or method to which one's name becomes eponymously attached. The very multiplicity, range, complexity, and contrariness of Sperry's contributions made it unlikely that he would be remembered for any clearly memorable or single final message. His truly pioneering work on localization and fixed wiring is overshadowed by the interpretations he later drew from the split-brain studies. His more philosophical final explorations (prescient as they have proved to be) are diplomatically ignored in his *Festschrift* volumes. It is hardly surprising that Sperry was awarded the Nobel Prize for his split-brain work though in terms of scientific importance his earlier work must be judged to be of at least equal merit. The inherent fascination of mind-brain issues inevitably means that the subject attracts wide and public attention, but it is doubtful that this enhanced Sperry's longer-term reputation.

According to Trevarthen, Sperry was "a taciturn, socially reticent person with a rich and creative private life. Something of a maverick, he shunned formalities and preferred to vacation in remote, wild places . . . he discovered record-breaking ammonites . . . caught big fish off the shores of Baja California . . . was an industrious artist and scientific illustrator, filling his home with sculptures and ceramics including busts of his family and life drawings."[45] Another colleague wrote, "Though a rather private person, preferring the quiet beauty of remote places to large crowds, he was known during the early 1960s for his delightful parties . . . and his special 'split-brain' punch."[46] Personality factors surely lie at the foundation of what will eventually lead to reputation creation. Being a private person can easily lead to being seen as swimming against the stream. On the other hand, he had twenty-three Ph.D. students, fifty-two postdoctoral fellows, and more than forty visiting research associates, published about two hundred papers, and won numerous prizes.[47] His high public standing is indicated by the inclusion of his 1981 opinion piece in a highly rated neuroscience review; this personal, impassioned article about values in science is unique in the publishing history of the *Annual Review of Neuroscience.* His motivation does not seem to have been religious in any strict sense; more likely it was an extension of his involvement with nature. Sperry had a barbed sense of humor. One senses that he was regarded with awe yet often also with affection. He was very generous about sharing authorship; he insisted on being listed as the second or third author in much of the work published with students.[48]

And yet the determination of how a particular scientist is remembered in the long term probably depends on quite different considerations. One key factor is the founding of a "school," because one's followers elaborate and perpetuate one's life's work, sometimes leading to a retrospective construction of a hero figure via myth creation.[49] Sherrington is a good example.[50] But more important is how in tune one's work is with the continuing trends and fashions in a given science.[51] A number of members of Sperry's team became vocal contributors to what became known as cognitive neuroscience.[52] Sperry was an early inspiration, but the subject (now dominated by artificial intelligence and computer modeling) has now moved well beyond his ideas. The focus of interest has shifted to "brain scanning" studies, which, paradoxically, have merely raised in a different guise—and certainly not resolved—the Humpty Dumpty dilemma that was so important for Sperry.[53]

Sperry was no rebel in any institutional or personal sense, and yet his position was often contrary to orthodoxy—but only on the basis of the high standards of experimental science that he exemplified.[54] Why is Sperry not remembered to an extent that matches the merits of his work? Perhaps, in his struggle to maintain a focus on the biggest and boldest questions, he took a step too far by venturing so openly into issues of consciousness and values. Still he has the image of a maverick: a brilliant but ultimately confusing scientific original. History may yet redefine that image.

NOTES

The author is grateful for information supplied by M. E. Glickstein, D. N. Robinson, C. Trevarthen, W. R. Uttal, and L. Weiskrantz.

1. C. Trevarthen, "Roger W. Sperry (1923–1994)," *Trends in Neuroscience* 17 (1994): 402–404.

2. R. W. Sperry, "Neurology and the Mind-Brain Problem," *American Scientist* 40 (1952): 291–312, a thoughtful paper written when Sperry was recovering from his first bout of tuberculosis; E. Erdmann and D. Stover, *Beyond a World Divided* (Boston: Shambhala, 1991), 52.

3. See C. Trevarthen, ed., *Brain Circuits and Functions of the Mind* (Cambridge: Cambridge University Press, 1990) (contains a complete bibliography); F. G. Worden, J. P. Swazey, and G. Adelman, eds., *The Neurosciences: Paths of Discovery* (Cambridge: MIT Press, 1975) (includes a useful flow chart of Sperry's work); D. F. Benson and E. Zaidel, eds., *The Dual Brain* (New York: Guilford, 1985); Erdmann and Stover, *Beyond a World Divided; Neuropsychologia,* special issue, 36 (1998): 953–1096. Sperry reviews his history in *Progress in Brain Research,* ed. M. A. Corner and D. F. Swaab (Amsterdam: Elsevier, 1976), 45:7–38; R. W. Sperry, "Embryogenesis of Behavioral Nerve Nets," in *Organogenesis,* ed. R. J. DeHaan and H. Ursprung (New York: Holt, Rinehart and Winston, 1965), 161–185; R. W. Sperry, "In Search of Psyche," in Worden, Swazey, and Adelman, *The Neurosciences,* 425–434; and in R. W. Sperry, "Some Effects of Disconnecting the Cerebral Hemispheres," in *Nobel Lectures: Physiology or Medicine, 1981–1990,* ed. J. E. Lindsten (Singapore: World Scientific, 1993).

4. T. J. Voneida, "Roger Wolcott Sperry, August 20, 1913–April 17, 1994," *Biographical Memoirs of the National Academy of Sciences* 71 (1995): 315–331; D. Hubel, "Roger W. Sperry (1913–1994)," *Nature* 369 (1994): 186.

5. T. J. Horder, "The Organizer Concept and Modern Embryology: Anglo-American Perspectives," *International Journal of Developmental Biology* 45 (2001): 97–132.

6. On Lashley and his holistic views see J. Orbach, *Neuropsychology After Lashley* (Hillsdale: Lawrence Erlbaum, 1982); K. S. Lashley, "Basic Neural Mechanisms in Behavior," *Psychological Reviews* 37 (1930): 1–24; K. S. Lashley, "Coalescence of Neurology and Psychology," *Proceedings of the American Philosophical Society* 84 (1941): 461–470 (outlining Lashley's views of connectionism, neural circuits [especially Lorente de No's reverberating circuits], endocrinology, intelligence, learning capacity and Gestalt theory). "Karl Lashley surmised that if it were feasible, a surgical rotation through 180 degrees of the cortical brain center for vision would probably not much disturb visual perception" (R. W. Sperry, "Problems Outstanding in the Evolution of Brain Function," in *Encyclopedia of Ignorance,* ed. R. Duncan and M. Weston-Smith [Oxford: Pergamon, 1977], 2:423–433. For background see K. Goldstein, *The*

Organism (New York: American Book, 1939); A. Harrington, *Medicine, Mind, and the Double Brain* (Princeton: Princeton University Press, 1987).

7. On contending views of localization see E. G. Boring, "The Problem of Originality in Science," *American Journal of Psychology* 39 (1927): 70–90; N. Geschwind, "The Neglect of Advances in the Neurology of Behaviour," in *Encyclopedia of Medical Ignorance,* ed. R. Duncan and M. Weston-Smith (Oxford: Pergamon, 1984), 1:9–15; Harrington, *Medicine, Mind, and the Double Brain;* T. Shallice, *From Neuroscience to Mental Structure* (Cambridge: Cambridge University Press, 1988); J. Horgan, *The Undiscovered Mind* (London: Weidenfeld, 1999); W. R. Uttal, *New Phrenology* (Cambridge: MIT Press, 2001).

8. On Weiss and his theories see P. A. Weiss, "Neural Specificity: Fifty Years of Vagaries," in Worden, Swazey and Adelman, eds., *The Neurosciences,* 77–100; D. J. Haraway, *Crystals, Fabrics, and Fields* (New Haven: Yale University Press, 1976); P. A. Weiss, "Neurobiology in *Statu Nascendi*," in *Progress in Brain Research,* ed. M. A. Corner and D. F. Swaab (Amsterdam: Elsevier, 1976), 45:7–38; B. Grafstein, "Bernice Grafstein," in *The History of Neuroscience in Autobiography,* ed. L. R. Squire (San Diego: Academic, 2001), 3:248–282. Weiss's notion of "neural modulation" (or "myospecificity") arose in order to explain the homologous (or "myotypic") response (his finding that amphibian limbs moved in normal sequences even after transplantation and rotation; given that this was maladaptive, such motor behavior could not have been learned). This work is reviewed in R. W. Sperry, "Neuronal Specificity," in *Genetic Neurology,* ed. P. A. Weiss (Chicago: University of Chicago Press, 1950), 232–239; R. M. Gaze, *The Formation of Nerve Connections* (London: Academic, 1970); Haraway, *Crystals, Fabrics, and Fields;* T. J. Horder, "Functional Adaptability and Morphogenetic Opportunism: The Only Rules for Limb Development?" *Zoon* 6 (1978): 181–192. Weiss repeatedly defended his theories with a variety of shifting positions; "resonance" gave way to "modulation" (or "impulse specificity theory," which gets close to the primitive but discredited notion of specific nerve energies) (Boring, "Problem of Originality").

9. Haraway, *Crystals, Fabrics, and Fields;* A. Harrington, *Reenchanted Science* (Princeton: Princeton University Press, 1996).

10. Sperry, "In Search of Psyche," 426.

11. The domains of embryology, neuroscience, and psychology shared many terms at the time (e.g., *fields, equipotentiality, individuation, organizers, gradients,* and *axial polarization*). Loeb, Coghill, Weiss, Whitman, and Child covered all these fields. See Lashley, "Basic Neural Mechanisms"; C. S. Sherrington. *Man on His Nature* (Cambridge: Cambridge University Press, 1940).

12. Horgan, *Undiscovered Mind.*

13. Quoted in Voneida, "Roger Wolcott Sperry," 319.

14. A number of Sperry's earliest studies were attempts to test Lashley's "law of mass action," a view of the brain that was strongly against localization; see especially R. W. Sperry, "Physiological Plasticity and Brain Circuit Theory," in *Biological and Biochemical Bases of Behavior,* ed. H. F. Harlow and C. N. Woolsey (Madison: University of Wisconsin Press, 1958), 401–421. Early experiments (1947) on the cerebral hemispheres in monkeys (some by Sperry involving multiple slicing of cortical circuits and some by Lashley in 1951 involving short-circuiting electrical forces within them with metal implants) seemed to support mass action because learning was unaffected. In 1955 Sperry implanted insulating strips in cats and made cuts into the deeper white matter; as a result, learning *was* defective. The involvement of the white matter pointed to the importance of connectivity. As for Lashley, the ultimate target for Sperry was the localization of memory (the "engram"); hence his use of tests of learning (rather than of perception, much favored by Gestaltists).

15. R. W. Sperry, "The Problem of Central Nervous Reorganisation After Nerve Regeneration and Muscle Transposition," *Quarterly Review of Biology* 2 (1945): 311–369.

16. Trevarthen, *Brain Circuits and Functions of the Mind,* 41.

17. R. W. Sperry, "The Eye and the Brain," *Scientific American* 194, no. 5 (1955): 48–52.

18. R. W. Sperry, "The Growth of Nerve Circuits," *Scientific American* 201, no. 5 (1959): 68–75. Sperry was still sympathetic to Weissian concepts of modulation; see also Sperry, "Physiological Plasticity and Brain Circuit Theory," 418.

19. See R. W. Sperry, "Embryogenesis of Behavioral Nerve Nets," in *Organogenesis,* ed. R. J. DeHaan and H. Ursprung (New York: Holt, Rinehart and Winston, 1965), 161–185; Grafstein, "Bernice Grafstein."

20. M. Jacobson, "Historical development of the concept of neuronal specificity," in *Formation and Regeneration of Nerve Connections,* ed. S. C. Sharma and J. W. Fawcett (Boston: Birkhauser, 1993), 1–11. On the history of Sperry's terminology see Trevarthen, *Brain Circuits and Functions of the Mind,* 5–9.

21. See R. L. Meyer and R. W. Sperry, "Retinotectal Specificity: Chemo-affinity Theory," in *Neural and Behavioral Specificity,* ed. G. Gottlieb (New York: Academic, 1976), 3:111–149; J. E. Cook and T. J. Horder, "The Multiple Factors Determining Retinotopic Order in the Growth of Optic Fibres into the Optic Tectum," *Philosophical Transactions of the Royal Society (London)* B278 (1977): 261–276; T. J. Horder and K. A. C. Martin, "Morphogenetics as an Alternative to Chemospecificity in the Formation of Nerve Connections," in *Cell-Cell Recognition,* ed. A. S. G. Curtis (Cambridge: Cambridge University Press, 1978), 275–358. By the mid-1970s it was becoming clear that chemo-affinity was not the only consideration; mechanics of nerve fiber growth

and correlations in nerve impulse patterns were also involved. Sperry was unconcerned about the nature of chemoaffinity as such (see Sperry, "Problems Outstanding in the Evolution of Brain Function," 424); this may have contributed to an initial unfortunate exaggeration in terms of the molecular and genetic elaboration implicitly required. For more recent reviews, some of which do not mention Sperry, see S. B. Udin and J. W. Fawcett, "Formation of Topographic Maps," *Annual Review of Neuroscience* 11 (1988): 289–327; F. M. Mann, W. A. Harris, and C. E. Holt, "New Views on Retinal Axon Development: A Navigation Guide," *International Journal of Developmental Biology* 48 (2004): 957–964; T. McLaughlin and D. D. M. O'Leary, "Molecular gradients and development of retinotectal maps," *Annual Review of Neuroscience* 28 (2005): 327–355. Even today many issues remain unresolved, e.g., the relative importance of chemical and impulse pattern matching factors. Strangely, the notion of modulation was never entirely shaken off among Sperry's collaborators (see Meyer and Sperry, "Retinotectal Specificity: Chemoaffinity Theory," 130). The explanation of the homologous response in urodele limbs still remains to be resolved to everybody's satisfaction.

22. D. H. Hubel, *Brain and Visual Perception* (Oxford: Oxford University Press, 2005).

23. After Sperry's cortical slicing experiments, section of the CC was a natural follow-on. Lashley had suggested such experiments (Trevarthen, *Brain Circuits and Functions of the Mind,* xx–xxi). The function of the CC had long been of interest owing to the apparent normality—despite the massive size of this intercerebral tract—of earlier cases of split-brain humans; for history see M. S. Gazzaniga, *The Social Brain* (New York: Basic, 1985); Harrington, *Medicine, Mind, and the Double Brain;* J. E. Bogen. "The Callosal Syndromes," in *Clinical Neuropsychology,* ed. K. M. Heilmann and E. Valenstein (New York: Oxford University Press, 1993), 337–407. Even today the number of nerve fibers it contains is uncertain; it may contain up to one billion. T. E. Feinberg, *Altered Egos* (Oxford: Oxford University Press, 2002), 90.

24. Voneida, "Roger Wolcott Sperry," 319.

25. Harrington, *Medicine, Mind, and the Double Brain,* 261.

26. Lindsten, *Nobel Lectures, Physiology or Medicine, 1981–1990,* 7–8.

27. Ibid., 9–19. In his work on animals, before the start of the human studies, Sperry was struck by the completeness and autonomy of the functions that each hemisphere could undertake. See, e.g., R. W. Sperry, "Cerebral Organization and Behavior," *Science* 133 (1961): 1749–1757. On the other hand, human brains may be uniquely lateralized. The prevailing assumption was that the CC was functionless (Geschwind, "Neglect of Advances in the Neurology of Behaviour")—this was the "Akelaitis-inspired perception of the corpus callosum as a big but essentially 'useless' structure" (Harrington, *Medicine, Mind, and the Double Brain,* 277)—so that it seemed to follow that

neither hemisphere needed communication with the other. It could therefore be argued that each was sufficient alone. But Sperry's tendency to emphasize this feature in humans, in the face of all the evidence for lateralization of functions such as speech, was a constant source of friction. Bogen went so far as to take his name off some of the early papers about language in order to dissociate himself from the contradictions (Trevarthen, *Brain Circuits and Functions of the Mind,* 306; Lindsten, *Nobel Lectures,* 11). Bogen saw Sperry as a "revisionist" (Bogen, "Callosal Syndromes," 371). But in a 1969 paper co-authored with Gazzaniga and Bogen, Sperry argued ingeniously that neurologists' evidence could be reconciled: "the unilateral lesion evidence . . . has been misleading" (Lindsten, *Nobel Lectures,* 11–12; Trevarthen, *Brain Circuits and Functions of the Mind,* 314–315). But it remained the case that "the answers . . . are still not entirely resolved" (Lindsten, *Nobel Lectures,* 11).

It appeared possible to some that, in the long term, the nondominant hemisphere could in some cases approximate the capacity for language. Gazzaniga wrote: "[O]ver the years it has become clear that our first three cases were unusual" (M. S. Gazzaniga, "The Split Brain Revisited," *Scientific American* 279 [1998]: 35–39, 37). Problems of interpretation have concerned preoperative compensation to pathology, adequacy of testing procedures, self-deception, confabulation, the role of intact alternative bilateral commissural pathways, and individual variability. Sperry developed a "respect for the inherent individuality in the structure of human intellect" (Lindsten, *Nobel Lectures,* 14). For critiques see M. S. Gazzaniga and J. E. LeDoux, *The Integrated Mind* (New York: Plenum, 1978); Shallice, *From Neuroscience to Mental Structure;* Bogen, "Callosal Syndromes," 376–379; R. Efron, *The Decline and Fall of Hemispheric Specialisation.* (Hillsdale, N.J.: Erlbaum, 1990); Erdmann and Stover, *Beyond a World Divided;* Feinberg, *Altered Egos.*

28. See Gazzaniga and LeDoux, *Integrated Mind.* Sperry's use of the word *consciousness* is merely a convenient portmanteau word for *mind* or *mentalism.* He consistently avoided matters of definition, philosophical treatments of the subject, the literature from introspectionist psychologists, and the fashionable work done in the 1950s and 1960s regarding sleep, attention, and the brain-stem "reticular activating system" (see Sperry, "In Search of Psyche," 429; Sperry, "Problems Outstanding in the Evolution of Brain Function," 429). He was initially hesitant about presenting his mentalist views as they began to emerge around 1959 (Voneida, "Roger Wolcott Sperry," 324): after he published in nonscience journals from 1965, *Psychological Reviews* (1969) published his formal account, and, as a response to critics, his best explication of the neurological foundation of his conclusions (R. W. Sperry, "An objective approach to subjective experience: Further explanation of a hypothesis," *Psychological Reviews* 77 [1970]: 585–590). For recent scientific approaches see B. J. Baar, W. P. Banks, and J. B. Newman, eds., *Essential Sources in the Scientific Study*

of Consciousness (Cambridge: MIT Press, 2003); C. Koch, The Quest for Consciousness (Reading, Pa.: Roberts, 2004).

29. R. W. Sperry, Science and Moral Priority (New York: Columbia University Press, 1983), 28.

30. Most unusually, Sperry gave equal attention to sensory and motor aspects of the behavioral system; thence, his consistent focus was the intermediary neural processes linking them (i.e., the complete causal chain that underlies integrated behavior). Sperry is credited (D. I. McCloskey, "Corollary Discharges: Motor Commands and Perception," in Handbook of Physiology: Neurophysiology, ed. V. B. Brooks [Bethesda: American Physiological Society, 1981], vol. 1, pt. 2, 1415–1447) with having pioneered the important concept of "reafferance" (Sperry termed it "corollary discharge"), i.e., the idea that the brain's state of predictive anticipation ("set") is actively involved in achieving coordinated motor behavior. See especially R. W. Sperry, "On the Neural Basis of the Conditioned Response," British Journal of Animal Behavior 3 (1955): 41–44, which parallels Lashley's treatment of plasticity and learning while emphasizing "set" (J. J. Gibson, "A Critical Review of the Concept of Set in Contemporary Experimental Psychology," Psychological Bulletin 38 [1953]: 781–817); he implies that this is incompatible with defined, fixed wiring of neuronal connections and ultimately with localization.

31. Sperry, "Retinotectal Specificity," 9.

32. R. W. Sperry, "Changing Priorities," Annual Review of Neuroscience 4 (1981): 1–15.

33. Sperry, Science and Moral Priority, 112.

34. Ibid., 108.

35. Ibid., 60. See also M. Jacobson, Foundations of Neuroscience (New York: Plenum, 1993); Erdmann and Stover, Beyond a World Divided.

36. Grafstein, "Bernice Grafstein."

37. "Curiously, the neural model for conditioning that I eventually settled on involved a rejection of connectivity in a sense. I concluded that it had been an error to search for newly formed sensory-motor connections, that we should think of the new sensory-motor linkages observed behaviorally as being effected instead by means of transient cerebral facilitating sets" (Sperry, "In Search of Psyche," 428).

38. Today the issues raised by the study of split brains are probably of greater interest to philosophers than to neuroscientists, as are the nature and meaning of consciousness and the person. The range of approaches is quite extraordinary and includes the claim that any attempt to relate such attributes to the brain is meaningless. For examples see P. S. Churchland. Neurophilosophy (Cambridge: MIT Press, 1986); J. Glover, I: The Philosophy and Psychology of Personal Identity (London: Penguin, 1988); J. Hymen, ed., Investigating Psychology (London: Routledge, 1991); D. C. Dennett, Consciousness Explained (London: Penguin, 1993); M. R. Bennett and P. M. S. Hacker,

Philosophical Foundations of Neuroscience (Oxford: Blackwell, 2003). See also exchanges in *British Journal of the Philosophy of Science,* 1973–1977. For an amusing and accurate account of these phenomena, see Horgan, *Undiscovered Mind.*

39. J. C. Eccles, *Facing Reality* (New York: Springer Verlag, 1970). For Sperry's rebuttal, see R. W. Sperry, "Mind-Brain Interaction: Mentalism, Yes; Dualism, No," *Neuroscience* 5, no. 2 (1980): 195–206. See also K. R. Popper and J. C. Eccles, *The Self and Its Brain* (London: Routledge and Kegan Paul, 1983).

40. Sperry, "Problems Outstanding in the Evolution of Brain Function," 424.

41. Trevarthen, *Brain Circuits and Functions of the Mind,* 381.

42. Some aspects are identified in Boring, "Problem of Originality"; D. L. Hull, "Scientific Bandwagon or Traveling Medicine Show?" in *Sociobiology and Human Nature,* ed. M. S. Gregory, A. Silvers, and D. Sutch (San Francisco: Jossey-Bass, 1978), 136–163; P. G. Abir-Am, "How Scientists View Their Heroes: Some Remarks on the Mechanisms of Myth Construction," *Journal of the History of Biology* 15 (1982): 281–315; N. Elias, H. Martins, and R. Whitley, eds., *Scientific Establishments and Hierarchies* (Dordrecht: Reidel, 1982); D. J. Coon, "Eponymy, Obscurity, Twitmyer, and Pavlov," *Journal of the History of the Behavioral Sciences* 18 (1982): 255–262; Harrington, *Medicine, Mind, and the Double Brain;* D. L. Hull, *Science as a Process* (Chicago: University of Chicago Press, 1988); A. G. Gross, *The Rhetoric of Science* (Cambridge: Harvard University Press, 1990); P. G. Abir-Am, ed., *Commemorative Practices in Science,* special issue of *Osiris,* 2d ser., 14 (2000); T. J. Horder, "Organizer Concept"; E. B. Hook, ed., *Prematurity and Scientific Discovery* (Berkeley: University of California Press, 2002); D. K. Simonton, *Great Psychologists and Their Times* (Washington, D.C.: American Psychological Association, 2002); M. Biagioli and P. Galison, *Scientific Authorship: Credit and Intellectual Property in Science* (New York: Routledge, 2003); F. L. Holmes, *Investigative Pathways: Patterns and Stages in the Careers of Experimental Scientists* (New Haven: Yale University Press, 2004).

43. See, e.g., J. Simmons, *The 100 Most Influential Scientists* (London: Robinson, 1997).

44. L. Weiskrantz, *Blindsight* (Oxford: Oxford University Press, 1986); W. Penfield, *The Mystery of the Mind* (Princeton: Princeton University Press, 1975); Uttal, *New Phrenology.*

45. Trevarthen, "Roger W. Sperry," 404.

46. Voneida, "Roger Wolcott Sperry," 327.

47. See Trevarthen, *Brain Circuits and Functions of the Mind.*

48. Voneida, "Roger Wolcott Sperry," 322.

49. On scientific "schools" and leadership see Jacobson, *Foundations of Neuroscience;* A. Bryman, *Charisma and Leadership in Organizations* (Lon-

don: Sage, 1993); G. L. Geison and F. L. Holmes, *Research Schools,* special issue of *Osiris,* 2d ser., 8 (1993); Abir-Am, "How scientists view their heroes"; Horder, "Organizer Concept."

50. M. Jacobson, *Foundations of Neuroscience.*

51. See, e.g., Geschwind, "Neglect of Advances in the Neurology of Behaviour."

52. On cognitive neuroscience see H. Gardner, *The Mind's New Science* (New York: Basic, 1987); Shallice, *From Neuroscience to Mental Structure;* M. S. Gazzaniga, *The New Cognitive Neurosciences* (Cambridge: MIT Press, 2000); M. A. Boden, *Mind as Machine* (Oxford: Clarendon, 2006); J. L. van Hemmen and T. J. Sejnowski, *23 Problems in Systems Neuroscience* (Oxford: Oxford University Press, 2006); L. T. Benjamin, *A Brief History of Modern Psychology* (Malden, MA: Blackwell, 2007).

53. Uttal, *New Phrenology.*

54. Sperry saw himself as a participant in a revolution (R. W. Sperry, "The Impact and Promise of the Cognitive Revolution," *American Psychologist* 48 [1993]: 878–885). For personal accounts of recent would-be rebels in psychology see R. J. Sternberg, ed., *Psychologists Defying the Crowd* (Washington, DC: American Psychological Association, 2003).

FURTHER READING

Harrington, A. *Medicine, Mind, and the Double Brain* (Princeton: Princeton University Press, 1987).

Sperry, R. W. *Science and Moral Priority* (New York: Columbia University Press, 1983).

Trevarthen, C., ed. *Brain Circuits and Functions of the Mind: Essays in Honor of Roger W. Sperry* (Cambridge: Cambridge University Press, 1990).

CHAPTER ELEVEN

Leon Croizat:
A Radical Biogeographer

DAVID L. HULL

The contributors to this volume deal primarily with rebels in biology whose iconoclastic views eventually became accepted, but not all iconoclasts succeed. In fact, very few do. I deal mainly with Leon Croizat, a biogeographer who worked in almost total isolation from other biogeographers of his day. "I do not like to share responsibilities, so I always do or die by myself," he wrote.[1] If anyone counts as a rebel, renegade, maverick, or iconoclast, it is certainly Leon Croizat. To what extent did he succeed? To what extent did he fail? And can we learn anything about science from studying Croizat's career?

From before Darwin's time to the present, scientists who studied the distribution of plants and animals around the world have disagreed with each other about the nature of the various mechanisms responsible for these biogeographic patterns and the methods of analysis to be used in discerning them. Were there centers of origin scattered across the continents, and once a center originated did new species disperse from it? Were continents stable, or did they drift, carrying along with them the species that inhabited them? Of equal importance, what methods are to be used to subdivide plants and animals into groups? On the basis of one classificatory method, certain patterns might emerge. If different methods are used, different patterns might be discernable. Which methods of classification are preferable?

For example, the distribution of species around the South Pole has fascinated biogeographers since the time of J. D. Hooker (1817–1911), Darwin's closest friend. Hooker used the striking resemblances between disjoint populations of the species inhabiting the landmasses that circle Antarctica to support Darwin's theory of evolution. Hooker explained the biogeography of the Antarctic region by means of migrations across land connections. Darwin agreed with Hooker that the

Figure 11.1. Leon Croizat. Courtesy of Jonathan N. Baskin.

biogeography of the Antarctic region supported his theory of evolution but thought that the distributions of plants and animals were best explained by means of long-distant dispersal over the intervening stretches of ocean. Put crudely, the contrast is between dispersal across land and dispersal across water. During the following century the "dispersalist" views of Darwin prevailed, culminating in the works of W. D. Matthew (1871–1930), G. G. Simpson (1902–1984), Ernst Mayr (1904–2005), and P. J. Darlington (1904–1983).

BIOGEOGRAPHY TODAY

Gareth Nelson and Norman Platnick, in their magnum opus *Systematics and Biogeography,* attempted to combine the views of Willi Hennig (1913–1976), Karl Popper (1902–1994), and Leon Croizat (1894–1982).[2] When Hennig first introduced his system of classification to the systematics community in 1966, he termed it "phylogenetic systematics" to emphasize the role of phylogenetic trees in classification.[3] Later it also came to be known as *cladism, clade* being the Greek word for "branch." Hennig's contemporaries considered his views to be nonstandard. The heterobathmy of synapomorphy indeed! Now, however, Hennig's system of classification is widely accepted, albeit in several different forms.

In his early years Popper considered himself to be a maverick, and his disciples have attempted to carry on in this tradition though Popper is one of the rare philosophers who not only influenced his fellow philosophers but also had an impact on scientists.[4] Because Popper's views mainly concerned philosophy, I do not discuss them in any great detail in this chapter except to mention that Croizat rejected Popper's views in no uncertain terms. He also rejected Hennig's principles of systematics, preferring the principles that were more "phenetic" in character. Hennig argued for a system of classification in which biological groups (taxa at all levels) form perfectly nested sets, and Croizat was willing to settle for statistical covariation.

Thus Nelson and Platnick were in a bind. They proposed to combine the systematics of Hennig, the philosophy of Popper, and the biogeography of Croizat when these three men either ignored or disagreed with each other. As Croizat put it in the final year of his life, "Hennig and Croizat have not found their work particularly compatible. Hennig never cited Croizat, and Croizat (1976) has published negative comments of Hennig, and neither one has indicated any interest in Popper's views or cited them as being compatible with his own."[5]

It is obvious that Croizat took Darwinian dispersalists who believed in centers of origin to be his enemies. Croizat referred to his own views as *vicariance/vicariism,* but when Nelson and Platnick began referring to their own views as vicariance biogeography, increasingly he began to characterize his own views as *panbiogeography.* As Nelson used the term *vicariance,* it referred to geographical speciation, that is, speciation that takes place when a population becomes geographically isolated for a sufficiently long time to become a separate species.

Nelson's biogeographic theory combined vicariance and cladistics, but Croizat rejected cladistics. What else Croizat intended is hard to say. Time and again other biogeographers begged him to clarify his terminology. The closest he came was to contrast Nelsonian vicariance biogeography as a "theory" with panbiogeography as a "method."[6] According to Croizat, panbiogeography consists in a "manner of (statistical) wholly factual investigation of living and fossil records of the geographic distributions of plants and animals, directed to establish the coordinates of *time* and *space* present in organic *evolution*." The term *track* might lead one to think in terms of wholesale movement of organisms, but that was not Croizat's intent at all. A track is "essentially a graph drawn to render visible and comparable the results of biographical investigation," a baseline in character.[7] If these terms seem strange, they should. They never became part of biogeographic terminology (see figure 11.2).

In this chapter I concentrate on a wide variety of issues, ranging from sociological and psychological matters to those that concern the ways in which scientific theories are individuated and the role that power plays in science. I only touch on truth—who was wrong, who was right, and why? Because a system of biogeography termed *vicariance biogeography* succeeded and panbiogeography failed, we are tempted to conclude that the former succeeded since it was correct and the latter failed because it was mistaken, but the story with respect to the content of these two theories is much more complicated than one might expect. One thing is certain. If Croizat was anything, he was a rebel.

A HARD LIFE

Croizat was born of French parents in Turin, Italy, near the French border. When Croizat was six, his parents separated, and by the time Croizat's father died in 1915, the family was destitute. For the next forty years, Croizat was never more than poor and was sometimes close to starvation. While he was in the Italian army from 1914 to 1919, he married and began a family. After the war he quickly earned a law degree and went to work in a textile mill owned by a friend, but the rise of Fascism soon forced him to immigrate to the United States. In 1923 he landed penniless in New York City with a wife and two children. For six years he made money any way he could, including the sale of his watercolors.

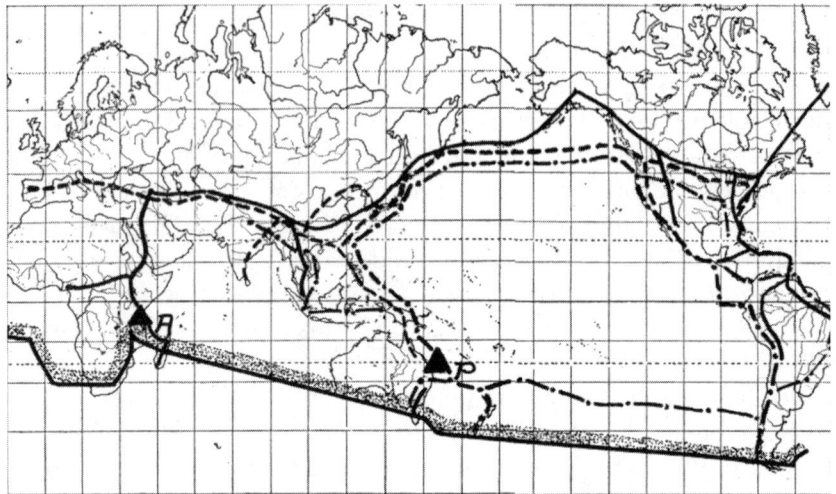

Figure 11.2. Vicariance biogeography: the dispersal of *Ericaceae*. The triangles marked A and P represent centers of angiosperm dispersal. The lines represent tracks and ranges of different groups. From Leon Croizat, *Space, Time, Form: The Biological Synthesis* (Caracas, Venezuela: Published by the Author, 1962), 296.

When the stock market crashed, the demand for original works of art crashed with it, and Croizat moved with his family to Paris, but the life of an unknown artist in Paris turned out to be even harder than it had been in New York, and Croizat returned to New York. Finally, he obtained a job identifying plants for a topographical survey of New York parks. During the Great Depression the federal government not only paid artists to cover the walls of public buildings with murals of rolling farmlands and muscular steel workers but also commissioned such projects as the production of floras and faunas. While working on his flora, Croizat met E. D. Merrill (1876–1956), who soon thereafter became director of the Arnold Arboretum of Harvard University and hired Croizat as a technical assistant. At long last he had a steady job.

While working for Merrill, Croizat wrote up a botanical paper and submitted it to the journal published by the arboretum. Generally such in-house journals automatically publish anything submitted by one of their own, but Croizat was only a technical assistant, and to make matters worse, in his paper he sharply criticized the work of one of the most powerful botanists on the staff. When the journal rejected the paper, Croizat published it elsewhere. In spite of this flap, Merrill kept Croizat on as an assistant, but when Merrill was replaced as director of the arboretum in 1946, Croizat was dismissed just a few months shy of achieving tenure.

Once again Croizat immigrated, this time to Venezuela, where he held a number of academic posts in botany between 1947 and 1952. Eventually he divorced his first wife and married a woman who had emigrated from Hungary and started a highly successful landscaping business in Caracas. For the rest of his life, Croizat was able to pursue his research unimpaired. He published a stream of books and papers that numbered, he was happy to declare, more than ten thousand printed pages, in part because he could publish his work at his own expense. No editor need be involved.

PANBIOGEOGRAPHY

In 1958 Croizat published his massive *Panbiogeography, or an Introductory Synthesis of Zoogeography, Phytogeography, and Geology,*[8] in which he set out his own views and took such authorities as Simpson to task for their mistakes. According to Croizat, the distribution of plants and animals must be plotted to see where these distributions consistently overlap. He termed these areas of overlap "standard tracks." To his surprise, one of the most general tracks marched right across the Atlantic Ocean between Africa and South America. That large a number of chance dispersals seemed very unlikely. Croizat argued that the discovery of standard tracks was the primary tool of biogeography and that equally general causes had to be found for these pervasive phenomena. The leftovers could then be explained in terms of chance dispersal. The converse line of reasoning struck Croizat as totally wrong-headed.

In 1959 Simpson happened upon Croizat's *Panbiogeography* in the library of the American Museum of Natural History and was dismayed by what he took to be the emotional venom lavished on him by its author. Simpson promptly wrote to Croizat to protest his treatment. Croizat was pleased that such an authority as Simpson had seen fit to communicate with him, but nothing came of this correspondence save alternations between irate indignation and assurances of good intentions on the part of both men. Croizat was prickly; so was Simpson.

In general, the strongest tool that well-established scientists have at their disposal is silence. Simpson chose not to refer to Croizat in print throughout the rest of his career; this slight did not go unnoticed. Simpson did complain in private that Croizat was a "member of the lunatic fringe."[9] Ernst Mayr joined with Simpson in ignoring Croizat, at least until 1982, when he reviewed Nelson and Rosen's *Vicariance Biogeography: A Critique*.[10] In this review Mayr complained of Croizat's "undisciplined verbiage."[11] Croizat's work, however, was not completely overlooked when he first published it. In his response to Bigelow's objections to phylogenetic systematics, Kiriakoff praised Croizat's panbiogeography, and later Brundin chastised Darlington for ignoring Croizat's "blazing sermon."[12]

GARETH NELSON AND VICARIANCE BIOGEOGRAPHY

In 1966, however, Croizat's fortunes took an unexpected turn. Lars Brundin published a monograph about the transarctic relationships of a certain group of midges in which he rehabilitated Hooker's view that Antarctica is a center of evolution.[13] According to Brundin, Hooker and other earlier biogeographers had misanalyzed the phylogenetic relationships among the species under investigation. In order to do biogeography properly, one had to adopt Hennig's method of phylogenetic analysis, especially his deviation rule. According to Hennig,[14] the most closely related groups of organisms frequently differ in their geographic distribution and, when they do, they replace each other geographically. He termed this phenomenon *vicariance.*

At this time a young ichthyologist, Gareth Nelson, took on the task of promulgating the views of Hennig and Brundin regarding classification. In this undertaking Nelson succeeded in convincing quite a few of his fellow systematists of the virtues of Hennig's system of classification and stepped on quite a few toes at the American Museum of Natural History in the process. Croizat was a rebel, a renegade, an iconoclast, a maverick. So was Nelson.

When Brundin was having trouble getting a paper published in *Systematic Zoology,* Nelson complained vehemently. In response, the editor suggested that Nelson himself might as well become editor. He did. As if Nelson had not already taken on enough in championing Hennig's principles, promoting Brundin's biogeographic extensions of Hennig, and assuming editorship of a major scientific journal, he chose this time to resurrect Croizat's theory of panbiogeography.

In 1973, while Nelson was editor of *Systematic Zoology,* he received a manuscript from Croizat. Instead of sending it off to two or three referees to be evaluated, the usual procedure, he sent it out to a total of twenty-nine referees—a practice that was anything but standard. Nineteen referees responded. Of these, only one referee found the manuscript ready for publication. Most thought that it needed major or minor revisions. Four urged outright rejection. It was in his review of Croizat's manuscript that Simpson referred to Croizat as belonging to the lunatic fringe.

One of the most common complaints about Croizat's manuscript concerned the writing style. As might be expected from his cosmopolitan background, he had some facility in several languages, including French (the language of his parents), Italian (the language of his birth), English, and Spanish, as well as Portuguese, Latin, Russian, German, and Greek. Although Croizat's flair for languages is intimidating, the manuscripts that he submitted for publication frequently departed significantly from standard academic prose. Many other scientists can be found using colorful prose, but Croizat was especially adept at it. For example, in response to Hennig's views concerning dispersal, Croizat declared that he stood on his record "with over 10,000 printed pages on the subject of dispersal, which allow me to laugh and titter when faced by, for instance, the 'zoogeography' of Hennig, poor soul."[15]

Nelson returned Croizat's manuscript to him for revision. On receipt of these revisions, Nelson sent Croizat's manuscript to be refereed again. The results this time were no different from those garnered by the original manuscript. In order to get this manuscript published, Nelson suggested to Croizat that he write a joint paper with Donn Rosen and himself.[16] Croizat agreed. Nelson and Rosen made two major changes to the manuscript. They introduced numerous lengthy footnotes clarifying Croizat's position and added Hennig's principles of phylogenetic analysis or, as it also came to be termed, "cladistic" analysis. As far as Croizat was concerned, however, the resulting paper was vicariance biogeography, not panbiogeography.

Because Nelson, Platnick, and Rosen considered Hennig, Croizat, and Popper to form three pillars of their views about classification, biogeography, and scientific method, respectively, they tried to obtain for them signs of scientific respectability. In 1973 Nelson, Rosen, and others launched a sustained effort to have the American Museum of Natural History award gold medals to these three men. They succeeded with respect to Hennig and Popper but failed with respect to Croizat. Several members of the museum staff considered Croizat no less a crackpot than Velikovsky. Nelson and company did succeed, however, in having Croizat elected a Corresponding Member of the Museum for a five-year period stretching from 1973 to 1978. When Croizat heard about Hennig's receiving a gold medal from the American Museum of Natural History, he was furious. For the rest of his life he complained of the injustice.

VICARIANCE BIOGEOGRAPHY AND THE NEW YORK SCHOOL

In 1979 Nelson and Rosen decided to sponsor a conference about vicariance biogeography. In response to a publicity release from the American Museum of Natural History, a reporter from the *New Yorker* appeared. Although Croizat was unable to attend the conference, a paper by him was read at the conclusion of the meeting, and Nelson was quoted as giving credit to a "wealthy Venezuelan amateur botanist named Leon Croizat."[17] When Croizat read this reference to himself he fired off a letter to the editors of the magazine in protest. He was neither wealthy nor an amateur. Nor was he a Venezuelan, either by birth or by tradition. No retraction was published.

Croizat's anger was only fueled when he read a review of the conference by Virginia Ferris in which she noted that Croizat's paper was read at the end of the three-day affair and that the discussions would have taken a very different turn had it been read earlier.[18] When Croizat read the review, he interpreted the placement of his paper as a means to preclude discussion of his work. To make matters worse, he was convinced that Nelson and Rosen continued to confuse their vicariance biogeography with his panbiogeography. This confusion was only exacerbated, Croizat complained, by placing his name on the paper with Nelson's and Rosen's.[19]

In October 1980 Croizat, at the tender age of eighty-six, submitted a paper to *Systematic Zoology* in which he took the advocates of vicariance biogeography to task. The associate editor who received the manuscript asked whether Croizat would mind if his paper were edited for style before being sent out to be refereed. Croizat agreed. Within the month Croizat received letters from Ferris, Nelson, Rosen, and Norm Platnick, a young specialist in spiders, apologizing for any unintentional slight that he may have perceived concerning the scheduling of his paper at the conference. Croizat accepted their apologies, withdrew his paper, and as a sign of his good will invited them to visit him and his wife if ever they happened to be in Venezuela.

Although Croizat claimed that he no longer harbored any hard feelings about the earlier misunderstanding about vicariance and panbiogeography, the appearance of Nelson and Platnick's *Systematics and Biogeography: Cladistics and Vicariance*[20] was too much for him to stomach. In a book of 567 pages Croizat received only nine citations, and once again Nelson and Platnick were confusing their system with his.

This book led Croizat to rewrite the manuscript that he had withdrawn earlier and resubmit it to *Systematic Zoology,* where it appeared in 1982.[21] This version of the paper turned out to be one long diatribe against Nelson. When Sadie Coats, Nelson's wife, wrote to Croizat to ask permission to visit him while they were in Venezuela collecting specimens, Croizat wrote to inform her that he did not care to receive Nelson, stating, "You are welcome in my house at any time and for as long as you please, but he will not be admitted."[22]

ROBIN CRAW AND PANBIOGEOGRAPHY

For several years, both Croizat and Nelson had been corresponding with a young biogeographer from New Zealand named Robin Craw. Although Croizat was eighty-six years old, blind in one eye, and able to type with only one finger on his left hand, he kept up an extensive correspondence with Craw, Nelson, and others including this author. Early on Croizat had developed a high opinion of Craw. As Croizat was to repeat frequently during the next half-dozen years, he found Craw to be his most intelligent interpreter. No matter how often others might misunderstand his ideas, Craw got them right.

In 1979 Craw published a paper in *Systematic Zoology* defending Croizat against the criticisms of R. M. McDowell,[23] in particular, the methodological views that Croizat attributed to Popper. In his acknowledgments he thanked Nelson and Rosen, among others. In the same year Craw submitted a manuscript to *Systematic Zoology* dealing with Cuvier, Darwin, and Kant, but the referees found it too historical for the journal, and the manuscript was rejected. In 1982 Craw published yet another paper defending Croizat that appeared immediately after the paper by Croizat attacking Nelson. In his acknowledgments Craw thanked both Nelson and Croizat.[24]

Craw did express one complaint to Croizat, however.[25] Craw had published two papers defending Croizat against his critics, but Croizat had not cited Craw's work in any of his subsequent publications. Croizat acknowledged that Craw was justified in feeling slighted. In his defense Croizat cited his ill health and the hostility that might be directed at the young man if Croizat were to boast of him as an ally.

Croizat was unable to read Craw's last two letters; he died of a heart attack on November 30, 1982. But before his death, Ernst Mayr at long last acknowledged his existence. Mayr wrote reviews of Nelson and Platnick's *Systematics and Biogeography*[26] and of Nelson and Rosen's *Vicariance Biogeography.*[27] He does not mention Croizat in his first review—justifiably so, because Croizat is hardly mentioned in this hefty tome—but *Vicariance Biogeography* is quite another matter. In this anthology Croizat is cited as frequently as are Nelson and Rosen and far more often than anyone else. Croizat, after being ignored by Mayr for so many years, was more than happy to use Mayr's review to defend himself. He crowed: MAYR MENTIONS CROIZAT.[28]

THE NEW ZEALAND CONTINGENT

With Croizat's death there was considerable danger that panbiogeography might die with him, but Craw, along with a half-dozen other young biogeographers based in New Zealand, strove to keep Croizat's views alive. Craw and G. W. Gibbs arranged to have a special issue of *Tuatara* devoted entirely to Croizat's work on panbiogeography.[29] It was in this publication that Croizat's posthumous response to Mayr appeared. The rest of the papers were by Craw, John R. Grehan, and Michael Heads. In the same year Gibbs published a short paper titled "Who Was Croizat?" noting that, though his subject had been publishing in biogeography since the 1950s, Croizat's views were "virtually unknown today."[30]

In 1989 a conference about panbiogeography was held at the New Zealand National Museum. The papers were published in a special issue of the *New Zealand Journal of Zoology.*[31] Ten years later Craw, Grehan, and Heads published *Panbiogeography: Tracking the History of Life.*[32] Between Croizat's death in 1982 and 1999, Croizat's defenders published extensively, but they did not do very well in obtaining academic positions. Grehan began his career at the Victoria University in Wellington, where he received his Ph.D. in 1987. Thereafter he worked at the Entomology Research Laboratory at the University of Vermont, in Wainuiomata, New Zealand, and finally moved to the Buffalo Museum of Science, in Buffalo, New York. Heads worked at the University of Otago in Dunedin, the University of Goroka, in Papua New Guinea, and finally at the University of South Pacific, in Suva, Fiji.

But it was Craw who paid the heaviest price for his spirited advocacy of Croizat. Craw received his Ph.D. in September 1983 from the Victoria University of Wellington and then obtained a position in the Entomological Division of the Department of Scientific and Industrial Research. At the time he was discussing the possibility of publishing an abridged version of Croizat's *Panbiogeography* with the University of Chicago Press. Even biogeographers who had their doubts about Croizat's work nevertheless would be happy to read a condensed, simplified version of his theory. Yet none of Croizat's descendants was willing to take on this formidable task.

In 1992, as part of a restructuring of science in New Zealand, Craw was notified that he was facing redundancy because he did not have any funding beyond 1993. In response to his reclassification, Craw decided to drop out of science and pursue a career elsewhere. As a result Croizat lost his strongest and most persuasive ally. During this period, the local panbiogeographers lost their jobs (Craw and F. Climo), kept their jobs but stopped publishing works on this subject (R. Gray and T. M. Henderson), or had to find employment overseas (Grehan, Heads, and R. Page). As Heads summarized the fate of panbiogeography in New Zealand after 1989, "no funding proposal for pangeographic research has succeeded, no panbiogeographers have been employed there, and no panbiogeographic work has been accepted for publication either in the government or the Royal Society journals."[33]

WHO WON? WHO LOST? AND WHY?

In disputes such as those surrounding vicariance and panbiogeography, the questions always arise, Who won? Who lost? And why? Given a superficial reading, answers to the first two questions seem obvious. Vicariance biogeography won and panbiogeography lost. One sign of the difference in fates between these two theories can be found in encyclopedia entries. For example, the *Oxford Encyclopedia of Evolution,* edited by Mark Pagel includes three entries for Biogeography: Island Biogeography, Human Influence on Biogeography, and Vicariance Biogeography.[34] Richard Mayden, a cladist, wrote the entry about the latter. According to Mayden, vicariance biogeography "has emanated from decades of dialogue in the history of biogeography as the predominate theory."[35] Some remnants of Croizat's work can be found in Mayden's entry (for example, reference to the coevolution of the earth and its organisms), but he largely ignores panbiogeography. Such was the case for Cox and Moore as well.[36]

One of the problems that confronts anyone who tries to decide the fate of scientific theories is that they are very difficult to individuate. If each of the biogeographic theories that I have treated had an essence that distinguished it from all others, deciding who won or who lost would be easier—easier but still not easy. But as Grehan observed, defining panbiogeography in terms of a "defining essence" would be misleading.[37] The same conclusion follows for other scientific research programs as well.

Matthew, Simpson, Mayr, and Darlington are lumped together as if they formed a single school, but anyone who reads their works soon discovers that they disagreed with each other on a variety of counts. The same can be said of vicariance biogeographers, but not panbiogeographers. Croizat went to great lengths to keep his theory from being contaminated by the views of opponents and disciples alike. The chief exception was the paper published with Nelson and Rosen in 1972,[38] but Croizat soon rejected it. When Heads and Craw compiled a bibliography of Croizat's scientific work, they omitted reference to this paper because it did not truly represent Croizat's views.[39]

The dispute between Croizat and Nelson concerned two main sets of issues—one surrounding biogeography, the other concerning taxonomic analysis. But the complexity does not stop there. Croizat favored conventional phenetic methods of taxonomic analysis, whereas cladists opted for the methods of cladistics analysis.[40] Hennig's descendants modified Hennig's views extensively, eventually splitting into two groups—pattern cladists and phylogenetic cladists. The contrast was between pattern and process. Pattern cladists strove to eliminate all reference to scientific theories from their classifications, at least initially, whereas phylogenetic cladists insisted that scientific theories must enter into classification right from the start. On this score, Popper agreed with the phylogenetic cladists. So did Croizat. According to Croizat, a "process is always more important than any of its byproducts."[41]

SCIENTISTS AND SOCIAL ORGANIZATION

Scientists are much more interested in the content of science than anything that might count as the social structure of science, but social structure does make a difference. The founders of the synthetic theory of evolution were not really all that much in agreement with each other about the fundamentals of the synthetic theory of evolution—in fact, synthetic theory was not very synthetic—but they agreed to mute their disagreements. They did not go after each other in print.

Nelson was also very good at gathering together a group of systematists and biogeographers to push his views. If anything sets Croizat apart from the other biogeographers of his day, it was his inability to form and maintain professional alliances. Early on, Croizat formed professional friendships with Lars Brundin and Søren Løvtrup, but these friendships did not last long. Croizat was interested in his own research program, not theirs.

When Nelson adopted Croizat's panbiogeography, Croizat should have been delighted, but very rapidly he came to reject Nelson and the "New York School." Craw, Heads, and Grehan came to Croizat's rescue. Initially, at least, members of the "New Zealand School" were content with trying to make Croizat's views comprehensible to a larger scientific community. They failed.

Croizat, Nelson, Craw, Heads, and Grehan joined in the controversy about biogeography and classification, but of equal importance, they themselves were rebels. Of these iconoclasts, Croizat started out cooperating with Nelson but soon became a dire enemy. At first Craw was on good terms with both Nelson and Croizat, but eventually he came to view Nelson as an enemy. Even so, Nelson continued to correspond with Croizat and Craw in an effort to stay on good terms with both of these contentious men.

In the 1969 correspondence between Simpson and Croizat, Simpson suggested that Croizat publish a shorter book about panbiogeography so that others could understand his views. No such book was forthcoming. Later, an editor at the University of Chicago Press sought to publish an abridged version of Croizat's massive *Panbiogeography*. None was forthcoming. Anyone who wanted to understand Croizat's views had to plod through several lengthy tomes. Not until 1982 did Croizat publish a short summary of his work and describe how it differed from that of his opponents. In this paper he acknowledged roles for both vicariance and dispersal, noting that a "biogeographer must be a vicariant in principle and a dispersalist in detail, case by case according to the merits of each case."[42]

Mavericks tend to oppose what they take to be censorship and the abuse of power. They see the refereeing process as a means for well-established scientists to reject new ideas. Croizat published most of his books at his own expense so that he could bypass the refereeing process. But that process is a valuable tool in science. We need more of it, not less. Referees and editors can point out places where the manuscript can be improved. They are not the enemies that Croizat thought that they were.

Croizat, Nelson, Rosen, Craw, Heads, and Grehan had one thing in common. They had no patience for authorities. When Croizat was a technical assistant at the Arnold Arboretum, he challenged one of the most powerful men on the staff. Nelson and other Hennigians became famous for playing according to the New York rules of conduct, rules that were rough and ready and showed no respect for the powerful men in their field. In fact, these scientists seemed to enjoy baiting authorities. Heads, in his "History and Philosophy of Panbiogeography," devotes an entire section to dogmatism, deliberate falsehood, oppression, and plagiarism.[43]

One cannot ignore the personalities of the scientists involved in the battles concerning biogeography. Croizat took offense easily but was unable to understand that others might take offense at him. When Simpson wrote Croizat in 1959, it was not to object to his biogeographic views but to complain about the "emotional venom" that Croizat directed at him. Croizat was even harsher in his private correspondence. In a letter to one of the associate editors of *Systematic Zoology* in October 1980, he declared that

> NELSON IS AN UNMITAGATED SCHEEMER, BETTER SAY A C R O O K, a being who inside or outside the field is dishonest . . .
>
> Nelson is very presumptuous when picking up a quarrel with me, more than presumptuous, indeed stupid, and you can freely tell him that much. I am not going to take all that rubbish and filth laying down, and when he tries to be astute, he risks being ridiculous.[44]

But Croizat could also be very charming.[45] He ended one of his last letters to Craw by declaring: "Cheer up! The sun is rising for you, even if it [is] setting for me, which is by now of little consequence. Do not worry about your work: *it is excellent.*" Before the year was out, Croizat was dead, and within a decade Craw had been ruled redundant.

THE ENTANGLED BANK OF SCIENCE

Thus far in this chapter I have concentrated on the sociology and psychology of science, but what really matters in science is content.

Why did Croizat fail? Did his hostility toward such a powerful bio-geographer as G. G. Simpson consign him to the lunatic fringe? Nelson and his colleagues were no less hostile to the men in power, and yet phylogenetic (that is, cladistic) classification took hold. So did vicari-ance biogeography. The members of the New Zealand School showed considerable hostility toward the big names in biogeography, but such hostility was not sufficient to get their views an extensive hearing. Diatribes against the powers that be can result in either a conspiracy of silence or irate responses in the literature. Both strategies can have some effect.

But, one might object, all of this psychology and sociology ignores the most fundamental reasons why certain views fail and others suc-ceed—reason, argument, and evidence. One might argue that the rea-son vicariance biogeography won out over panbiogeography was its greater ability to handle biogeographic data. Perhaps Croizat could have had a more extensive and fairer hearing for his views if only he had presented them in a more palatable form, but ultimately what matters is evidence and coherence. If we look at the data available at the time, we see that both sides of the dispute held views that we now take to be mistaken. Certainly the classical school (for example, Simpson and Mayr) initially rejected continental drift. A similar story can be told for classification but with the opposite conclusion. The early Hennigians did not present their views in a very conventional fashion, and yet they succeeded.

Darwin ended his *On the Origin of Species* by observing that nature forms an "entangled bank."[46] One of the goals of scientists is to un-tangle these banks. That is what the biogeographers and systematists that I have discussed in this chapter do. But science itself also forms an entangled bank, as all the chapters in this volume attest.

NOTES

1. Leon Croizat, "Vicariance/Vicariism, Panbiogeography, 'Vicariance Biogeography,' etc.: A Clarification," *Systematic Zoology* 31, no. 3 (1982): 298.

2. Gareth Nelson and Norman Platnick, *Systematics and Biogeography: Cladistics and Vicariance* (New York: Columbia University Press, 1981).

3. Willi Hennig, *Phylogenetic Systematics* (Chicago: University of Illinois Press, 1966).

4. Karl R. Popper, *The Logic of Scientific Discovery* (New York: Basic, 1959); Michael Mulkay and Nigel Gilbert, "Putting Philosophy Back to Work:

Karl Popper's Influence on Scientific Practice," *Philosophy of the Social Sciences* 12 (1981): 389–407; W. W. Bartley, "A Popperian Harvest," in *Pursuit of Truth*, ed. L. Levinson, 249–289 (New York: Humanities, 1982); J. R. Skoyles, "Popper's Success or Failure," *Nature* (1992) 359: 100, and David L. Hull, "The Use and Abuse of Sir Karl Popper," *Biology and Philosophy* 14, no. 4 (1999): 481–504.

5. Croizat, "Vicariance/Vicariism," 295. See also Croizat, *Biografía Analítica y Sintética ("Panbiogeografia") de las Américas* (Caracas: La Biblioteca de la Cademia de Ciencias Físicas, Matémáticas y Naturales de Venezueal, 1976).

6. Croizat, "Vicariance/Vicariism"; for a retrospective history, see John Grehan, "Panbiogeography, 1981–91: Development of an Earth/Life Synthesis," *Progress in Physical Geography* 15, no. 4 (1991), 331–363.

7. Croizat, "Vicariance/Vicariism."

8. Leon Croizat, *Panbiogeography, or an Introductory Synthesis of Zoogeography, Phytogeography, and Geology; with Notes on Evolution, Systematics, Ecology, Anthropology, etc.* (published by the author). As if the title of Croizat's 1958 trilogy were not long enough, he ended it with "etc."

9. Gareth Nelson, "Leon Croizat, Biografía Analítica y Sintética ('Panbiogeografia') de las Américas," *Systematic Zoology* 26, no. 4 (1977): 451.

10. Gareth Nelson and Donn E. Rosen, *Vicariance Biogeography: A Critique* (New York: Columbia University Press, 1981).

11. Ernst Mayr, "Gareth Nelson and Donn Rosen (eds.), 'Vicariance Biogeography,'" *Auk* 99, no. 3 (1982): 618.

12. Lars Brundin, "Transantarctic Relationships and Their Significance, as Evidenced by Chrinonomid Midges," *Kungliga Svenska Vetenskapsakademiens Handlingar* 11, no. 4 (1966): 1–472. See also R. S. Bigelow, "Classification and Phylogeny," *Systematic Zoology* 7, no. 2 (1958): 49–59; S. G. Kiriakoff, "Phylogenetic Systematics versus Typology," *Systematic Zoology* 8, no. 2 (1956): 117–118.

13. Brundin, "Transantarctic Relationships."

14. Hennig, *Phylogenetic Systematics,* 169.

15. Leon Croizat, "Résumé," *Boletin de la Academia de Ciencias Fisicas Matematicas y Naturales* 38, no. 116 (1978): 124.

16. Leon Croizat, Gareth Nelson, and Donn Eric Rosen, "Centers of Origin and Related Concepts," *Systematic Zoology* 23, no. 2 (1974): 265–287.

17. "The Talk of the Town," *New Yorker,* September 30, 1979, 38.

18. Virginia Ferris, "A Science in Search of a Paradigm? Review of the Symposium, 'Vicariance Biogeography: A Critique,'" *Systematic Zoology* 29, no. 1 (1980): 73.

19. Croizat, Nelson, and Rosen, "Centers of Origin."

20. Gareth Nelson and Norman Platnick, *Systematics and Biogeography: Cladistics and Vicariance* (New York: Columbia University Press, 1981).

21. Croizat, "Vicariance/Vicariism."

22. Leon Croizat to Sadie Coats, June 29, 1981, Special Collections, Room 363 Hillman Library, University of Pittsburgh, Pittsburgh, PA.

23. Robin Craw, "Generalized Tracks and Dispersal in Biogeography: A Response to R. M. McDowall," *Systematic Zoology* 28, no. 1 (1979): 99–107; R. M. McDowell, "Generalized Tracks and Dispersal in Biogeography," *Systematic Zoology* 27, no. 1 (1978): 88–104.

24. Robin Craw, "Phylogenetics, Areas, Geology and the Biogeography of Croizat: A Radical View," *Systematic Zoology* 31, no. 3 (1982): 304–316.

25. Robin Craw to Leon Croizat, May 27, 1982, Special Collections, Room 363 Hillman Library, University of Pittsburgh, Pittsburgh, PA.

26. Ernst Mayr, "Gareth Nelson and Norman Platnick's *Systematics and Biogeography*," *Auk* 99, no. 3 (1982): 621–622.

27. Ernst Mayr, review of "Vicariance Biogeography."

28. Leon Croizat, "Mayr vs. Croizat: Croizat vs. Mayr—an Enquiry," *Tuatara* 27, no. 1 (1984): 54.

29. Robin Craw and G. W. Gibbs, *Croizat's Panbiogeography & Principis Botanica: Search for a Novel Biological Synthesis* (Timaru, N.Z.: Timaru Herald Print, 1984).

30. G. W. Gibbs, "Who Was Croizat?" *Weta* 7, no. 1 (1984): 2–4.

31. C. Matthews, ed., "Panbiogeography Special Issue." *New Zealand Journal of Zoology* 16, no. 4 (1989).

32. Robin Craw, John Grehan, and Michael Heads, *Panbiogeography: Tracking the History of Life* (Oxford: Oxford Biogeography Series, 1999).

33. Michael Heads, *The History and Philosophy of Panbiogeography* (Mexico City: Regionalización Biogeográfica en Iberoamerica y Tópicos Afines, 2005), 67–123.

34. Mark Pagel, ed., *Oxford Encyclopedia of Evolution,* 2 vols. (Oxford: Oxford University Press, 2002).

35. Richard Mayden, *Encyclopedia of Evolution,* s.v. "Vicariance Biogeography," 1:97–101 (Oxford: Oxford University Press, 2002).

36. C. B. Cox and P. D. Moore, *Biogeography: An Ecological and Evolutionary Approach* (Oxford: Blackwell, 1993).

37. Grehan, "Panbiogeography, 1981–91," 333.

38. Croizat, Nelson and Rosen, "Centers of Origin."

39. Michael Heads and Robin Craw, "Bibliography of the Scientific Work of Leon Croizat, 1932–1982," in *Croizat's Panbiogeography and Principia Botanica: Search for a Novel Biological Synthesis,* ed. Robin Craw and G. W. Gibbs (Timaru, N.Z.: Timaru Herald Print, 1984), 67–75.

40. A. Hallam, "Relative Importance of Plate Movements, Eustasy, and Climate in Controlling Major Biogeographical Changes Since the Early Mesozoic," in *Vicariance Biogeography: A Critique,* ed. G. Nelson and D. Rosen, 303–330 (New York: Columbia University Press, 1981).

41. Croizat, "Résumé," 59.

42. Croizat, "Vicariance/Vicariism," 297. See also Cox and Moore, *Biogeography;* Craw, Grehan, and Heads, *Panbiogeography.*

43. Heads, "History and Philosophy of Panbiogeography."

44. Croizat to David Hull, October 1980, Special Collections, Room 363 Hillman Library, University of Pittsburgh.

45. For more extensive coverage of Croizat's personality, see an interview conducted by Jon Baskin in Caracas, Venezuela, August 1974. The tapes are housed in the archives of the American Philosophical Society, 105 South Fifth Street, Philadelphia.

46. Charles Darwin, *On the Origin of Species* (London: Murray, 1859), 480.

FURTHER READING

Craw, R. "Panbiogeography: Method and Synthesis in Biogeography," in *Analytical Biogeography,* ed. A. A. Myers and P. S. Giller, 405–435 (1988).

Hull, David L. *Science as a Process: An Evolutionary Account of the Social and Conceptual Development of Science* (Chicago: University of Chicago Press, 1988).

Wiley, E. O. "Vicariance Biogeography," *Annual Review of Ecology and Systematics* 19 (1988): 513–542.

CHAPTER TWELVE

Dogma, Heresy, and Conversion: Vero Copner Wynne-Edwards's Crusade and the Levels-of-Selection Debate

MARK BORRELLO

The idea of the rebel carries with it a whiff of the romantic and often an element of tragedy; this chapter contains a bit of both. I describe the life and career of Vero Copner Wynne-Edwards (1906–1997) and his infamous contribution to the biological literature of the twentieth century, the theory of group selection. As David Hull did for Leon Croizat (see Chapter 11), I present the work of a rebel who was ultimately unsuccessful. This, I think, provides the historian with an interesting opportunity to delve into the nature of success in science. Though his theory was never accepted into the mainstream of Darwinian thinking, Wynne-Edwards was the focus of a debate that spurred some of the most important developments in evolutionary theory in the latter half of the twentieth century. Indeed, it is a debate that has continued into the present and therefore serves to illuminate the nature and structure of evolutionary theory. From the first presentation of his theory in the mid-1950s, Wynne-Edwards was fully aware that his idea fell outside the recently developed neo-Darwinian paradigm. By the end of his career, he bemoaned his status as heretic and outsider, but his dogged conviction regarding the importance of group selection to the evolutionary process indicates that he wouldn't have had it any other way.

The possibility that selection acted on groups of organisms as well as individuals was a component of Darwin's thinking beginning in the *Origin* and more fully developed in *The Descent of Man.* In the *Origin*, Darwin looked to the social insects and argued that "we can perhaps understand how it is that the use of the sting should so often cause the insect's own death: for if on the whole the power of stinging

Figure 12.1. Vero Copner Wynne-Edwards (left) with students,
Baffin Island Expedition, 1953. Courtesy Vero C. Wynne-Edwards
fonds, Queen's University Archives—5137.1-8-43.

be useful to the community, it will fulfill all the requirements of natural selection, though it may cause the death of some few members."[1]
Darwin would later argue in *The Descent of Man* that the inheritance
of social instincts was of the utmost importance to the later development of human society and that the development of these instincts
was for the good of the community over and above the advantage of
the individual. He wrote: "Finally, the social instincts which no doubt
were acquired by man, as by the lower animals, for the good of the
community, will from the first have given him some wish to aid his
fellows, and some feeling of sympathy."[2] These passages set the stage
for a proliferation of "good of the species" arguments during the
eclipse of Darwinism—from the late 1880s to the mid-1920s.[3] With the
rediscovery of Mendel, the development of classical genetics, and
the rise of theoretical population genetics, the focus of evolutionary
theory shifted more exclusively to the individual and subsequently
the gene.

Wynne-Edwards began his training at Oxford University in the
midst of these developments. He was immediately intrigued by the attempts of the ecologist Charles Elton to understand population dynamics in terms of evolutionary theory. How could the biologist explain

the countervailing demands of the individual and the group? How could evolutionary theory account for the maintenance of populations below the exploitation level of the environment if each individual was striving to maximize the number of its offspring? For Wynne-Edwards, the answer required an embrace of the good-of-the-community arguments of Darwin and the development of a theory of group selection that accounted for phenomena that were not conducive to individual-level explanation.

GROUP SELECTION TODAY

The theory of group selection remains controversial among evolutionary biologists today.[4] It does not engender the kind of vehement rejection that Wynne-Edwards's theory received in the 1960s and 1970s, but neither has it been seamlessly incorporated into contemporary evolutionary biology. My analysis of his contribution demonstrates that Wynne-Edwards's legacy is a complicated one that is too often oversimplified and underappreciated. The story of his development and advocacy of group selection is the story of an insider who becomes an outsider and, to some extent, revels in that status.

EARLY LIFE AND NATURAL HISTORY

Wynne-Edwards's early life experiences and schooling might lead one to expect a productive and conventional scientific career; there is nothing particularly rebellious or iconoclastic about him except for his scientific work. Born the fifth of six children in 1906 to the Reverend Canon John Roslindale Wynne-Edwards and his wife Lillian Agnes, Wynne-Edwards was brought up in a manner reminiscent of the stories of another iconoclastic naturalist—Charles Darwin. Though his family did not have the means of the Darwins, his youth in the Yorkshire Dales was filled with many of the activities familiar to Darwin. From early on, Wynne, as he was called, was a collector and list maker. His fondness for natural history found expression in long bike rides and collecting expeditions with family and friends. Wynne was precocious; at the age of eleven he won the junior botany prize from Leeds Grammar School for his collection of named wildflowers and ferns.[5] In 1920, at the age of thirteen, Wynne was sent to the prestigious Rugby School. There, he also became interested in astronomy and furthered his passion for natural history in the school's herbarium and natural history society. While at Rugby, Wynne also came into contact with some scientific luminaries of the day; Sir Ernest Shackleton spoke to the students on the eve of his final Antarctic expedition, and Julian Huxley "lectured awfully well" on the Oxford expedition to Spitzbergen.[6]

After graduating from Rugby Wynne-Edwards entered New College, Oxford, to read zoology with Julian Huxley. Although Huxley took the chair of the Zoology Department at King's College London after Wynne-Edwards's first year, Oxford remained an excellent choice. His new tutor, Charles Elton, sparked an interest in the ecology of animal populations that burned for the remainder of his lengthy career. Wynne-Edwards was trained by an unparalleled group of biologists. He read comparative anatomy with the Linacre Professor, E. S. Goodrich, embryology and experimental zoology with Gavin de Beer, and genetics with E. B. Ford. Wynne-Edwards graduated from Oxford with first-class honors in 1927 well prepared to begin his professional career. The influence of Elton, and his interest in animal populations, would dominate his professional life and led to his most significant and controversial contribution: the theory of group selection.

FROM THE CENTER TO THE PERIPHERY

If Wynne-Edwards's early life was a model of traditional British education and professionalism, his first academic position, at McGill University in Montreal, Canada, marked the beginning of his move from the center. He immigrated to Canada in 1930 with his new wife and former classmate at Oxford, Jeannie Morris. Though removed from the hub of the intellectual universe, Canada proved an excellent spot for Wynne-Edwards to begin his career. Indeed, he wasted no time turning their trans-Atlantic voyage to their new home into an opportunity for research. On the trip from Portsmouth to Halifax, he kept numerical logs of sea bird species. On arrival he secured funding for three more crossings and subsequently established a pattern of sea bird distribution. He published an article describing this pattern of inshore (coastal), offshore (to the edge of the continental shelf) and pelagic (deep-water) zones of sea bird distribution. This article, "On the Habits and Distribution of Birds on the North Atlantic," won the Walker Prize of the Boston Society of Natural History and hailed the arrival of a new first-rate zoologist at McGill.[7]

Wynne-Edwards remained there through the end of World War II.

During his fifteen years in Canada he participated in a number of field expeditions to Newfoundland, Labrador, and the Arctic. On these expeditions he developed a view of nature that influenced his later development of group selection theory. From his early interest in the roosting behavior of starlings to his survey of the distribution patterns of birds on the North Atlantic to his analysis and explanation of non-breeding behavior in sexually mature fulmars, Wynne-Edwards was fascinated by the structure of populations and the mechanisms that regulated their size and density. His 1937 paper "Intermittent Breeding of the Fulmar with Some General Observations on Non-Breeding in Sea Birds" marked a shift in his work from a descriptive, observational approach based on natural history to one more concerned with theory. In this paper we can see the seeds of his later work concerning group selection.[8] Wynne-Edwards wrote that based on previous observations it was estimated that only between one-third and two-fifths of the fulmars (both male and female) present at a particular colony appeared to be engaged in reproduction. This surprising fact, he wrote, required explanation. If, consistent with Darwinian theory, individuals are constantly striving to increase their representation in subsequent generations, why were so many of these fulmars not engaged in reproductive activity?

From the 1930s through the 1950s Wynne-Edwards participated in multiple expeditions in northern Canada. As the naturalist on the Baird Expedition to Baffin Island in 1950 he devoted twelve weeks to the collection of plants and of freshwater and terrestrial animals, and above all, to a study of the breeding birds. In his report, "Zoology of the Baird Expedition" published in 1952, he observed: "Competition between individuals for space and nourishment seems commonly to be reduced to a low level among members of the arctic flora and fauna; they live somewhat like weeds, the secret of whose success lies in their ability to exploit transient conditions while they last, in the absence of serious competition. In the Arctic the struggle for existence is overwhelmingly against the physical world, now sufficiently benign, now below the threshold for successful reproduction, and now so violent that life is swept away, after which recolonization alone can restore it."[9] By this point in his career, he had come to recognize that in order to account for the regulation of populations below the threshold of exploitation, a mechanism other than individual-level selection was necessary.[10]

Wynne-Edwards was in the Arctic when he learned that the Regius

Chair for Natural History had become available at Aberdeen University in Scotland. His return to the United Kingdom led to his appointment to a number of scientific advisory committees. As he related in his autobiographical memoir, this committee work for the National Environmental Research Council and the Nature Conservancy, among others, "draws one's attention to useful scientific knowledge that would otherwise have gone unnoticed . . . and kept me in close contact with scientists working at sea and at the bench, and added very opportunely to my intellectual stock in trade in the important decade of 1952 to 1962."[11] This decade was, indeed, an important one for Wynne-Edwards. It began with the publication of David Lack's book *Natural Regulation of Animal Numbers* and concluded with Wynne-Edwards's response—a book-length argument for group selection titled *Animal Dispersion in Relation to Social Behavior* (figure 12.2 shows a photo of a cliffside gannetry used to buttress his argument).

A CHALLENGE TO ORTHODOXY?

By the early 1950s what we now know as the Modern Synthesis was fairly well established. The population genetics of Fisher, Haldane, and Wright was the common currency of evolutionary theory. Dobzhansky, Simpson, Mayr, Huxley, and Stebbins had published their accounts of the "neo-Darwinian" genetics, paleontology, systematics, general biology, and botany, respectively, in a series of books that would set the standard for the study of biology in a Darwinian framework. Historians have recently begun to look more carefully at what was actually agreed upon in the synthesis and what remained unresolved; not surprisingly, there is a fairly wide range of opinion.[12] For the purposes of this chapter, I suggest, consistent with Stephen Jay Gould, that for most biologists the Modern Synthesis represented a hardening of biological theory around the analysis of adaptation and the mechanism of individual-level selection. Of course, this interpretation of the synthesis is not universally accepted (neither by the participants nor by historians of biology). Nevertheless, there is sufficient evidence that for many biologists of the period, the developments represented in the aforementioned texts would lead to a new and unified biology.

Wynne-Edwards's first direct challenge to the individual-level selectionist accounts came in the form of a review of Lack's *Natural Regulation of Animal Numbers*. In a letter in which he agreed to write

Figure 12.2. The gannetry at Cape St. Mary, Newfoundland. Wynne-Edwards shot this photo in 1935 and used it in his 1962 book to demonstrate that group selection maintained bird populations below the threshold of resource exploitation. Above there is ample space available for expansion of the colony; however, the birds do not expand into those areas to breed. Courtesy Vero C. Wynne-Edwards fonds, Queen's University Archives—5137.1-8-69-55.

the review, Wynne-Edwards lamented that he was busy with other projects but felt compelled to do it because "Lack has failed to penetrate the first principles of the subject, and in order to demonstrate this one must go very deep oneself."[13] In his review he criticized Lack with regard to a number of fundamental assumptions. He suggested that Lack's view that natality was the independent variable for population regulation and that mortality adjusts itself (through density-dependent effects) to match it was insufficiently supported. Wynne-Edwards sketched Lack's argument as follows: "Any individual which contributes more than its share to posterity must be favored by selection, and any which contributes less than its share (i.e. has a smaller egg-number than average) is doomed to decline. One may well suspect a fallacy here. As the March Hare said to Alice, 'Why you might just as well say that "I see what I eat" is the same thing as "I eat what I see!"'"[14] He also pointed out that if the most fecund are not the most fit, the stock might deteriorate and be replaced by some stock of more moderate fecundity. Recall his earlier work concerning the nonbreeding behavior of sea birds as important support for his criticism of Lack's thesis.[15] Lack's explanation for the regulation of animal numbers fit perfectly within the newly minted paradigm of the modern synthesis: population size was the result of natural selection acting on individuals and limiting broods to the maximum size that would result in the highest number of successful fledglings, thereby maximizing the fitness of each breeding pair.

In the same year, Wynne-Edwards presented a paper to the Eleventh International Ornithological Congress in Basel, Switzerland, that directly challenged Lack. This paper marked his first clear presentation of his theory of group selection. Though claims about the good of the species did not originate with Wynne-Edwards, his presentation of a particular mechanism (that is, group selection) raised the interest and ire of many biologists and generated an immediate and sustained response. Wynne-Edwards had shared the draft of his Basel paper with some colleagues at Oxford, most notably his mentor Charles Elton. In a brief letter, Elton wrote that he found the paper interesting and was sure it would generate a great deal of discussion. Wynne-Edwards was disappointed by the less-than-enthusiastic response.[16]

In his presentation, Wynne-Edwards introduced his developing theory of group selection in qualified terms. He wrote, "It is theoretically possible to regulate the numbers in the population by density dependent control of the recruitment rate alone." He continued, "Control of this sort could be largely intrinsic, that is, depending for its operation on the behavior-responses of the members of the population themselves." Later in the discussion section he added: "A collective response by a social group to general conditions of food productivity does not appear so very much more abstract and improbable than the corresponding individual responses by male birds in claiming territory."[17] The concluding paragraph of the paper makes the most clear and unwavering statement of his new theory, his challenge to the neo-Darwinian focus on individual-level selection. "The theory that slowly breeding birds have evolved a series of interrelated adaptations, giving them a great measure of autonomic control of their numbers, permits, at any rate, a rational explanation to be offered of many hitherto unconsidered or anomalous features of their breeding biology. It shows that if they were adapted to impose their own limit on the number and size of their breeding colonies . . . they could combine optimum feeding conditions with maximum numbers."[18] The paper presented in Basel and the review of Lack's book marked the first clear steps of Wynne-Edwards's long journey into apostasy. He had staked his claim on the notion that selection could and did act on groups of individuals in order to maintain populations below the threshold of environmental exploitation. Where Lack saw the population as the result of selection acting on the breeding pair to maximize reproduction in particular environmental conditions, Wynne-Edwards saw selection acting on groups to preserve the breeding habitat for the long-term survival of the colony. Between 1955 and 1962 Wynne-Edwards did not publish much; he continued with his administrative duties and worked to establish the Culterty Field Station (the site of the red grouse studies that would feature prominently in his second book about group selection). Most important, he worked on his book *Animal Dispersion in Relation to Social Behavior.*

GROUP SELECTION IS EVERYWHERE

Whereas Wynne-Edwards's early formulation of the theory of group selection was applied mostly to colonial sea birds, *Animal Dispersion* would demonstrate that group selection acted on populations across the animal kingdom from plankton to pachyderms. The 650 pages of *Animal Dispersion* formed an exhaustive (and exhausting) catalog of the effects of group selection. From the pronking behavior of springboks and the swarming of whirligig beetles to the overwintering of monarch butterflies, Wynne-Edwards surveyed the behavioral literature and reinterpreted much of what had gone before in terms of group selection—his idea that populations had been selected to maintain themselves below the exploitation level of the environment.

The preface states his goal as the linking of the subjects of population and behavior to provide a "rational, single explanation of the origin of social behavior." He writes that this new theory "has provided me with a novel and, it has often seemed, commanding viewpoint from which to survey the everyday events of animal behaviour; and some of the most familiar activities of animals, the purpose of which has never properly been understood, have readily been seen to have important and obvious functions."[19] In the following paragraph he provides further evidence as to why his theory would become so controversial. "One of the interesting corollaries," he writes "has been to interpret the zoological background of conventional behaviour in man. This is not without its philosophical implications . . . seldom, as a matter of fact, could one find any clearer indication than emerges here of the closeness of man's kinship with his fellow animals."[20] Thus it is clear from the outset that this is meant as a fundamental challenge to the recently minted neo-Darwinian paradigm built on individual-level natural selection as the primary if not the exclusive explanation for the development of social behavior.

The first chapter of *Animal Dispersion* presents an analogy between natural population regulation and the problem of overfishing in the North Atlantic (something Wynne-Edwards was quite familiar with). He suggests that the best approach to the subject of optimum density was to study man's own experience exploiting natural resources. Following the presentation of the overfishing example, he asserts that something must constantly restrain populations, while in the midst of plenty, from overexploiting their resources. He rejected the application of such terms as *free enterprise* and *unchecked competition* to natural populations, as individual-level selectionists often did, and invoked instead the concept of the balance of nature. According to Wynne-Edwards a society is, in its most primitive function, merely an organization capable of providing for conventional competition. The existence of this conventional competition precludes direct competition for food or other resources and thereby assures the persistence of the social group through the avoidance of overexploitation.

Wynne-Edwards also made clear that his own theory was distinct from the traditional "Darwinian heritage" (read: neo-Darwinism). He cited the standard interpretation of natural selection, which occurs at two levels, the individual (intraspecific) and the species (interspecific) level, and argued that neither of these covered the social adaptations that are of interest. From his perspective, it takes a group of individuals to maintain social conventions, and this in turn requires the selection of groups. He spelled out the function of group selection as follows: "Evolution at this level can be ascribed, therefore, to what is here termed group-selection—still an intra-specific process, and for everything concerning population dynamics, much more important than selection at the individual level. The latter is concerned with the physiology and attainments of the individual as such, the former with the viability and survival of the stock or race as a whole. Where the two conflict, as they do when the short-term advantage of the individual undermines the future safety of the race, group selection is bound to win, because the race will suffer and decline, and be supplanted by another in which antisocial advancement of the individual is more rigidly inhibited."[21] If for the first half of the twentieth century biologists allowed for explanations of adaptations at various levels, where individual-level explanations of morphological traits were published side by side with group- and species-level explanations of social behavior, Wynne-Edwards's theory of group selection provided a clear target for the neo-Darwinists.

A DEVASTATING RESPONSE

Animal Dispersion sold well and led to a round-the-world lecture tour. In 1965 Wynne-Edwards published a précis of the book in *Scientific American* that generated 350,000 offprint requests. The seed of his theory, it seemed, had landed on fertile ground. Indeed, *Animal Dispersion* remains a citation classic.[22] The response from the professional community was, however, not so welcoming. Wynne-Edwards's formulation of group selection was seen as a return to the kind of good-of-the-group explanations that had been supplanted by individual-level explanations modeled by the population geneticists. The intuitive appeal of group selection seemed to rally an immediate response. Of course, David Lack took issue with Wynne-Edwards's interpretation of his own work and field data.[23] The mathematical biologists William Hamilton and John Maynard Smith developed a new model of kin selection that could be applied to many of the social behaviors that Wynne-Edwards had insisted required group selection. In 1966, George C. Williams, an American evolutionary biologist, published what would become one of the most influential books in evolutionary biology of the twentieth century, *Adaptation and Natural Selection.*[24] Williams's book set a new standard for evolutionary biology and focused the research at the level of the gene. The trend toward molecularization had been under way in biology for more than a decade, and Williams's sweeping critique of group selection theory marked the beginning of a new age of genic selection. The decade following the critiques by Lack and Williams was one of continuing decline for group selection. Hamilton, who had initially published his theory of kin selection in 1964, continued to hone his ideas and drive the focus of evolutionary biology to the level of the gene. Smith also engaged in an ongoing criticism of Wynne-Edwards's work and supplied perhaps the most crucial challenge to his theory: that of the possibility of invasion of an altruistic population by an opportunistic cheater. In 1971, Robert Trivers published his paper concerning reciprocal altruism, which again thwarted the group-selected explanations for population regulation that Wynne-Edwards had proffered for various seemingly self-sacrificial behaviors. Essentially, Trivers argued that behaviors that seemed altruistic (for example, breeding restraint) were best interpreted as individual strategies to encourage cooperation, which benefited the individuals so engaged, not the group as a unit of selection.[25] The culmination of the gene-centered view came with the 1976 publication of Richard Dawkins's book *The Selfish Gene*.

Dawkins took the gene-centered view and presented it, incredibly persuasively, to both popular and professional audiences. In the preface to the 1989 edition, Dawkins wrote of his personal goal: "I would write a book extolling the gene's-eye view of evolution. It should concentrate its examples on social behaviour, to help correct the unconscious group-selectionism that then pervaded popular Darwinism."[26] In Dawkins's recollection, the intellectual bogeyman of group selection was still quite prevalent, at least when he began writing in 1972. He went on to describe *The Selfish Gene* as merely the logical outgrowth of the orthodox neo-Darwinism of R. A. Fisher and perhaps a more "full-throated account" to accentuate the "laconic expressions" of G. C. Williams and W. D. Hamilton that had so inspired him. In the broader context, *The Selfish Gene* served as the nail in group selection's coffin, which had been built ten years earlier by G. C. Williams's *Adaptation and Natural Selection*.

SECOND TIME AROUND: *EVOLUTION THROUGH GROUP SELECTION* (1986)

By the mid-1970s it seemed that Wynne-Edwards had nearly given up on group selection. In 1977 the Royal Geographical Society held a symposium titled "Population Control by Social Behavior" in order to "examine how far animal populations are controlled by intraspecific mechanisms, and in particular, to debate the status of Wynne-Edwards' massive hypothesis that such control involves contests for social status, territory or other conventional prizes rather than direct competition for food."[27] In his opening paper Wynne-Edwards discussed research on the red grouse in Scotland and beaver populations in Canada that supported his theory but ultimately allowed that "as a hypothesis [group selection] cannot be tested experimentally because of the prohibitive length of time required; but in the last 15 years many theoreticians have wrestled with it, and in particular with the specific problem of the evolution of altruism. The general consensus of theoretical biologists is that credible models cannot be devised, by which the slow march of group selection could overtake the much faster spread of selfish genes that bring gains in individual fitness. I therefore accept their opinion."[28] This capitulation was, however, short-lived. Wynne-Edwards was soon to begin another book-length treatment of group selection that he was certain would turn the tide.

Through the end of the 1970s and into the 1980s Wynne-Edwards wrestled with criticisms of group selection in a series of notebooks. He was all too familiar with the challenges presented by Hamilton's kin-selection model, John Maynard Smith's development of the haystack model, which demonstrated that any self-regulating population was vulnerable to invasion by selfish individuals, and Robert Trivers's theory of reciprocal altruism. In trying to answer these critics, Wynne-Edwards looked back to the evolutionary genetics of Theodosius Dobzhansky and Sewall Wright for support. He interpreted Dobzhansky's work as not supportive of kin selection theory, because the populations he described were not necessarily comprised of closely related individuals. The benefit of genetic variation (which was so important in Dobzhansky's account) could not be ascribed to the individual because it was a characteristic of the population. Sewall Wright's shifting balance model was equally important to Wynne-Edwards's second attempt to establish the fundamental importance of group selection.

His notebooks demonstrate an increasing level of interest in Wright throughout the 1970s that culminated in a chapter in *Evolution Through Group Selection* about evolution in structured populations that drew heavily on Wright's model. He wrote: "My own ideas about both functions [ecological and evolutionary] have of course been reached by following ecological and sociobiological paths; but there is an older and more powerfully and persistently argued theory that has led to very much the same conclusions, with regard to the evolutionary conclusions alone, in the work of Sewall Wright (e.g. 1931, 1940, 1945, 1949, 1968–78, 1980)."[29] After describing the importance of Wright's early work, Wynne-Edwards turned to the most recent paper and pointed out that, much like his own theory, Wright's had been frequently misrepresented and misinterpreted. Wynne-Edwards's new book was meant to set things straight. In Wright's attempts in a 1980 paper to clarify his own ideas for the community of evolutionary biologists, Wynne-Edwards saw an opportunity to demonstrate the consistency of his own thought with Wright's and thereby salvage his theory from oblivion. He was particularly keen to show that his theory, like Wright's, need not always be working in opposition to individual-level selection. Given the earlier critiques of Williams and Maynard-Smith, this was an especially important point of clarification.

Another important development for Wynne-Edwards was the new research concerning group selection theory, in particular, the experimental work of Michael J. Wade and the theoretical work of David Sloan Wilson.[30] Wade and Sloan Wilson had corresponded with Wynne-Edwards, but the connection between their work and his theory was never simple. The "new" group selection theorists had to distinguish themselves from what Sloan Wilson characterized as the "naïve" group selection of Wynne-Edwards, and Wynne-Edwards had concerns about priority and credit. In his initial manuscript of *Evolution Through Group Selection,* he made only brief reference to Wade's or Sloan Wilson's work. It was only at the urging of the publisher that Wynne-Edwards incorporated their work to a more significant extent.

The reception of *Evolution Through Group Selection* was largely one of silence. Wynne-Edwards had hoped that his book would re-ignite the controversy he had initiated in the 1960s; this was not to be. The reviewers dismissed the book as a work of advocacy more than of science. In *Ethology* the Oxford zoologist Mark Ridley wrote: "[W]hat a strange man Wynne-Edwards is, to precipitate perhaps that most interesting controversy in evolutionary biology of the past quarter century, and then to write another book on the same subject, but which ignores the controversy completely." Although Ridley gave Wynne-Edwards credit for his synopsis of the recent work of Sloan Wilson and Wade, he pointed out that ultimately the book "contained no argument at all. It instead describes a point of view."[31] This period must have been a particularly difficult one for Wynne-Edwards; nevertheless, he soldiered on.

THE DEATH OF WYNNE-EDWARDS AND THE LIFE OF AN IDEA

Wynne-Edwards spent the last years of his life in the quest for a fair hearing from a professional audience. By this time, he was convinced that the system was set against him and his theory, that editors would not review (let alone publish) his work, that instructors misrepresented his theory to their students, that senior biologists would not change their minds, and that junior biologists would never be given the opportunity. He continued his correspondence during his eighties and worked doggedly to get a précis of the book published in a journal with a broad audience. He was rejected by *Nature* and *Trends in Ecology and Evolution* among other leading journals. Finally, in 1993 the *Journal of Theoretical Biology* published his final article, "A Rationale for Group Selection." He was eighty-six years old.[32] The article occupied the first twenty-two pages of the journal and reinvigorated Wynne-Edwards—now he needed surrogates. In a letter to the Royal Society Professor at Oxford, Robert May, he made a final attempt to usher his theory into the next generation. After a brief introduction he wrote: "There is of course massive objection to accepting [group selection theory] from the majority of biologists who have been brought up to regard it as a heresy, although on dogmatic not experimental grounds. They seem unlikely to change their minds unless they are given the true message in a hurry by authorities whose competence they trust." It is quite striking to see how Wynne-Edwards presents his predicament. His theory is "heresy" disqualified from consideration not on empirical grounds but by the fact that it doesn't conform to the "dogma" of individual-level selection. He went on to lament: "I am eighty-seven years old and can play no further part. In the last eight years I have grown accustomed to having articles rejected by editors who, being unqualified to make independent judgments, just assume I am a heretic." The lamentation was followed by an entreaty: "You and your Oxford colleagues constitute a centre of excellence in this field of knowledge. Though it might involve some back-tracking, you could, if you saw fit, take over the conversion task as your own, with lasting benefits to scholarship at large, and great delight from this Oxonian son. But perhaps I presume too much."[33] Indeed, he presumed too much. May demurred with a letter that acknowledged the significance of the idea but pointed out that he was busy with his own work and therefore "reluctant to embrace any new crusades." The use of language in this exchange is fascinating; Wynne-Edwards writing of dogma, heresy and conversion, May describing the invitation as an embarkation on a crusade. Perhaps more than anything else this characterizes the rebellion of Wynne-Edwards.

The iconoclasm of Wynne-Edwards is largely recognized and acknowledged. He represents, for most who have heard of him, the wrong-headedness of the "good-of-the-group" arguments of old-fashioned evolutionary theorizing. What I think we can surmise from this account is a slightly more nuanced understanding not only of the role of group selection theory in the development of evolutionary biology but also of the role of iconoclasm in science more generally.

Wynne-Edwards initially presented his theory at a time of increasing professional and intellectual focus for evolutionary biology. His work challenged one of the basic tenets of that focus: the power of the mechanism of natural selection acting on the individual. The architects of the modern synthesis had labored mightily to find the points of commonality for the varied subdisciplines of biology, and Wynne-Edwards's assertion of the importance of group selection only complicated that concordance.

In the past ten years we have seen a move toward a more hierarchical approach to evolutionary theory. Supporters of this approach argue that selection acts not only on genes and individuals but also groups, species, and clades. Though Wynne-Edwards was never successful in convincing biologists that group selection was fundamentally important to the evolutionary process, his work focused efforts on the levels-of-selection question. His iconoclasm was born of his view of nature, his experiences in Canada and the Arctic, his career in institutions outside the centers of research and authority, and his dogged commitment to his theory. Though he might have been frustrated in life, Wynne-Edwards would certainly revel in the notion that the conversation, if not the controversy, continues.

NOTES

1. C. Darwin, *On the Origin of Species.* facsimile 1st ed. (Cambridge: Harvard University Press, 1859), 203.

2. C. Darwin, *The Descent of Man, and Selection in Relation to Sex* (Princeton: Princeton University Press, 1871), 103.

3. P. J. Bowler, *The Eclipse of Darwinism: Anti-Darwinian Evolution Theories in the Decades Around 1900* (Baltimore: Johns Hopkins University Press, 1983).

4. M. Borrello, "The Rise, Fall and Resurrection of Group Selection," *Endeavour* 29, no. 1 (2005): 43–47; L. Dicks, "All for One," *New Scientist* 167, no. 2246 (2000): 30–35; R. Lewin, "Evolution's New Heretics," *Natural History* 105, no. 5 (1996): 12–17.

5. V. C. Wynne-Edwards, "Backstage and Upstage with 'Animal Dispersion,'" in *Leaders in the Study of Animal Behavior: Autobiographical Perspectives,* ed. D. Dewsbury, 486–512.

6. Ibid., 488.

7. V. C. Wynne-Edwards, "On the Habits and Distribution of Birds on the North Atlantic," *Proceedings of the Boston Society of Natural History* 40, no. 4 (1935): 233–346.

8. V. C. Wynne-Edwards, "Intermittent Breeding of the Fulmar, with Some General Observations on Non-Breeding in Sea Birds," *Proceedings of the Zoological Society of London* A109 (1939): 127–32.

9. V. C. Wynne-Edwards, "Zoology of the Baird Expedition," *Auk* 69, no. 4 (1935): 353–91, 384.

10. M. Borrello, "Mutual Aid and Animal Dispersion: An Historical Analysis of Alternatives to Darwin," *Perspectives in Biology and Medicine* 47, no. 1 (2004): 15–31.

11. Wynne-Edwards, "Backstage and Upstage," 501.

12. J. Cain, "Common Problems and Cooperative Solutions: Organizational Activity in Evolutionary Studies," *Isis* 84, no. 1 (1993): 1–25; M. Dietrich, "Paradox and Persuasion: Negotiating the Place of Molecular Evolution Within Evolutionary Biology," *Journal of the History of Biology* 31 (1998): 85–111; E. Mayr and W. Provine, eds., *The Evolutionary Synthesis: Perspectives on the Unification of Biology* (Cambridge: Harvard University Press, 1980); V. B. Simocovitis, *Unifying Biology: The Evolutionary Synthesis and Evolutionary Biology* (Princeton: Princeton University Press, 1996).

13. Wynne-Edwards Collection, Queen's University Archive, box 2, file 11.

14. V. C. Wynne-Edwards, "The Dynamics of Animal Populations." *Discovery,* October 1955, 434.

15. M. Borrello, "Synthesis and Selection: Wynne-Edwards' Challenge to David Lack," *Journal of the History of Biology* 36 (2003): 531–66.

16. Wynne-Edwards Collection, Queen's University Archive, box 2 file 2.

17. V. C. Wynne-Edwards, "Low Reproductive Rates in Birds, Especially Sea-Birds," *Acta of the XI International Congress of Ornithology* (1955): 545–46.

18. Ibid., 547.

19. V. C. Wynne-Edwards, *Animal Dispersion in Relation to Social Behavior* (Edinburgh: Oliver and Boyd, 1962), v.

20. Ibid.

21. Ibid., 20.

22. R. P. McIntosh, "Citation Classics of Ecology," *Quarterly Review of Biology* 64, no. 1 (1989): 31–49.

23. D. Lack, *Population Studies of Birds* (Oxford: Clarendon, 1966). See esp. appendix 3.

24. G. C. Williams, *Adaptation and Natural Selection: A Critique of Some Current Evolutionary Thought* (Princeton: Princeton University Press, 1966).

25. W. D. Hamilton, "The Genetical Evolution of Social Behavior, I and II," *Journal of Theoretical Biology* 7 (1964): 1–52; J. M. Smith, "Group Selection and Kin Selection," *Nature* 201 (1964): 1145–47; R. Trivers, "The Evolution of Reciprocal Altruism," *Quarterly Review of Biology* 46 (1971): 33–57.

26. R. Dawkins, *The Selfish Gene,* new ed. (London: Oxford University Press, 1989), ix.

27. F. J. Ebling and D. M. Stoddart, eds. *Population Control by Social Behaviour.* Symposia of the Institute of Biology (London: Institute of Biology, 1978), vi.

28. V. C. Wynne-Edwards, "Intrinsic Population Control: An Introduction," in *Population Control by Social Behavior,* ed. F. J. Ebling, D. M. Stoddart (London: Institute of Biology, 1978), 1–22, 19.

29. V. C. Wynne-Edwards, *Evolution Through Group Selection* (Oxford: Blackwell, 1986), 203.

30. M. J. Wade, "Group Selection Among Laboratory Populations of *Tribolium,*" *Proceedings of the National Academy of Sciences* 73 (1976): 4604–7; M. J. Wade, "Experimental Study of Group Selection," *Evolution* 31 (1977): 134–53; M. J. Wade, "A Critical Review of the Models of Group Selection," *Quarterly Review of Biology* 53 (1978): 101–14; D. S. Wilson, "A Theory of Group Selection," *Proceedings of the National Academy of Sciences* 72 (1975): 143–46; D. S. Wilson, "Evolution on the Level of Communities," *Science* 192 (1976): 1358–60; D. S. Wilson, *The Natural Selection of Populations and Communities* (Menlo Park, Calif.: Benjamin/Cummings, 1980).

31. M. Ridley, "Evolution Through Group Selection—Review," *Ethology* 74, no. 3 (1987): 260–61.

32. V. C. Wynne-Edwards, "A Rationale for Group Selection," *Journal of Theoretical Biology* 162 (1993): 1–22.

33. Wynne-Edwards Collection, Queen's University Archive, box 4, file 1.

FURTHER READING

Borrello, Mark E. "Synthesis and Selection: Wynne-Edwards' Challenge to David Lack," *Journal of the History of Biology* 36, no. 3 (2003): 531–566.
———. "Mutual Aid and Animal Dispersion: An Historical Analysis of Alternatives to Darwin," *Perspectives in Biology and Medicine* 47, no. 1 (2004): 15–31.
———. "The Rise, Fall and Resurrection of Group Selection," *Endeavour* 29, no. 1 (2005): 43–47.
Pollock, Gregory B., "Suspending Disbelief—of Wynne-Edwards and His Reception," *Journal of Evolutionary Biology* 2 (1989): 205–221.
Sober, Elliott, and D. S. Wilson, *Unto Others: The Evolution and Psychology of Unselfish Behavior* (Cambridge: Harvard University Press, 1998).
Williams, George C., ed., *Group Selection* (Chicago: Aldine Atherton, 1971).
Wilson, D. S., "The Group Selection Controversy: History and Current Status," *Annual Review of Ecology and Systematics* 14 (1983): 159–187.

CHAPTER THIRTEEN

Peter Mitchell:
Changing the Face of Bioenergetics

JOHN PREBBLE and BRUCE WEBER

The systematic study of the way chemical energy is provided from the breakdown of foodstuff in living cells started early in the twentieth century. The initial concepts developed more or less logically as experimental evidence provided insights into the chemical processes involved. Yet one central mechanism, that for the final oxidation by oxygen, known as oxidative phosphorylation, began to emerge as a major problem when provisional ideas could not be underpinned by experiment. Thus a new approach was clearly needed, and several were forthcoming. The problem was significant because oxidative phosphorylation is the process that provides most of the cell's energy. Peter Mitchell's ideas, initially published in 1961, provided a truly original basis for a new theoretical approach to the problem and directed others in the field toward a totally novel way of thinking about chemical oxidative processes. The radically new proposals of Mitchell were not readily accepted, and the publication of his theory in a revised form in 1966 led to a period of turmoil. Some have described this period as the "ox phos wars." Initial concepts to which researchers had made a personal commitment were not readily abandoned, and Mitchell's contribution consisted of the formulation and elaboration of a bold and imaginative hypothesis together with battling to persuade others of its veracity. It is remarkable that even as controversy raged about the mechanistic details of his proposals, Peter Mitchell was awarded the 1978 Nobel Prize in Chemistry.

INITIAL CONCEPTS OF OXIDATIVE PHOSPHORYLATION

The first metabolic pathways to be studied were concerned mainly with glucose breakdown and were worked out in the first half of the

Figure 13.1. Peter Mitchell. Courtesy of John Prebble.

twentieth century. These pathways represented some of the early achievements of the relatively new science of biochemistry, but many of the underlying skills involved were those of the organic chemist. In fact, organic chemistry was being successfully applied to the study of cellular processes generally and to understanding the role of enzymic catalysts that made chemical reactions possible at physiological pH and temperature. Thus the way was open for the explosion of biochemical knowledge in the second half of the twentieth century. A more problematic aspect of the organic chemistry approach to biological systems was the treatment of the cell as an unorganized "bag of enzymes" in which the structural and vectorial (directional) aspects of cellular activity were treated as either irrelevant or as secondary phenomena. Nevertheless, the early achievements of metabolic biochemistry not only promoted confidence in the successful application of organic chemistry but also dictated the approach to one of the major puzzles presented by these studies: the mechanism of oxidative phosphorylation, the process whereby oxygen is used to "burn" foodstuffs to provide energy for the cell.

Starting with the studies of the Russian biochemist Vladimir Egelhardt and in particular the Danish biochemist Herman Kalckar in the 1930s, it became clear that a substantial amount of cellular phosphorylation was linked to cell oxidations. Progressively it emerged that there were phosphorylations linked to the respiratory chain (electron transport chain) found in mitochondria. The respiratory chain had been described by the Cambridge biochemist David Keilin and was later shown to be linked to the synthesis of adenosine triphosphate (ATP), the energy currency of the cell. A similar process driven by light energy operates in chloroplasts. Thus the energy from oxidation of cellular molecules in mitochondria could be conserved in the synthesis of ATP. At the heart of this process, however, was a puzzle. In the laboratory the oxidations of the electron transport chain could be demonstrated to occur without ATP synthesis under some circumstances, whereas in others, particularly in fresh preparations of mitochondria, the process of oxidation and ATP synthesis were tightly coupled. The essence of the puzzle was the mechanism by which the two processes were linked. How did the use of oxygen lead to ATP production?

Biochemists, with their confidence in the successful chemical approach to biochemistry, proposed "high-energy" chemical intermediates that were formed during oxidation and consumed during the synthesis of ATP. The first proposal was made by Fritz Lipmann in 1946, but a more detailed model, the chemical theory, was formulated by E. C. (Bill) Slater in 1953 based on two sequential steps in the glucose breakdown pathway where ATP was synthesized.[1] This formulation predicted one or more unknown chemical intermediates between the respiratory chain and ATP synthesis. Enormous effort, both personal and financial, was expended during the 1950s and 1960s and on into the early 1970s in order to define the properties of such high-energy intermediate substances and to establish their identity.

There were two observations for which the chemical theory could not provide a convincing explanation. First, there were several chemical substances (not necessarily structurally similar) that could break the link between the respiratory chain and ATP synthesis. It was suggested that these "uncouplers" in some way destroyed the high-energy intermediate substance, but explanations for the role of uncouplers remained problematic. Second, all known oxidative phosphorylation and photophosphorylation systems were found in membranes, but the chemical theory had no necessity for a membrane. As time passed, however, the overriding objection to the theory became the failure to find the putative chemical intermediates despite extensive experimentation. Inevitably there were claims that it had been identified, but usually the suggested intermediate did not possess all of the necessary properties.[2]

Thus by the early 1960s the field of cell bioenergetics was progressively moving into a crisis in which its central theory lacked experimental support. It was against this background that in 1961 Peter Mitchell (1920–1992) provided an entirely new approach to the puzzle of how oxidation-reduction in the respiratory chain and the synthesis of ATP were linked. Without experimental evidence but with a number of theoretical ideas from the field of membrane permeability, he proposed that the link was a pH (or proton) gradient across the membrane, in effect an electrochemical gradient. The energy from respiration was used to create the gradient (which thus stored the energy), and the gradient drove the synthesis of ATP.

PETER MITCHELL

Peter Mitchell is best described as an individualist who had a strong personality and confidence in his own intellectual ability.[3] His fellow scientists tended to react strongly to him. His rugged individualism stood him in good stead when others were too easily inclined to dismiss his ideas. These aspects of his character were combined with great charm and genuine concern for people. In many respects he had the characteristics of an artist rather than a scientist. He had the confidence to allow his intuition to guide him. His theorizing had a strong visual component in its early stages, though later he converted his models of nature into mathematical language.

Mitchell was born at Mitcham, Surrey, England, in 1920, the younger of two brothers. His father was a successful civil servant who was awarded an OBE for his services but later committed suicide. Mitchell appeared to seek father figures, whom he found first in the headmaster of his school, Queen's College Taunton, and later in David Keilin, the distinguished Cambridge biochemist who rediscovered cytochromes and should, in the opinion of many, have received a Nobel Prize. His father's brother, Sir Godfrey Mitchell, had significant influence on Mitchell's life. Godfrey had acquired a small insolvent construction company, George Wimpey, at the end of World War I. He built it up to become one of the premier construction companies, and shares in the company were a source of Peter's affluence. But his mother had the dominant influence on Peter. She was well educated for the time and imparted to Peter a keen interest in classical music that he retained for the rest of his life. She instilled in him a strong sense of personal responsibility and also encouraged him in a number of pursuits, which included a well-equipped workshop that allowed Peter to develop his mechanical interests. Although he was educated at a Methodist school, that reinforced the strong moral upbringing he had received from his mother, Queen's College failed to interest him in organized religion.

His academic performance does not seem to have been brilliant. He failed the scholarship examination for Cambridge, but his headmaster, who recognized his abilities, persuaded Jesus College, Cambridge to accept him in 1939. There he read biochemistry, obtaining a degree with upper second-class honors. At Cambridge he pursued his interest in music, adopted a flamboyant lifestyle, and read widely. He tended to seek the company of artists and writers rather than scientists. He developed a strong interest in philosophical questions applied to biology, especially the work of Joseph Woodger, a philosopher of biology, and of Frederick Gowland Hopkins, head of the Cambridge Biochemistry Department until 1943. His first thesis for the Ph.D. contained a substantial element of his philosophical approach and was referred (rejected). He did not obtain his Ph.D. until 1950.

Mitchell's background was not in the field to which he brought his revolutionary ideas. Although he had graduated in biochemistry, he became interested in microorganisms and in transport across microbial membranes. Initially he worked with Jim Danielli, whose interests were in membranes and membrane transport. After Danielli's departure to King's College, London, Mitchell became increasingly involved with the microbiology group and, following his troublesome Ph.D. studies, his work centered on the uptake of phosphate by bacteria. In 1955, he moved to the University of Edinburgh, where he published his most original work. During the mid-1950s he developed a theoretical approach to transport in which he suggested that membrane-bound enzymes functioned vectorially (directionally) so that the substrate approached the enzymes' active site from one side of the membrane and the product left on the opposite side.[4] Thus the process of membrane transport and enzyme catalysis would be almost the same thing and metabolism and transport were dual aspects of an underlying vectorial process. In his scheme for phosphate uptake into bacterial cells, uptake was seen as involving phosphorylation of a substance that was released into the cell as the sugar phosphate. Thus transport and the formation of an organic phosphate were catalyzed by a single integral membrane protein.[5] Mitchell described such a system in 1959 as "chemiosmotic," implying a process carrying out both chemical and osmotic work; such a proposition could be seen as applicable to mitochondrial membranes.[6] These ideas were applied to a sugar phosphorylation system believed to be membrane-bound and tentatively to ATP synthesis in oxidative phosphorylation in 1960 at a symposium in Stockholm, but the full initial version was not published until the summer of 1961.

THE CHEMIOSMOTIC HYPOTHESIS FOR OXIDATIVE PHOSPHORYLATION

In 1961 two theories of oxidative phosphorylation, both based on protons, were published. The first was by R. J. P. Williams, who saw protons, produced by the respiratory chain, driving ATP synthesis in the hydrophobic (water-avoiding) environment of the membrane.[7] It was based on sound chemical principles of polyphosphate synthesis in water-free systems, but its influence on the field of bioenergetics was limited. The relationship between this theory and Mitchell's proposals is complex and has been explored elsewhere.[8]

The second theory, Mitchell's chemiosmotic hypothesis (figure 13.2), was based on the possibility that a pH (proton) gradient across the membrane was formed by respiration and that this gradient would drive the synthesis of ATP. The chemistry of the initial proposal was dubious, relying on extraction of protons and hydroxyl ions from the membrane, but it was reformulated in 1966 in a form that has survived. The principle that respiration would lead to the formation of a proton gradient was based on theoretical arguments in a review published the previous year by Rutherford Robertson.[9] The principal fields where most of these proposals had developed were ion uptake by plant roots and acid secretion in the stomach. The proposition that respiration would form a pH gradient had virtually no experimental support. The novel proposal was that the proton gradient would drive ATP synthesis; it owed much to Mitchell's earlier thoughts about "vectorial" functioning of membrane-bound enzymes. A concomitant requirement was that the membrane itself should be impermeable to protons. The theory was published in *Nature* in July 1961,[10] but the absence of experimental evidence led a later commentator to describe the theory as "counter intuitive."[11] Note, however, that the basic idea

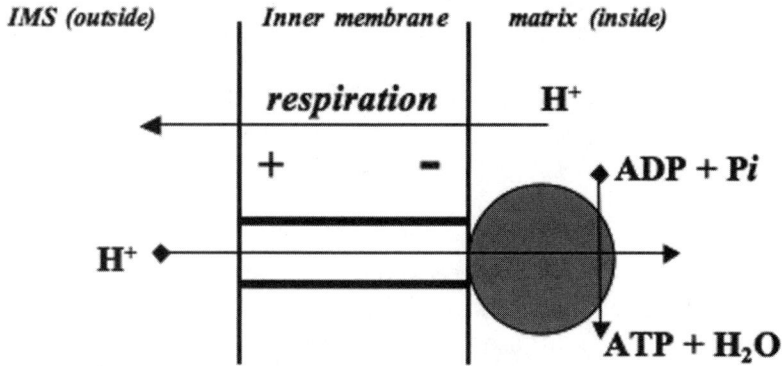

Figure 13.2. Diagram representing the principles of the chemiosmotic theory. Respiration is located in the inner mitochondrial membrane, where it pumps protons (H⁺) outwards as oxygen is reduced to water. This movement of protons creates a greater concentration outside than inside and a membrane potential that is positive outside and negative inside. Both the concentration difference and the membrane potential encourage protons to return inwards through the ATP synthase, which synthesizes ATP.

underlying Mitchell's proposals had been foreseen by Robert Davies in a paper written with Krebs. Davies's comments were restricted to a very few lines in the paper and had no influence on the field. Mitchell maintained that such ideas were not those of the vectorial chemiosmotic theory, although others have tended to disagree.[12]

SEEKING EVIDENCE

Between 1961 and 1965 little attention was paid to Mitchell's proposals. He himself was ill. He resigned his post at Edinburgh and, after a period of recuperation, set up a private research laboratory at Glynn, Cornwall, England, financed by the family inheritance of shares in George Wimpey and Company. In 1960, before leaving Edinburgh, Mitchell had started experimental work on oxidative phosphorylation and soon discovered that there were errors in his initial proposals. In particular, the polarity of the membrane had to be reversed so that protons giving a low pH were moved outwards, not inwards. Later he realized that the number of protons transported per oxygen atom reduced had to be doubled from three to six;[13] these quantities were to require further revision in the 1980s. The feasibility of Mitchell's proposals was judged in part on thermodynamic calculations based on proton numbers, and the higher values tended to remove one of the criticisms of the hypothesis. In addition, Mitchell gave further consideration to the mechanisms at the heart of his theory. Hydroxyl ions were abandoned, and the respiratory chain, together with the enzyme synthesizing ATP (the ATP synthase), was now clearly seen as transporting protons across the membrane. All these developments were summarized in two works published in 1966: a medium-length statement in *Biological Reviews* and an extensive account published by the Glynn Research Institute that became known as the first "Grey Book."[14] The ATP synthase was now a proton-translocating enzyme that provided a fundamental redirection of the field. Nevertheless, Mitchell never contributed successfully to the further understanding of this enzyme.

With his own ideas now much more clearly formulated, Mitchell was free to develop evidence for the theory, which became one of the two roles of the Glynn Research Institute. The other was to test whether it was possible to carry out world-class science in a small private research establishment; Glynn never employed more than about twelve including its director, Mitchell, and his life-long collaborator, Jennifer Moyle. Moyle had originally worked with Mitchell at Cambridge and had moved with him to Edinburgh. She was now his key experimental collaborator and his close associate in the refurbishing and setting up of Glynn as a research institute that also provided a home for Mitchell, his wife Helen, and their family.

Somewhat surprisingly, the first independent evidence for the chemiosmotic theory came not from the mitochondrial field, where the major proposals for oxidative phosphorylation had been developed, but rather from the study of chloroplast biochemistry. André Jagendorf and his co-workers in the McCollum Institute at Johns Hopkins University initially interpreted their results in terms of the chemical theory, but he began to realize that they were consistent with the build-up of a proton gradient as proposed by the chemiosmotic hypothesis.[15] In May 1964, Jagendorf wrote to Mitchell saying that his experiments were in agreement with a proton gradient's being the intermediate between chloroplast electron transport and phosphorylation. More important, in 1965 Jagendorf showed that a proton gradient could drive ATP synthesis in the dark.[16] Such results had a major influence on the field of bioenergetics, although they did not convince workers that the hypothesis was correct.

Initial results from the Glynn laboratory included a demonstration that six protons were ejected from mitochondria for each oxygen atom that was converted to water. This number eventually proved contentious, in part because Mitchell and Moyle had not adequately controlled the movements of other ions, particularly phosphate, although they believed they had done so. In 1986, after a long argument in which he stuck to his original numbers, Mitchell admitted that "at least 8" protons were ejected for each oxygen atom.[17] In the mid-1960s Mitchell and Moyle also measured the protons transported for each molecule of ATP hydrolyzed and obtained the number 2. This gave them a ratio of ATP synthesized to atoms of oxygen used of 3.[18] At that time the latter figure was widely accepted and gave credence to Mitchell's theory and his experiments.

So far as the validity of Mitchell's theory was concerned, a key question was whether a proton gradient could drive the synthesis of ATP in mitochondria in the same way as Jagendorf had demonstrated for chloroplasts. Mitchell and Moyle published results showing that the application of a proton gradient to mitochondria could induce the synthesis of ATP. But the amounts synthesized were small, and several other leading laboratories failed to reproduce the results, causing Bill Slater to write to Mitchell and others reporting that ATP formation could not be demonstrated in mitochondria by creating an artificial proton gradient. Part of the problem arose because one key aspect of Mitchell's proposals had not been appreciated. This was the electrical potential across the membrane that would accompany the transport of protons by respiration and would constitute an important part of the energy conserved from oxidation and utilized for ATP synthesis. In mitochondria (but not in chloroplasts) most of the energy is in the electrical potential.

Such problems made the theory suspect in the eyes of many. The major problem, however, was the physical nature of Mitchell's proposals. Many believed that the answer must lie in some chemical solution involving metabolic intermediate substances or chemical interactions between molecules. The notion that the link between oxidation processes and ATP synthesis could be a physical ion gradient seemed implausible.

COMPETING THEORIES

In the mid-1960s, opinions about oxidative phosphorylation became more diverse; the era of the "ox phos" wars had dawned. Mitchell's views were now being seriously discussed though not widely accepted. The decade was also a period when the understanding and significance of conformational changes in the structure of proteins began to be appreciated, and this introduced new and intriguing possibilities in the study of biochemistry. Among these was the idea that the energy of respiration might be transferred to the ATP synthase by means of conformational changes in the proteins involved. Thus the conformational hypothesis of oxidative phosphorylation was first proposed by Paul Boyer in 1965.[19] As the popularity of the chemical theory began to wane in the late 1960s and early 1970s, the conformational hypothesis proved attractive to a number of biochemists who could not accept the proton gradient ideas of Mitchell. Eventually it emerged that the value of the conformational approach lay not in defining the core events of oxidative phosphorylation but in establishing the mechanism of the ATP synthase enzyme itself.

During the 1960s the chemical theory retained its attraction as a proposal firmly based in classical biochemistry, particularly to those who could not accept the idea of ion gradients as intermediates in oxidative phosphorylation because of their expectation that chemical intermediates would be involved. But the continued failure to find any of the predicted high-energy chemical intermediates left the proposal in a vulnerable position. In the Biochemical Society's Sixth Keilin Memorial Lecture in 1974, Slater announced the demise of the chemical theory that he himself had proposed a little more than twenty years earlier. Slater remained skeptical about the chemiosmotic theory for a long time and felt that there were good experimental reasons for considering the conformational theory seriously.[20]

By the early 1970s Mitchell had shown to his satisfaction, if not everybody else's, that the basic tenets of his hypothesis were true. The membrane was impermeable to protons (necessary to create a gradient), the operation of the respiratory chain would generate a pH gradient, and such a gradient could drive ATP synthesis. Note, however, that experimental results were frequently open to alternative interpretations, and this made agreement among researchers difficult. Nevertheless, Mitchell's proposals began to gain credibility as a number of experimental results by various workers pointed to a chemiosmotic interpretation in both mitochondria and chloroplasts. Identifying the molecular mechanisms by which the proton gradient was maintained remained a problem, but in 1975 Mitchell proposed a mechanism for transporting protons (the Q cycle) in the middle protein complex of the respiratory chain.[21] This proposal, based on an imaginative application of Mitchell's theoretical concepts, has so far stood the test of time.

MOVING TOWARDS FINAL ACCEPTANCE OF THE CHEMIOSMOTIC THEORY

Apart from Mitchell's continued advocacy of his theory and the work of Jagendorf, Efraim Racker at Cornell University did much to persuade senior members of the bioenergetic community to take the chemiosmotic theory seriously. True, many of the younger generation, such as Peter Hinkle, also at Cornell, found Mitchell's views very attractive.

Racker showed that the three principal complexes of the mitochondrial respiratory chain and also the mitochondrial ATP synthase were able to transport protons as part of their function. One elegant experiment that did much to persuade the biochemical community of the validity of the chemiosmotic theory neatly demonstrated the synthesis of ATP by a biologically generated proton gradient. Walther Stoeckenius had shown that a purple bacterial membrane protein would transport protons across the membrane when illuminated. Racker and Stoeckenius set up a purified vesicular system containing the purple protein (bacteriorhodopsin) and the mitochondrial ATP synthase. When activated by light this system would synthesize ATP, thereby demonstrating a simple chemiosmotic system.[22] Mitchell, however, felt that the experiment did little to persuade those working on mitochondrial oxidative phosphorylation.

Racker was also responsible for a very different development that probably did much to establish the work of Mitchell as central to the field of bioenergetics. He was concerned about the poor image of the field, especially in the United States, where funding agencies felt that despite much investment relatively little progress had been made. In 1974 he endeavored to persuade his colleagues that they should all work toward the publication of a joint statement, an approach rather similar to that tried earlier by Lars Ernster of Stockholm. After about three years of what can only be described as squabbling among the main participants, a review of oxidative phosphorylation was published in the *Annual Review of Biochemistry*.[23] Unlike probably all other reviews published in that periodical, this one consisted of six separate contributions, reflecting the lack of agreement and trust in the field. Paul Boyer, Britton Chance, Lars Ernster, Peter Mitchell, Efraim Racker, and E. C. Slater contributed one "mini-review" apiece.[24] Many have felt that this marked the end of the dispute about the mechanism of oxidative phosphorylation, but a careful reading of the texts shows that some were by no means willing to accept the full validity of the chemiosmotic theory. Nevertheless, the publication of the review probably paved the way for the Chemistry Committee of the Nobel Foundation to choose Mitchell for the 1978 Nobel Prize. The citation reads slightly oddly, being rather general: "For his contribution to the understanding of biological energy transfer through the formulation of the chemiosmotic theory."[25]

Although Mitchell's Nobel Prize seemed to settle the argument, it would be wrong to pretend that the debate about the theory had totally ceased. Indeed, the ox phos wars continued far into the 1980s with arguments about mechanisms for proton transport by the respiratory chain.[26] Such issues might have been regarded as mere details in the understanding of the theory, but to Mitchell they formed a key part of the theory itself.

REBELLIOUSNESS AND COMMITMENT

Mitchell was a natural rebel in that he instinctively ploughed his own furrow. As a young man, he preferred to deduce scientific understanding from first principles rather than gaining insight from textbooks. His thesis for his Ph.D. at Cambridge departed from the normal pattern for a thesis in biochemistry and a significant part was taken up with his own philosophical ideas, with the result that it had to be resubmitted along orthodox lines. He found working in a university restrictive and preferred the freedom of his own research institute, Glynn. He also disliked interference by editors and published his revised theory independently as the Grey Books. His approach to biochemistry was that of a theorist, experimental work serving to test preexisting theories. In contrast, most of his contemporaries viewed experimental work as the route of new knowledge and theory formulation. In particular, he considered the overall processes associated with a single cell (he worked on microorganisms) and deduced his understanding of biochemistry from that consideration. Cells were like flames; they took in substances from the environment and engaged in chemical reactions, and the products left. This led Mitchell to pay particular attention to membranes that controlled the process of entry and exit. It also led him to think about the vectorial (directional) aspects of living systems implied by these views, and he felt that contemporary biochemistry had an essentially scalar (nondirectional) view of cellular reactions.

One of the few examples he could find of a truly vectorial process was the respiratory chain described by David Keilin. Thus the respiratory chain was an example of a vectorial system whereby electrons moved along the chain in one direction in the membrane. Keilin had befriended Mitchell in Cambridge, and Mitchell's Nobel lecture was titled "David Keilin's Respiratory Chain Concept and Its Chemiosmotic Consequences."[27] In order to develop his vectorial ideas he considered that metabolic events associated with membranes would also be directional—the substrate might arrive at a membrane-bound enzyme from one side, and the product might be released on the other. This idea was the ground from which the chemiosmotic theory grew.

Mitchell's own perspective drew little from the biochemists who were concentrating on trying to understand oxidative phosphorylation. It was the field of transport (which at that time belonged primarily to physiology rather than biochemistry) to which his ideas had a natural affinity. He drew on ideas from transport to advance his hypothesis. In 1959 he wrote, "The belief in the necessity for coupling separate membrane-transport systems and metabolic systems has partly been a reflection of the necessity for coupling physiologists to biochemists."[28] Thus Mitchell's innate rebelliousness arose in his approach to biochemistry itself rather than in relation to the field where he made his key contribution. He felt that most biochemists had lost sight of a holistic view and in particular were ignoring the vectorial quality of living systems; this rebelliousness led him along the unorthodox path to the chemiosmotic theory.

This rebelliousness cannot be appreciated without also understanding the importance of personal commitment of scientists to particular theories. It was summarized by the German physicist Max Planck (1858–1947), who wrote, "A new scientific truth does not triumph by convincing its opponents and making them see the light, but rather because its opponents eventually die, and a new generation grows up that is familiar with it."[29] In Mitchell's case, the commitment was complex. Initially he proposed his theory as an interesting idea and carried out tests to determine whether it was sensible. When he saw

that there were good experimental reasons for taking the chemiosmotic theory seriously, however, he became personally committed
to it, despite his attempts at maintaining a detached approach. From
about 1970 onward he became interested in the falsification ideas of
the philosopher Karl Popper, who subsequently became a personal
friend. This led him to view his theory as provisional and to carry out
tests as a means of falsifying the proposal. This approach remained
somewhat at odds with his personal passion for and commitment to
the theory.

The fierce commitment of Mitchell and his opponents to their respective views created distrust between researchers and made open
discussion of problems more difficult. The disciplinary background of
most of those in the field is a crucial part of understanding the early
stages of the chemiosmotic debate of the 1960s. The chemical theory,
as noted above, was born of a hitherto highly successful chemical
approach to the processes taking place in living cells. Workers in bioenergetics had interpreted their experiments in terms of that theory
and had committed a major part of their lives to endeavoring to establish its mechanisms and the identity of its components. More particularly for science in the second half of the twentieth century, the need
to obtain funding and the views of the funding agencies were crucial
to scientific success. To seek funding and to engage in research in
support of a theory were in themselves elements of personal commitment. Even before Mitchell became deeply involved in the debate, it
was observed that people went to the federation meetings concerned
with oxidative phosphorylation because they knew in advance that
there would be a good "punch up." Indeed, the situation in the early
1960s, when Mitchell put his theory forward, could be summarized
by the now well-known dictum of Efraim Racker: "Anyone who is
not thoroughly confused, just does not understand the situation."[30]
The frustration that workers felt in not being able to make progress
with the core issue of oxidative phosphorylation, the personal risk
to their reputations that workers felt, and the difficulty of designing
good experiments resulted in a community of scientists who became
well known for their acrimonious disputes. Against this background
Mitchell was seen as a rebel.

The chemiosmotic approach led to an understanding of the core
mechanism of the biochemical processes known as oxidative phosphorylation and photosynthetic phosphorylation. It also showed why
a membrane was necessary for the process and gave an immediate

entry into the understanding of ion movements into and out of the mitochondrion and in bacteria and chloroplasts. This in turn opened up a new era for the study of "active" transport across membranes, transport that was energy-dependent. It also provided a basis for building an understanding of the proton-translocating ATP synthetases, although the work on the mechanisms involved depended on other insights, particularly those of Paul Boyer (for which he received the Nobel prize in 1997). Mitchell's work also contributed to diverse areas of biology such as the mechanism of flagella action and theories of the origins of mitochondria. In brief, Mitchell's formulation and advocacy of the chemiosmotic theory gave a totally new direction to the field of biochemistry known as bioenergetics.

NOTES

1. E. C. Slater, "Mechanism of Phosphorylation in the Respiratory Chain," *Nature* 172 (1953): 975–978.

2. See Douglas Allchin, "A Twentieth Century Phlogiston: Constructing Error and Differentiating Domains," *Perspectives in Science* 5 (1997): 81–127.

3. John Prebble and Bruce Weber, *Wandering in the Gardens of the Mind: Peter Mitchell and the Making of Glynn* (New York: Oxford University Press, 2003).

4. Peter Mitchell and Jennifer Moyle, "Group-Translocation: A Consequence of Enzyme-Catalysed Group-Transfer," *Nature* 182 (1958): 372–373.

5. Peter Mitchell, "A General Theory of Membrane Transport from Studies of Bacteria," *Nature* 180 (1957): 134–136.

6. Peter Mitchell, "Structure and Function in Microorganisms," *Biochemical Society Symposia* 16 (1959): 73–93.

7. Robert J. P. Williams, "Possible Functions of Chains of Catalysts," *Journal of Theoretical Biology* 1 (1961): 1–17; Robert J. P. Williams, "Possible Functions of Chains of Catalysts II," *Journal of Theoretical Biology* 3 (1962): 209–229.

8. Bruce Weber and John Prebble, "An Issue of Originality and Priority: The Correspondence and Theories of Oxidative Phosphorylation of Peter Mitchell and Robert J. P. Williams, 1961–1980," *Journal of the History of Biology* 39 (2006): 125–163.

9. Rutherford Robertson, "Ion Transport and Respiration," *Biological Reviews* 35 (1960): 231–264.

10. Peter Mitchell, "Coupling of Phosphorylation to Electron and Hydrogen Transfer by a Chemi-osmotic Type of Mechanism," *Nature* 191 (1961): 144–148.

11. Leslie E. Orgel, "Are You Serious, Dr. Mitchell?" *Nature* 402 (2000): 17.

12. Robert E. Davies and Hans A. Krebs, "The Biochemical Aspects of the Transport of Ions by Nervous Tissue," *Biochemical Society Symposia* 8 (1952): 77–92. For a discussion of the issues raised by the work of Davies see John Prebble, "Successful Theory Development in Biology: A Consideration of the Theories of Oxidative Phosphorylation Proposed by Davies and Krebs, Williams, and Mitchell," *Bioscience Reports* 16 (1996): 207–215.

13. Peter Mitchell and Jennifer Moyle, "Stoichiometry of Proton Translocation Through the Respiratory Chain and Adenosine Triphosphatase Systems of Rat Liver Mitochondria," *Nature* 208 (1965): 147–151.

14. Peter Mitchell, *Chemiosmotic Coupling in Oxidative and Photosynthetic Phosphorylation* (Bodmin, Cornwall: Glynn Research, 1966). This Grey Book was an amplification of Peter Mitchell, "Chemiosmotic Coupling in Oxidative and Photosynthetic Phosphorylation," *Biological Reviews* 41 (1966): 445–502. Another Grey Book was produced in 1968 to expand on the implications of the theory: Peter Mitchell, *Chemisomotic Coupling and Energy Transduction* (Bodmin, Cornwall: Glynn Research, 1968). These books were circulated widely to the bioenergetic community and helped explain Mitchell's ideas to those who found them opaque.

15. See Geoffrey Hind and André Jagendorf, "Separation of Light and Dark Stages in Photophosphorylation," *Proceedings of the National Academy of Sciences U.S.A.* 49 (1963): 715–722; J. Neumann and André Jagendorf, "Light Induced pH Changes Related to Photophosphorylation by Chloroplasts," *Archives of Biochemistry and Biophysics* 107 (1964): 109–119.

16. André Jagendorf and Ernest Uribe, "ATP Formation Caused by Acid-Base Transition of Spinach Chloroplasts," *Proceedings of the National Academy of Sciences U.S.A.* 55 (1966): 170–177.

17. Roy Mitchell, Ian C. West, A. John Moody, and Peter Mitchell, "Measurement of the Proton-Motive Stoichiometry of the Respiratory Chain of Rat Liver Mitochondria: The Effect of N-ethylmaleimide," *Biochimica Biophysica Acta* 849 (1986): 229–235.

18. Mitchell and Moyle, "Stoichiometry of Proton Translocation."

19. Paul Boyer, "Carboxyl Activation as a Possible Reaction in Substrate-Level and Oxidative Phosphorylation and in Muscle Contraction," in *Oxidases and Related Redox Systems,* ed. Tsoo E. King et al. (New York: John Wiley, 1965), 994–1017. The conformational theory was developed further; see, e.g., Boyer's critique of Mitchell's chemiosmotic theory in "A Model for Conformational Coupling of Membrane Potential and Proton Translocation to ATP Synthesis and to Active Transport," *FEBS Letters* 58 (1975): 1–6.

20. E. C. Slater, "From Cytochrome to Adenosine Triphosphate and Back," *Biochemical Society Transactions* 2 (1974): 1149–1163.

21. Peter Mitchell, "Protonmotive Redox Mechanism of the Cytochrome

b-c₁ Complex in the Respiratory Chain: Protonmotive Ubiquinone Cycle," *FEBS Letters* 56 (1975): 1–6.

22. Efraim Racker and Walther Stoeckenius, "Reconstitution of Purple Membrane Vesicles Catalyzing Light-Driven Proton Uptake and Adenosine Triphosphate Formation," *Journal of Biological Chemistry* (1974): 249, 662–663.

23. For an account of the steps leading to the publication of the review, see Prebble and Weber, *Wandering in the Gardens of the Mind.*

24. Paul D. Boyer, Britton Chance, Lars Ernster, Peter Mitchell, Efraim Racker, and E. C. Slater, "Oxidative Phosphorylation and Photophosphorylation," *Annual Review of Biochemistry* 46 (1977): 955–1026.

25. For the comment of the chairman of the Nobel Chemistry Committee, see Bo Malmström, "Mitchell Saw the New Vista, if not the Details," *Nature* 403 (2000): 356. Malmström does, however, overlook the significance of the Q cycle. See John Prebble, "The Lasting Value of Mitchell's Mechanisms," *Nature* 404 (2000): 330.

26. John Prebble, "Peter Mitchell and the Ox Phos Wars," *Trends in Biochemical Sciences* 27 (2002): 209–212.

27. Peter Mitchell, "David Keilin's Respiratory Chain Concept and Its Chemiosmotic Consequences," *Science* 206 (1979): 1148–1159.

28. Peter Mitchell and Jennifer Moyle, "Coupling of Metabolism and Transport by Enzymic Translocation of Substrates Through Membranes," *Proceedings of the Royal Physical Society Edinburgh* 28 (1959): 19–27.

29. Max Planck, *A Scientific Autobiography,* trans. F. Gaynor (New York: Philosophical Library, 1949). See p. 33.

30. Efraim Racker, "Multiple Coupling Factors in Oxidative Phosphorylation," *Federation Proceedings* 22 (1963): 1088–1091.

FURTHER READING

Gilbert, G. Nigel. Michael Mulkay, *Opening Pandora's Box: A Sociological Analysis of Scientists' Discourse* (Cambridge: Cambridge University Press, 1984).

Lane, Nick. *Power, Sex, Suicide: Mitochondria and the Meaning of Life* (Oxford: Oxford University Press, 2005).

Mitchell, Peter. "Bioenergetic Aspects of Unity in Biochemistry: Evolution of the Concept of Ligand Conduction in Chemical, Osmotic and Chemiosmotic Reaction Mechanisms," in *Of Oxygen, Fuels and Living Matter,* pt. 1, ed. G. Semenza (New York, John Wiley, 1981), 1–160.

Prebble, John, and Bruce Weber. *Wandering in the Gardens of the Mind: Peter Mitchell and the Making of Glynn* (New York: Oxford University Press, 2003).

CHAPTER FOURTEEN

Howard Temin:
Rebel of Evidence and Reason

DANIEL J. KEVLES

At the Nobel Prize ceremony banquet held in Stockholm's Town Hall, in December 1975, Howard Temin rose before the assemblage of Swedish royalty and twelve hundred distinguished guests to respond to the toast offered by the Prime Minister to the year's laureates in physiology or medicine—David Baltimore, Renato Dulbecco, and himself. Each had won his share of the prize for research in the interactions of viruses and cells related to the genesis of cancer. Temin's recognition was rooted, as the Nobel citation read, in his advancement of "a rather unorthodox idea [in the 1960s], which was not well received by the scientific establishment." The banquet was ordinarily an occasion for decorous celebration, but Temin was no more bound by convention in his social views than he was in his scientific ideas. At a press conference held earlier, pointing to what the shared prize honored, he had insisted on removal of the ashtrays from the heavy walnut table where the laureates were seated. Now, his tall, thin figure amplified in silhouette behind the brightly lit dais, he indecorously scourged the smokers in the audience, including the Queen of Denmark, declaring that, having journeyed to Stockholm to receive a prize for cancer research, he was "outraged that the one major measure available to prevent much cancer, namely the cessation of smoking, had not been more widely adopted."[1]

THE YOUNG REBEL

Temin grew up in Philadelphia, Pennsylvania, one of three sons of Jewish parents who belonged to the professional middle class and whose family values included active devotion to social justice and encouragement of independent thinking. At the Henry School, where he

Figure 14.1. Howard Temin, 1986. Courtesy of the
University of Wisconsin at Madison Archives, Image
#11236 H1-4.

edited the student newspaper, Temin published editorials advocating
social conscience. "What does charity mean to you?" he asked, admon-
ishing that it might mean giving up ice cream cones or movies in
favor of contributions to alleviate hunger. When he celebrated his bar
mitzvah in 1947, the family bypassed the standard celebratory party
in order to donate the funds that would have been spent to a camp
for displaced persons. At his graduation from Central High School, in
June 1951, he devoted his valedictorian's address to the issues raised
by the recent explosion of the hydrogen bomb and the possibility of
sending a man to the moon.[2]

While at Central, Temin was one of a handful of students selected
to attend a summer school at one of the nation's leading biology laboratories, in Bar Harbor, Maine. He found the experience thrilling,

and it sealed his ambition to become a scientist. The director of the summer program wrote to Temin's parents that he was "unquestionably the finest scientist of the fifty-seven students who have attended the program since its beginning. . . . I can't help but feel this boy is destined to become a really great man in the field of science."[3] As an undergraduate at Swarthmore College he excelled in biology, but he fell into a disagreement with the external examiners in the zoology honors program (people said that he "fought his examiners to a draw"). Temin refused to wear a cap and gown or march in the processional at commencement, insisting instead on sitting with his family at the ceremony and hearing his name read in absentia.[4]

AN ASSAY FOR THE ROUS SARCOMA VIRUS

Temin embarked on his Nobel Prize–winning research in virology while a graduate student at the California Institute of Technology, in Pasadena, where he matriculated in the fall of 1955. At the time, most virology, at least in the United States, was concerned with the study of viral diseases as such—with identifying the viral causes of disease, isolating the viruses, and attempting to find vaccines or drugs to deal with them. Virology at Caltech, however, was not directly concerned with human disease. It was rooted in the study of bacteriophage—viruses that prey on bacteria—and its dominant figure was Max Delbrück. In the late 1930s, Delbrück had helped establish the phage group, a loose collective of scientists in the United States that had pioneered the study of phage as a model controllable system for the analysis of genetics. The phage group's scientific hallmarks were an emphasis on the use of simple, uniform biological systems—notably bacteria and phage isolated and bred to have standard characteristics—and the study of these systems with quantitative experimental techniques. Their fundamental experimental tool was a flat culture of bacteria in a Petri dish that revealed the action of infective phage by the presence of countable clear holes of dead cells, called plaques, readily observable on the culture's opaque surface.[5]

The Caltech program expanded into animal virology not long after the arrival, in 1948, of Renato Dulbecco, who had been trained in his native Italy in mathematics, biology, and medicine. At the time, most research in animal viruses was carried out in live animals, which was time-consuming, cumbersome, and expensive. Some of it was done with animal cell cultures, but few viruses had been cultivated in such cultures, and even the most advanced methods lacked exactitude and specificity.[6] Beginning in 1950, with Delbrück's encouragement, Dulbecco began to extend to animal virology the precise, quantitative methods of the phage group. By early 1952, he had managed to obtain a flat culture of chicken embryo cells in which he could use the plaque technique to detect and measure the activity of the equine encephalitis virus. His results demonstrated, as he put it, that "the plaque count is an efficient assay technique." Dulbecco successfully used the technique to investigate the infective and genetic properties of both the Western equine encephalitis virus and, working with a collaborator, the polio virus.[7]

In 1955, when Temin arrived at Caltech, Dulbecco was urging virologists to apply the plaque technique to all research in animal virology, including tumor viruses.[8] In 1911, Peyton Rous, a biologist at the Rockefeller Institute of Medical Research, in New York City, had discovered that a virus causes sarcomas in chickens, but the field of tumor virology had languished for many years, largely because scientists had been unable to stimulate tumors in other animals by viral infection. It had begun picking up during the 1930s, when it was found that viruses were implicated in cancerous papillomas in rabbits as well as in mammary tumors in mice. Now viral oncogenesis was increasingly drawing the attention of basic oncological researchers because since the early 1950s viruses had been shown to be responsible for a variety of mouse leukemias. Dulbecco saw in quantitative analysis with the plaque technique a valuable tool for probing the way in which viruses might stimulate the transformation of normal cells into cancerous ones.[9]

Temin came to Caltech intending to work in embryology, but he was early drawn into Dulbecco's group and research on tumor viruses as the result of a chance meeting with Harry Rubin. Trained in veterinary medicine, Rubin had come to Caltech in 1953 as one of Dulbecco's postdoctoral fellows and was particularly interested in the Rous sarcoma virus. Temin recalled that he hardly ever saw Dulbecco, "a kind of hands-off professor." He found stimulation and support from his fellow graduate students, sharing a house with several devotees of the quantitative, analytical methods that, in keeping with the spirit of the phage school, pervaded biology at Caltech. And like many graduate students there, he became part of the extended intellectual and social family of Max Delbrück, often joining the Delbrücks on weekend camping trips into the desert.[10] But Temin's most immediate influence was Rubin. The two formed an odd but effective pair—Rubin, a big strapping man with large hands looking like the ex-football player he was, and Temin, tall and gangly, his speech marked by an irrepressible giggle.[11] They collaborated productively in research on the Rous sarcoma virus until Rubin left Caltech in 1959.

The Rous virus posed a special challenge for quantitative investigations with the plaque technique. Unlike the cells infected with the kind of animal viruses with which Dulbecco had originally worked, cells infected with the Rous virus did not die, producing dark spaces against the light-colored surface of the culture. Rather, they continued to multiply and to produce virus. They thus did not form plaques that could be counted and analyzed. In an effort to overcome this problem, Rubin had developed quantitative assay for the Rous virus by infecting a membrane in developing chicken embryos, but it was not reliable for precise and consistent analysis. Temin used the method in his first experiments with the Rous virus, but he and Rubin changed experimental direction when, in 1956, Robert Manaker and Vincent Groupé, at Rutgers University, reported that they had observed the development of discrete foci—that is, piles of tumorous cells—in a culture of chicken embryo cells in vitro that had been infected with the Rous virus (see figure 14.2).[12] The number of foci was proportional to the concentration of infecting virus, but Manaker and Groupé's method was not suitable for a quantitative assay because, among other reasons, it did not prevent newly released virus from infecting cells elsewhere in the ensemble.[13]

By 1958, after some struggle, Temin and Rubin had successfully improved the method, packing a single layer of cells on a Petri-dish plate about four inches in diameter at a density high enough that they would grow but low enough that they would divide and produce virus. Appropriating a technique of Dulbecco's, they also added an overlay of agar, a moderately hardened mixture of nutrient, to prevent the virus produced in one focus of infection from migrating to and infecting other cells on the plate. Within six days, foci containing about eleven tumor cells would appear that were round, growing in grape-like clusters. They stood out against the background of normal cells and were easily visible in a standard microscope.[14] Their method was a critical step forward for investigations of the Rous sarcoma virus. Though using the technique might require instruction, it enabled detection of the viral transformation of cells into cancerous ones by the foci they formed and use of the foci for analysis in the manner of plaques.

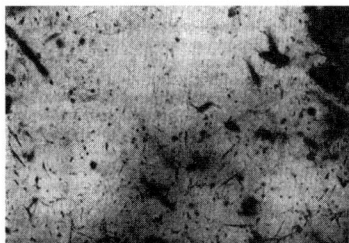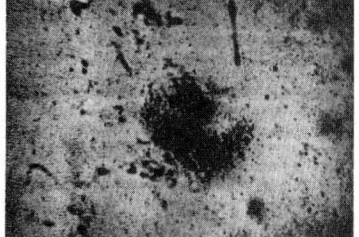

Figure 14.2. Left: A culture of embryonic chicken cells before infection by the Rous sarcoma virus. Right: A similar culture seven days after infection showing a focus of cells transformed by the Rous virus. *Engineering and Science,* California Institute of Technology, January 1960, 23.

THE PUZZLE OF THE ROUS VIRUS

The method revealed very quickly that chicken cells infected with the Rous sarcoma virus in vitro had remarkable properties. Most significant was that the infected cells were not quickly killed; rather, they were able to divide and produce virus for many months.[15] Temin and Rubin were guided in trying to understand this phenomenon by knowledge of the behavior of certain types of phage called "moderate" or "temperate" phage. On infecting the bacterial cell, moderate phage did not immediately kill it. Instead, the phage's genetic material appeared to integrate itself into the host cell's genetic machinery, where it lay in wait while the cell multiplied; it expressed itself later by producing virus in abundance, swelling the cell with progeny until it burst and died. Cells infected with the Rous virus also multiplied for a time without producing virus. Pointing to this similarity in behavior to temperate phage, Temin and Rubin suggested that "the genetic material of the virus is integrated into the genome of the cell, and is therefore transmitted to daughter cells as an inheritable property of the cell."[16] The integration might account for the transformation of the cell into a cancerous one, and the transmission would allow the daughter cells to produce the virus.

But Temin and Rubin emphasized that they had found important differences between the Rous virus and bacteriophage, enough to convince them that in dealing with the Rous virus "we must expect to encounter even more unique situations in this system and be prepared to deal with them in a unique and original way." One key difference was precisely that when cells infected with the Rous virus did produce new virus, the fresh production was, unlike the case with phage-infected cells, not lethal; the cells continued to live and divide. Another difference—and a particularly confounding one—was that, whereas the familiar temperate phage contained only DNA, the Rous virus's genetic material seemed to comprise RNA.[17]

Deoxyribonucleic acid (DNA) is, of course, the famed molecule that is built as a double helix. The two outer strands of the helix are joined at periodic intervals by rungs fashioned of one of two pairs of chemical bases that, in their ordering along the helix, form the information carried by the genetic code. This information becomes operative in a cell through the agency of RNA (ribonucleic acid). This is a single-stranded molecule, but it resembles DNA chemically, containing bases that are complementary to the bases in a stretch of DNA. A gene's information is transferred from DNA by the formation of a strand of RNA that is complementary to the DNA's coding bases.

It was plausible, as in the case with phage, that viral DNA could integrate into the DNA of a cell, but such integration was thought to be physically impossible for RNA from any organism. Nor was RNA thought to be capable of generating DNA complementary to itself that could be integrated into the DNA of a cell. According to the central dogma of molecular biology at the time, DNA could generate RNA that was complementary to it, but the reverse did not occur. It thus seemed implausible that the genetic material of the Rous virus could be integrated into the genome of the chicken cell and become a heritable property of it.

Temin completed his doctorate in 1959, working independently on the Rous virus after Rubin's departure. His experiments yielded additional evidence—for example, mutations in the virus changed the structural characteristics of the infected cell—that integration into the genome of the cell did occur. In his doctoral thesis he declared that his research had established that the Rous sarcoma virus has "some kind of close relationship with the genome of the infected cell," adding that it was thus "reasonable to assume that this relationship causes the cell to become malignant." He proposed that this transformation might result from the virus's contributing "information to the cell necessary for the cell to become malignant" or activating "a pre-existing potential for malignancy."[18]

Dulbecco recalled that when Temin defended his doctoral thesis, Delbrück took issue with his claim that the Rous virus's genes somehow associated themselves with the host cell's genes, insisting that Temin had not proved the association. Dulbecco, who had been inspired by

Temin and Rubin's research to begin his own work on DNA tumor viruses, defended Temin. He acknowledged that, strictly speaking, Delbrück was right, but contended that Temin had advanced a plausible interpretation of the behavior of the Rous virus, given what was known about temperate phage. Delbrück acknowledged that Temin's research was very good and approved the thesis with enthusiasm, but he remained skeptical about Temin's hypothetical conclusion.[19]

The skepticism of Delbrück, one of the world's leading biologists, might have dissuaded a less strong-minded young scientist to discard the idea of integration into the cellular genome, but not Temin. During a postdoctoral year at Caltech, his commitment to the idea was reinforced by further experiments showing that cells infected with the Rous virus could become tumor cells without producing virus and that the capacity for the production of virus was transmitted to daughter cells. Both lines of experimentation strongly suggested that the virus's genetic information was somehow incorporated into the host cell. In a popular article, titled "Cancer and Viruses," that he published in the Caltech alumni magazine in 1960, Temin noted that Rous tumor virology could unite two seemingly distinct theories of oncogenesis: those of cellular mutations and viral infection. "From a functional point of view," he wrote, "there's little difference between chromosomal or gene mutation and infection by Rous sarcoma. Both sets of events cause genetic changes in the cell." He added that "the virus, in some structural sense, as well as functional sense becomes a part of the genome of the cell," suggesting, "[p]robably it does not attach to a chromosome and may not even be in the nucleus, but becomes part of the general apparatus of the cell which controls what a cell is."[20]

CONTESTING ORTHODOXY

In the fall of 1960, Temin joined the faculty of the University of Wisconsin with an appointment in Madison at the McArdle Laboratory for Cancer Research. He later recalled that he did not have many job offers because in the world of modern biology interest in Rous tumor virology was limited. There was essentially no molecular biology or virology at McArdle, either, but Temin felt confident, even cocky, in his broad knowledge and abilities in both fields, and he liked the idea that he would be able to continue working independently. Between 1960 and 1970, he had only one graduate student and one or two postdoctoral fellows, and he was the sole author of all but one of the papers he published during that decade. Few if any of his colleagues in Madison understood his work, and some of the older faculty members disparaged it. His principal sounding board was perhaps Rayla Greenberg, a graduate student at Wisconsin in the population genetics of fruit flies, whom he met in 1961 and married in 1962.[21]

Temin's first grant proposal to the National Cancer Institute was rejected at the peer-review stage because, at age twenty-five, he was con-sidered too young to have his own laboratory, but the grant was awarded when a senior member of the NCI council intervened on his behalf, saying that Temin must be all right because he admired the chair at McArdle who had hired him.[22] Temin, working in a basement laboratory in a dingy building, soon acquired all-right status on his own. During his first two years at Wisconsin, he came increasingly to think of the genetic information that the Rous virus introduced into the cell as a "provirus." The term mimicked *prophage,* which designated the genetic material that infection with moderate phage introduced into bacterial cells and that lay latent until, activated, it produced new phage. But what was the Rous sarcoma provirus—RNA or DNA? An RNA provirus could account for the cell's further production of the Rous virus but not obviously for the transformation of the cell into a cancerous one. A DNA provirus could account for both phenomena, but it defied the central dogma that RNA could not generate DNA. Temin later recalled, "I was agnostic, I think probably still, as to RNA or DNA, and spoke of it as a genetic concept."[23]

In 1962, returning from their honeymoon, the Temins attended a symposium at the Cold Spring Harbor Laboratory, on Long Island, and there Temin learned about an antibiotic that he thought might help determine the nature of the provirus. Called actinomycin D, the antibiotic inhibits the expression of DNA. Temin reasoned that actinomycin D would not block the production of Rous virus in infected cells if the provirus consisted of RNA, but it would if it consisted of DNA. In Temin's experiments with the antibiotic, conducted in 1963, the cells did not generate any RNA, including viral RNA.[24]

To Temin, the results meant that the provirus was DNA or was at least located on the cell's DNA. He obtained additional experimental evidence that indirectly supported this view, demonstrating that infected cells contained what uninfected cells did not—new DNA complementary to the viral RNA.[25] Lurking in these results was the implication that the infecting Rous RNA virus, contrary to the central

dogma, somehow generated double-stranded DNA that was complementary to it. If that was true, here was the mechanism of the virus's
reproduction: from viral RNA to proviral DNA back to viral RNA. Here
also was a likely mechanism of cellular transformation: the proviral
DNA integrating itself into the cell's DNA, becoming part of—and
changing—its genetic endowment.

What Temin regarded as necessary, albeit exceptional, to account
for his evidence, however—namely, that RNA could synthesize DNA—
a number of other biologists regarded as scientifically bizarre and
wrongheaded. They greeted his presentations on the subject with skepticism.[26] Some ridiculed him behind his back. Temin heard about the
ridicule and about how, for example, at a meeting Harry Rubin himself
said in a review of Rous sarcoma work, "I'll give Howard's idea the
amount of time it's worth—none."[27]

The opposition and ridicule failed to demoralize Temin. Alice
Huang, a biomedical scientist, remembers asking him during a meeting if it was "very difficult to give a talk when no one seemed to believe
in any of it," noting, "He basically replied that he was used to it."[28]
Temin was convinced that he was correct and that ultimately he would
be vindicated. In a conversation in 1991, he told me, "Intellectually
I felt that the central dogma was true, but that it didn't explain my
results," adding, "Since this is biology, I didn't have any philosophical
problems with my results being an exception—biology doesn't have
the force of physics." He was sustained by his self-confidence, his
happiness in his marriage, his two children, and the environment at
Wisconsin, where the faculty, being somewhat isolated, tended to be
mutually supportive. His friend and colleague Bill Sugden has written
that Temin's "singular dedication to his research, coupled with his
exacting thought and speech, made him a formidable figure on the
Madison campus." His theory of how the Rous virus worked may not
have been believed or understood, Temin recalled, but he was known
to be bright, knowledgeable, and critical. "No one here accepted the
idea, but they accepted *me*," he later said.[29]

Despite the ridicule, the larger world of biology accepted him, too.
David Baltimore recalls that he and other biologists "did not think
[Temin] nuts; his ideas were imaginative and even compelling. The
problem was the lack of evidence. Howard kept plugging away over
the ten years, but the more experiments he did, and published, the less
persuaded were those who were staying tuned in. The experiments
were just not convincing. But the experiments were intelligent tests of

his notion and he did come up with some striking phenomena. So there was no difficulty funding his work because it was smart, thoughtful and even well-done. And it turned up information people wanted to know. But it never became unequivocal support for his ideas."[30]

Alice Huang, Baltimore's wife, recalls the existence of greater tolerance for Temin's ideas, saying that Temin "was considered to be somewhat out of the mainstream of virology, but nonetheless, a brilliant person who seemed to have the only sensible, consistent theory regarding this group of viruses. There weren't many people working in the field, but no one had contrary data, and he did have some support from a few other investigators."[31] Whatever the attitudes toward his unorthodox ideas, first-rate experimental work counted among biologists, perhaps more than theories, and if only for that reason Temin clearly commanded high respect in his field. Although he was not in high demand as a speaker, receiving perhaps only one invitation per year, he never lacked for grant money. The granting agencies recognized that he was using the same methods, the same terms, the same processes as everyone else, he later said. He was "just getting a discordant answer, which was unacceptable. But the techniques and the logic could not be criticized."[32]

REBEL TO THE END

The ridicule, however, did shape the focus of Temin's research. He felt that he had to prove conclusively that a DNA complement to the Rous viral RNA did integrate into the host cell's genome. From 1964 through 1969, that was his primary research target, rather than attempting to explain how RNA could be made into DNA (which was the critical step that ultimately brought him to Stockholm). He told me in 1991, "Our discussion was to prove that integration into the cellular DNA existed, not how it worked," elaborating, "Because no one else in the virology community took this seriously, the question of how the RNA might become DNA was never raised; there never was any intellectual discussion of the provirus hypothesis because it was treated with such ridicule. I was just defending it and establishing it, not taking the next step of trying to find a mechanism." Temin noted that he was not attentive to the particle, the virus, as such, let alone its capacity to generate the RNA-DNA switch. "I wish I could tell you that I was cleverer. I wish I could tell you that I was smarter than I was," he told me with a laugh.[33]

Temin was smart enough, nevertheless, to begin wondering by 1969 how the DNA provirus could be constructed from the Rous virus's RNA template. The turn in this direction evidently was prompted by discoveries of enzymes in other RNA viruses that catalyzed the synthesis of RNA from its constituent parts. He supervised a new postdoctoral fellow named Satoshi Mizutani in the conduct of an experiment designed to test whether an enzyme might be present in the host cell that catalyzed the creation of the provirus. The experiment demonstrated that, if there was such an enzyme, it existed in the virus, not in the cell. In the preceding two years, other biologists had identified enzymes in RNA viruses that catalyzed the production of RNA. Temin and Mizutani decided to look for an enzyme in the Rous virus that would catalyze the production of DNA. After several months, they found it.[34]

The same discovery was made simultaneously and independently by David Baltimore, a member of the M.I.T. faculty who had first met Temin during a high-school summer at the Jackson Laboratory. His own investigations with RNA viruses other than the Rous virus had led him to wonder whether an enzyme might account for how they multiply. He was well aware of Temin's provirus theory. In his Nobel Prize address, Baltimore recalled that although Temin's logic "was persuasive, and seems in retrospect to have been flawless, in 1970 there were few advocates and many suspects," adding, "Luckily, I had no experience in the field and so no axe to grind—I also had enormous respect for Howard dating back to when he had been the guru of the summer school I attended at Jackson Laboratory." Baltimore found evidence of an enzyme that could catalyze the creation of DNA in the leukemia virus of a mouse; then he found the same enzyme in the Rous sarcoma virus.[35]

In 1970, Temin and Baltimore reported the discovery of the enzyme in separate articles published in the same issue of *Nature*. It was promptly dubbed "reverse transcriptase," which signified its ability to transcribe RNA back into DNA. Temin noted at the end of his paper that confirmation of the discovery of reverse transcriptase would constitute "strong evidence" for the correctness of the provirus hypothesis and raise "strong implications for theories of viral carcinogenesis."[36]

The results were quickly and handsomely upheld. Baltimore recalls that "within weeks, the whole field reoriented itself and within months the stream of new developments hit the literature," adding that the speed of the changeover strongly suggests that the idea that RNA could generate DNA was not so "repugnant," that the "'central dogma' was not so fixed in people's minds."[37] Although the confirmation of reverse transcriptase added force to Temin's proviral theory of the way in which cells might be heritably transformed, however, the theory was commonly understood to be a hypothesis, not a finished explanation. Temin noted in 1971 that it did not illuminate how the integration of new genetic information from the virus made the host cells tumorous. He turned increasingly to that question, having been drawn by the Rous sarcoma work fully into cancer research, but it was the kind of cancer research that merged molecular virology with cellular transformation and that would lead to the age of oncogenes.

Temin recalled that after his discovery of reverse transcriptase he "went from rebel to establishment." The trappings of success did indeed appear, including invitations to speak and requests from students to work with him, but he remained ready when necessary to violate establishment norms.[38] In October 1976, less than a year after the award of the Nobel Prize, he traveled to Moscow and Leningrad to lecture at the invitation of the Academy of Medical Sciences of the Soviet Union. In the evening, he sought out dissidents and refuseniks (people denied exit visas for emigration to Israel), nervously looking for their apartment numbers in darkened corridors with a pocket flashlight. He brought them copies of *Science* and *Nature* as well as Hebrew books. Some nights he skipped dinner to pursue his clandestine missions, eating nuts and raisins stashed in his suitcase.[39]

Once Temin gave a lecture in someone's home to a group of refusenik scientists, meeting in what was known as the Sunday Seminar. Returning to give a second lecture on Monday, he learned that most of the seminar members had been arrested that morning by the KGB. Later that evening Temin visited Andrei Sakharov, a fellow 1975 Nobel laureate. Word finally came that the scientists had been released after being held twelve hours without charges. One of them came to Sakharov's home and tape-recorded an account of the events that he asked Temin to publicize. On returning to the United States, Temin gave the newspapers his tapes and his diary of the trip. Both were chilling. He advocated the cause of the beleaguered Soviet scientists to several political and scientific organizations and provided information about Soviet prisoners to groups such as Amnesty International.[40]

Temin's mission to the refuseniks was likely rooted in the same solid core of principle and conviction that his parents had encouraged and that had sustained him through the years of his unorthodox pursuit of the Rous sarcoma virus. It had been the spirit of Caltech biology, embodied in Delbrück, Dulbecco, and Rubin, that had brought him to quantitative studies of the virus and the exposure of its puzzling behavior. But his long, tenacious struggle to resolve the Rous conundrum was his own iconoclastic triumph, the product of a subtle brilliance and inner strength that were warmed by family and Wisconsin friends and of a deftness in the laboratory that won the support of most virologists, even if some scoffed at his heretical interpretation of the data. After winning the Nobel Prize, Temin used his laureate's pulpit to speak out, not only about smoking but about nuclear weapons, human rights, and AIDS. To the end of his life—he died of cancer in 1994 at age fifty-nine—he was a scientifically and socially engaged rebel whose purposes were grounded in reason, evidence, and principle.

ACKNOWLEDGMENTS

I am grateful to Horace Judson for a copy of the transcript of his interview with Howard Temin, to Bill Sugden for a copy of his biographical memoir of Temin, to Rayla Greenberg Temin for information, to Bill Meier for a copy of the tapes of Temin's oral history interview in the University of Wisconsin at Madison Oral History Project, and to David Baltimore and Alice Huang for their reflections on aspects of Temin's career.

NOTES

1. Rayla Greenberg Temin, "Foreword," in Geoffrey M. Cooper, Rayla Greenberg Temin, and Bill Sugden, *The DNA Provirus: Howard Temin's Scientific Legacy* (Washington, D.C.: ASM Press, 1995), xxi; Renato Dulbecco, *L'Aventurier du Vivant* (Paris: Plon, 1990), 248, 250; "The Nobel Prize in Physiology or Medicine 1975," Presentation Speech by Professor Peter Reichard of the Karolinska Medico-Chirurgical Institute, accessed at http://nobelprize.org/medicine/laureates/1975/presentation-speech.html, May 1, 2006. Temin recalled that after he spoke, everyone put out their cigarettes, but in David Baltimore's recollection, the Swedish king's uncle ostentatiously lit up in support of the Danish queen. Howard M. Temin, Oral History Interview with Margaret Andreasen and Barry Teicher, July–August 1993, University of Wisconsin-Madison Oral History Project, University of Wisconsin Archives, Madison, Wisconsin (hereafter Oral History, UW); David Baltimore, email to the author, May 17, 2006.

2. Rayla Greenberg Temin, "Foreword," xv–xvi.

3. Ibid., xvi–xviii.

4. Ibid., xvi. As David Baltimore remembers hearing the story, "Howard merely thought this a rare opportunity to discuss science with some very accomplished people so he sat down on the edge of a desk and started asking them questions." Baltimore, email to the author, May 17, 2006.

5. See Ernst P. Fischer and Carol Lipson, *Thinking About Science: Max Delbrück and the Origins of Molecular Biology* (New York: Norton, 1988).

6. H. Rubin, "Quantitative Tumor Virology," in *Phage and the Origins of Molecular Biology,* ed. John Cairns, Gunther S. Stent, and James D. Watson (Cold Spring Harbor, N.Y.: Cold Spring Harbor Laboratory of Quantitative Biology, 1966), 292–93.

7. Daniel J. Kevles, "Renato Dulbecco and the New Animal Virology: Medicine, Methods, and Molecules," *Journal of the History of Biology* 26 (Fall 1993): 409–42.

8. Renato Dulbecco, "Interaction of Viruses and Animal Cells: A Study of Facts and Interpretations," *Physiological Reviews* 35 (April 1955), 328.

9. Ibid., 328–31; Kevles, "Dulbecco."

10. Author's interview with Howard Temin, Jan. 28 and 29, 1993; Temin, Oral History UW. Temin's housemates included Frank Stahl, Matthew Meselson, and John Cairns, all soon to become leading figures in molecular biology. Rayla Greenberg Temin, "Foreword," xvii.

11. Dulbecco, *L'Aventurier du Vivant,* 179–80.

12. Rubin, "Quantitative Tumor Virology," 293–94. This was the chorioallantoic membrane. Temin later thought that the membrane infected with the Rous virus might in fact have been used for quantitative purposes but that it did not occur to him to try to use it for such analysis, perhaps because he had acculturated fully to the plaque religion at Caltech. Author's interview with Temin.

13. Howard M. Temin and Harry Rubin, "A Kinetic Study of Infection of Chick Embryo Cells *in Vitro* by Rous Sarcoma Virus," *Virology* 8 (1959): 209–22; Harry Rubin and Howard Temin, "Infection with the Rous Sarcoma Virus," *Federation Proceedings* 17 (December 1958): 994.

14. Rubin and Temin, "Infection with the Rous Sarcoma Virus," 995.

15. Temin and Rubin, "Kinetic Study," 219–20.

16. Rubin and Temin, "Infection with the Rous Sarcoma Virus," 994. See also Temin's doctoral thesis, Howard M. Temin, "The Interaction of Rous Sarcoma Virus and Cells *in Vitro*" (Ph.D. thesis, California Institute of Technology, 1960), 33–34.

17. Rubin and Temin, "Infection with the Rous Sarcoma Virus," 1003.

18. Temin, "Interaction of Rous Sarcoma Virus and Cells *in Vitro*," 46–47, 49.

19. Dulbecco, *L'Aventurier du Vivant,* 180–81, 190–93.

20. Rayla Greenberg Temin, "Foreword," xix; Howard M. Temin, "Cancer

and Viruses," *Engineering and Science* 23 (January 1960): 21–24; Howard M. Temin, "Recent Studies on Rous Sarcoma Virus," *Proceedings of the Avian Leukosis Conference* (East Lansing, Mich.: USDA Regional Poultry Research Laboratory, 1962), 30.

21. Judson, Interview with Temin, transcript, March 15–16, 1993, 6–7, 9–10; Bill Sugden, "Howard M. Temin, 1934–1994: A Biographical Memoir," National Academy of Sciences, *Biographical Memoirs* (Washington, D.C.: National Academy Press, 2001), 79:16; Rayla Greenberg Temin, "Foreword," xix; Temin, Oral History, UW.

22. Judson, Interview with Temin, transcript, 6–7; Rayla Greenberg Temin, "Foreword," xix.

23. Judson, Interview with Temin, transcript, 5; author's interview with Temin.

24. Howard M. Temin, "The Effects of Actinomycin D. on Growth of Rous Sarcoma Virus *in vitro*," *Virology* 20 (1963): 577–82; Howard M. Temin, "RNA-Directed DNA Synthesis," *Scientific American* 226 (1972): 27.

25. Temin, "Effects of Actinomycin D"; Howard M. Temin, "Homology Between RNA from Rous Sarcoma Virus and DNA from Rous Sarcoma Virus–Infected Cells," *Proceedings of the National Academy of Sciences* 52 (1964), 323–29.

26. Rayla Greenberg Temin, "Foreword," xx.

27. Author's interview with Temin; Temin, Oral History UW.

28. Alice Huang, email to the author, December 23, 2005; author's interview with Temin.

29. Sugden, *Temin,* 16; Judson, Interview with Temin, transcript, 17.

30. David Baltimore, email to the author, December 26, 2005. Baltimore notes, for example, that the data were simply "not convincing" in the paper that Temin published in 1964 (see note 25) claiming that infected cells contained new DNA that was complementary to the viral RNA. Baltimore, email to the author, May 17, 2006.

31. Huang, email to the author, December 23, 2005.

32. Judson, interview with Temin, transcript, 10, 17.

33. Author's interview with Temin.

34. Temin, "RNA-Directed DNA Synthesis," 29.

35. David Baltimore, "Viruses, Polymerases and Cancer," Nobel Lecture, December 12, 1975, in *Les Prix Nobel en 1975* (Stockholm: Nobel Foundation, 1976), 159–60.

36. Howard M. Temin and Satoshi Mizutani, "RNA-Dependent DNA Polymerase in Virions of Rous Sarcoma Virus," *Nature* 226 (1970): 1211–1213.

37. Baltimore, email to the author, December 26, 2005.

38. Temin, Oral History, UW.

39. Rayla Greenberg Temin, "Foreword," xxiii–xxiv.

40. Ibid.

FURTHER READING

Cairns, John, Gunther S. Stent, and James D. Watson, eds. *Phage and the Origins of Molecular Biology* (Cold Spring Harbor, N.Y.: Cold Spring Harbor Laboratory of Quantitative Biology, 1966).

Cooper, Geoffrey M., Rayla Greenberg Temin, and Bill Sugden. *The DNA Provirus: Howard Temin's Scientific Legacy* (Washington, D.C.: ASM Press, 1995).

Fischer, Ernst P., and Carol Lipson. *Thinking About Science: Max Delbrück and the Origins of Molecular Biology* (New York: Norton, 1988).

Kevles, Daniel J. "Courage, Viruses, and Cancer," in *Hidden Histories of Science,* ed. Robert B. Silvers (New York: New York Review of Books, 1995), 69–114.

Bill Sugden. "Howard M. Temin, 1934–1994: A Biographical Memoir," *Biographical Memoirs of the National Academy of Sciences,* vol. 79 (Washington, D.C.: National Academy Press, 2001).

Temin, Howard M. "RNA-Directed DNA Synthesis," *Scientific American* 226 (1972), 25–33.

CHAPTER FIFTEEN

Motoo Kimura and the Rise of Neutralism

JAMES F. CROW

Sometimes a member of the scientific establishment makes a discovery that is so surprising and so contrary to conventional wisdom that this person automatically becomes a maverick. An outstanding example, Barbara McClintock, is discussed in Chapter 8 of this book. Early in her career, she was recognized as *the* leading maize cytogeneticist, admired by all who were familiar with her work. Then she discovered genetic entities that jumped around the genome. At first she simply was not taken seriously. But eventually, thanks to similar discoveries in microorganisms, which were more accessible to molecular analysis and therefore more convincing to the genetics community, her findings were confirmed. She then became a household word, a feminist role model, and a Nobel Prize winner.[1]

Motoo Kimura is another example. His astonishing aptitude for formulating and solving stochastic equations in population genetics caused him to be widely regarded as the successor to the holy trinity—Wright, Fisher, and Haldane. Then in his early forties he found that newer data concerning molecular evolution seemed to imply that the bulk of evolutionary changes were selectively neutral. Rather than being the result of natural selection, they were driven by mutation and random drift. Those acquainted with the standard selectionist dogma regarded Kimura's idea as patently ridiculous. But he persisted; the neutral theory eventually found general acceptance, especially among molecular evolutionists, and he became world-famous.

Having achieved maverick status, both of these people became more iconoclastic. McClintock increasingly felt that she was not understood. She became more reclusive and did not publish in standard journals. She insisted that her transposable elements were important for normal development. Although transposability is no longer in doubt, this developmental aspect was and remains controversial because jumping genes seem too erratic to play a role in orderly development.

Figure 15.1. Motoo Kimura. Courtesy of the
University of Wisconsin News Service.

Kimura, after his discovery, became more argumentative, more obsessive about his views, and more determined to champion them.
Until his death, the neutral theory became the driving force in his life.

This chapter is based in part on publications and other written
records and in part on personal reminiscences.[2] I knew Kimura well
because he was my graduate student and we continued as working
colleagues until his death in 1994. We took advantage of every opportunity to meet, either in the United States or in Japan. And we remained close personal friends.

A BRIEF BIOGRAPHY

Motoo Kimura was born on November 13, 1924, in Okazaki, Japan.[3] He
came from a family with several generations of metal workers; they
made large bells for temples. His father loved flowers and had a garden
of ornamental plants. Because Motoo loved botany, his father gave him
a microscope, which brought young Motoo many happy hours. Visitors
at the Neishi elementary school in Okazaki can see the microscope
among other memorabilia of Kimura's elementary school days.

He retained his love of plants throughout his life.[4] Yet at an early age
he also displayed unusual mathematical ability. While he was in middle

school, his family was stricken with food poisoning. One of his younger brothers died, and Motoo was absent from school for several weeks. He used this time to study Euclidean geometry and worked all the problems in his text. His surprised teacher urged him to become a mathematician, and he might have done so had he realized that there were important mathematical problems in biology. Nevertheless, his mathematical skill served him well; very likely it was responsible for his being admitted to the highly competitive National High School in Nagoya.

Fortunately, this high school had a distinguished botanist, M. Kumazawa, on its faculty. His specialty was microscopic plant anatomy. Chromosome cytology was a popular subject in pre-war Japan, and Motoo joined the large community of cytologists studying lily chromosomes. He also enjoyed physics. His hero was Japan's most famous physicist, Hideki Yukawa, who later won the Nobel Prize for predicting the meson. Of special significance for Kimura's future, Kumazawa also taught a course in biometry. For the first time, Kimura realized that his mathematical skills could find a place in biology.

Kimura entered high school in 1942 when, because of World War II, the normal three-year course was shortened to two years. In 1944 he was admitted to the prestigious Imperial University in Kyoto. Japan's best-known geneticist at the time was Hitoshi Kihara, whose work on the cytology and evolution of wheat had brought him worldwide fame. Normally Kimura would have entered Kihara's lab in the Faculty of Agriculture, but Kihara was aware of a peculiarity in the wartime regulations. Students in the Faculty of Science were excused from military service, whereas those in the Faculty of Agriculture were not. So, on Kihara's advice, Kimura enrolled in botany. By the time he graduated, the war was over.

Life in Japan during the war years was hard. There was never enough good food, and there were military drills, which Kimura hated. If anything, conditions became worse after the war ended. Motoo had a cousin who was a quantum physicist and lived in the Kyoto suburbs, where better food was available. On Sundays Motoo visited his cousin for both gustatory and intellectual refreshment. Later he attributed much of his knowledge of the physical sciences and the philosophy of science to these visits.

After graduation, Kimura moved into Kihara's laboratory. Kihara's attitude was remarkable; it was just right for Motoo and for the future of population genetics. Recognizing Kimura's talent, Kihara assigned him no specific duties, leaving him free to study.[5] Kimura immediately started reading the works of Haldane, Fisher, and Wright. He was particularly interested in Wright and studied his papers in minutest detail. In those pre-photocopy days, he laboriously copied verbatim the papers of interest. Years later, I saw his handwritten copy of Wright's famous 1931 paper. It contained not only the full text in Kimura's neat hand but also extensive notes. Sometimes he had found alternative, improved derivations. Clearly he was destined to make a mark.

In 1949 the National Institute of Genetics was established in Mishima, and Kimura joined the staff. He remained there for the rest of his life. The institute was located in a wooden building that had been a wartime aircraft factory. It was hot in summer and cold in winter. Furthermore, at that time Mishima was a small provincial city lacking the cultural and intellectual attractions of Kyoto and Tokyo. He made frequent trips to these cities for library facilities and for intellectual refreshment. No one in Mishima understood or cared about Kimura's work, which increased his sense of isolation. Undeterred, he began writing papers; the first annual report of the Genetics Institute contains five of his reports, some startlingly original. It is interesting to read these early reports, for they foreshadowed some of the later work for which Kimura was to become famous.

One Japanese scientist who did show an interest in Kimura's work was Taku Komai, who had studied with T. H. Morgan at Columbia University. Kimura's work was also recognized by two American geneticists at the Atomic Bomb Casualty Commission, Duncan McDonald and my student Newton Morton. Together with Komai they were able to scrape together funds from various sources, making it possible for Kimura to come to the United States. He wanted to work with Sewall Wright, but by this time Wright was getting ready to retire from the University of Chicago and was not taking students. He recommended Iowa State University, where Kimura could work with America's best-known animal breeder, the Wright acolyte Jay L. Lush.

I first learned of Kimura through Newton Morton, who sent me some Kimura reprints. I realized that his was remarkable work and assumed that the author was a mature mathematician. I first encountered Kimura at a meeting of the Genetics Society of America, which met in Madison, Wisconsin, in the fall of 1953. Recognizing the name, I was surprised to find a young student, not the older person I had pictured.[6] He was on his way to Iowa State College.

Kimura had brought with him a manuscript written on the ship as it crossed the Pacific Ocean. The paper dealt with the effects of fluctuating selection, which can mimic random drift. Kimura had cleverly found a transformation that converted a nasty partial differential equation into a simple heat-diffusion formula known to every physics student. Wright reviewed the paper with unusual enthusiasm, and it was soon published.[7]

After entering Iowa State College, Kimura became dissatisfied with the research program, which was concerned with quantitative traits and emphasized subdivision of epistatic variance (that is, the variance component caused by gene interaction). Kimura understood this, but he really wanted to work on stochastic processes. Furthermore, he developed a strong dislike of Lush. During the winter he asked if he could transfer to Wisconsin and study with me. I knew I could add nothing to his mathematical knowledge, but I was reasonably sure that Sewall Wright would soon be moving to Wisconsin, so I accepted Kimura as a student. Having Wright as a colleague and two such outstanding students as Morton and Kimura made this an exciting time for a beginning graduate instructor.

Kimura spent two years, 1954–1956, obtaining his Ph.D. in Wisconsin. During this time he wrote several papers that are milestones in population genetics and are now textbook material. Invited to speak at the 1955 Cold Spring Harbor Symposium, he gave a talk that was notable for being almost impossible to understand, both because of the mathematics and because of his English pronunciation. Equally notable is what Wright said at the meeting. After the talk, Wright rose to say that only those who had struggled with these problems could appreciate Kimura's enormous accomplishment. By the time Kimura graduated, he was already a recognized leader in theoretical population genetics.

Kimura then returned to Japan. Except for occasional stays abroad, usually a year or less, he spent the rest of his life in Mishima. He died on November 13, 1994, his seventieth birthday.

POPULATION GENETICS THEORY

By the 1950s there was a well-developed population genetics theory, thanks mainly to the work of R. A. Fisher, Sewall Wright, and J. B. S. Haldane. The unit of study was the allele frequency, the proportion of a particular allele among the allelic forms at a gene locus. Although selection acts on genotypes, these are not stable, for they are scrambled by the Mendelian shuffle every sexual generation. Except for rare mutations, genes are stable through generations. Given assumptions about the strength of selection among the genotypes at this locus and the mating system (usually taken to be random), the theory predicts how rapidly the allele frequency changes or whether it reaches an equilibrium. A weakness in the application of the theory is that the strength of selection is almost always unknown. An exception was the replacement in industrialized England of standard light forms of the peppered moth by a dark form as tree trunks darkened with soot, a phenomenon called "industrial melanism." Because this moth reproduces annually and the original and later frequencies were approximately known, along with the time needed for the change to occur, Haldane was able to infer that the melanistic form enjoyed a selective advantage of about 50 percent. Haldane produced tables giving the number of generations required for a specified change, given the selective advantage or disadvantage. The theory was also extended to include the effects of mutation, migration, and departures from random mating for different modes of inheritance.

The problem becomes much more difficult when the effects of random processes are included. This was done mainly by Fisher and Wright. A typical problem studied by Wright is this: What is the steady state distribution of allele frequencies with a given intensity of selection balanced by recurrent mutation, along with random changes caused by a small population number? He presented the results in the form of a graph in which the abscissa is the allele frequency and the ordinate is the frequency of that frequency. By various clever tricks, Wright was able to obtain approximations of these distributions. Over a period of years he continued to develop these equations with increasing mathematical rigor.[8]

The difficulty was that the theory, beautiful as it was, had had very few successful applications. This changed with the coming of molecular methods, which supplied an abundance of data. It is at this point that Kimura's work changed the picture.

KIMURA'S CONTRIBUTIONS TO POPULATION GENETICS THEORY

Before coming to the United States, Kimura had discovered the two Kolmogorov equations. These are partial differential equations, one known as "forward" and the other as "backward," used to describe random processes such as Brownian motion and more general diffusion processes. Wright had used the forward equation—in fact, he rediscovered it himself—but Kimura was the first geneticist to employ the backward equation. He realized while still in Wisconsin that this equation was especially useful for some previously unsolved problems. Later he used it to study the age of a mutant allele in a population.

Soon after arriving in Wisconsin, Kimura obtained the complete distribution of allele frequencies under conditions of neutral random drift, at any time from any arbitrary starting frequency.[9] He soon extended this to three alleles, then to an indefinite number. He then included the effects of mutation, migration, and selection. These results were published in the Cold Spring Symposium mentioned above.[10] It was of great significance for his later work, however, that Kimura derived the probability of ultimate fixation of a mutant gene under various models of selection. His early work, mostly done as a graduate student, was summarized in a review article.[11]

Returning to Japan, he continued to develop equations for stochastic genetic models of greater generality. He introduced the "infinite allele" and "infinite site" models, widely used for evolutionary studies many years later, after the coming of molecular techniques. With his colleague Takeo Maruyama, he found a method for investigating several problems such as the number of individuals or the number of heterozygotes in the path to fixation or loss. They later determined the moments for the sum of any arbitrary function along the path. They also found a surprising paradoxical result: the conditional time to fixation from an arbitrary starting point is the same for a given degree of selection, whether the allele is selected for or against. Although the probability of fixation of a deleterious mutant is very low, if it does go to fixation, it does it in a hurry.[12]

Although Kimura's best-known work involved stochastic processes, he also did a number of other studies. He wrote two seminal papers about the rate of change of population mean fitness. The first was a more explicit formulation of Fisher's Fundamental Theorem of Natural Selection. The second, a quite remarkable paper, epitomizes Kimura's originality. He showed that, under continued directional selection for loosely linked loci, the epistatic (gene interaction) variance is almost exactly canceled out by the linkage disequilibrium variance. The result is that the results of selection can be predicted more accurately by ignoring epistasis than by including it.[13]

In another paper he and I showed, much to Wright's surprise, that mating the least closely related individuals is not the way to minimize the ultimate rate of decrease in heterozygosity.[14] Later, we showed that a crude approximation to truncation selection is almost as effective as true truncation selection in removing mutations from the population with minimum genetic load.[15]

As I emphasized above, in the 1960s population genetics had the most beautiful theory in biology, but there were few opportunities to apply it. Molecular biology changed everything. Data about the rates of molecular evolution were appearing and were awaiting analysis.[16]

The stage was set for Kimura's blockbuster.

THE NEUTRAL THEORY

At this time there was a growing body of data about evolutionary rates as inferred from amino acid substitutions in various proteins. Kimura noticed, following Zuckerkandl and Pauling, that for a given protein the rates in different mammals were remarkably uniform. The rates implied an enormous number of substitutions. If one made reasonable assumptions about the size of the genome, it looked as if there was about one mutation every two years. Kimura realized that to make this many substitutions by selection would impose a tremendous cost on the population. But the cost approaches zero as the selective difference becomes smaller. Hence Kimura concluded that the great majority of the detected mutations must be selectively neutral.

Kimura prepared a paper presenting these results. He sent a copy of the manuscript to Wright, who was very cool to the idea.[17] Kimura remained convinced, however, and the paper was published.[18] Soon afterward, King and Jukes published a paper with the provocative title "Non-Darwinian Evolution."[19] Their arguments were more biochemical in nature, emphasizing such things as synonymous mutations and the relation between the frequency of amino acids and the genetic code. But they supported Kimura's claim.

I spent several weeks in Mishima in the summer of 1968. I had seen the King and Jukes paper in manuscript, and Kimura and I, together with his colleague Tomoko Ohta, spent many hours in discussions of this paper and the new ideas that Kimura had produced in the meantime. Although initially skeptical, I became convinced that the neutral theory, though not solidly established, was certainly worthy of serious attention. I spoke at the final plenary session of the International Congress of Genetics in Japan late in the summer and endorsed Kimura's ideas.[20] Later, Kimura and Ohta also suggested that polymorphism was a phase of molecular evolution.

One of the most attractive features of the neutral theory was that it provided a basis for a molecular clock, previously suggested by Zuckerkandl and Pauling. Mutation-driven evolution is as constant as the mutation rate itself. Hence the number of changes between two species could be used to infer the time since they split from a common ancestor—a great boon to systematic biologists.

Immediately after the papers by Kimura and by King and Jukes appeared, the fur began to fly. Ever since Darwin's time, biologists had assumed that the driving force in evolution was natural selection, usually acting on minute differences. For many traditional evolutionists, the heretical view that mutation and random survival, rather than natural selection, were responsible for genetic change was totally unreasonable. Many of the arguments by classical evolutionists were emotional rather than logical. Two of the more reasoned arguments were from Bryan Clarke and Rollin Richmond.[21] Others were less restrained in their criticisms, especially in conversation and letters. Especially vehement were people from the world of classical evolution, who usually were not impressed by Kimura's mathematical arguments.

The main reason why the argument was so heated was that the two groups were firing past each other. Traditional evolutionists were thinking of such things as body structures, physiological processes, and behavioral differences. Kimura was thinking of changes at the DNA level. Although it was not obvious at the time, it soon became clear that, especially in vertebrates, the bulk of DNA changes in evolution were not in genes but in noncoding regions with no obvious function. And, within genes, a number of the DNA changes led to no important functional change, such as synonymous changes that did not change the amino acids. The crucial issue, still not settled, is the relative importance of selection and drift for amino acid changes. The answers are appearing on a case-by-case basis.

Kimura found many other arguments in the next few years. Amino acids in regions of less importance for the function of a polypeptide evolved faster than those important for the function. Particularly revealing was the insulin molecule, which has three regions, one of which is discarded and not used. The unused part evolved fastest. Within codons, synonymous changes were faster than nonsynonymous ones. To Kimura, slow evolution of some nucleotides was caused by "selective constraint." These regions already functioned well, and therefore most mutations were harmful. One of Kimura's most striking arguments came from the fact that the number of amino acid differences between the alpha and the beta hemoglobins in humans was about the same as that between human beta and carp alpha. The first two hemoglobins have been in the same cell for four hundred million years, while the latter two have been in fish and in the line leading to humans. The difference in selective forces could hardly be greater. If the amino acid changes were due to selection the two sequences should be enormously different from each other; but they weren't. Kimura summarized his views in a widely quoted book.[22] He devoted much of his energy for the rest of his active life to finding more evidence and arguing the case.[23]

The most telling arguments against the theory came from Hirotsugu Matsuda in Japan and John Gillespie in the United States, and these caused Kimura the greatest distress. These mathematical geneticists pointed out that fluctuating selection coefficients could mimic the long-time constant rate of evolutionary change.[24] Gillespie's paper was titled "The Molecular Clock May Be an Episodic Clock." According to him, the irregularities in the evolutionary rates were too great to be explained by the neutral theory. Kimura believed that Gillespie overemphasized the departures from constancy and, given the uncertainties of measurement, that the approximate constancy was more important.

Over the course of the following decades the debate died down. The neutral theory, if not accepted in full, nevertheless became a standard part of evolutionary theory. Whether neutrality is as widespread as Kimura thought is doubtful. Yet in many organisms there is a great deal of noncoding DNA. At the same time, the theory is here to stay for practical reasons. It is the natural null hypothesis for studies of selection. And, as I stated above, it provides a mechanistic basis for a molecular clock. The latter follows from the simple idea that neutral evolution is driven by the mutation rate and, when the random process is taken into account, the long-time average rate of nucleotide substitution is given simply by the mutation rate.[25]

Another reason for the neutral theory's popularity is that it is so simple. There is much to say for Gillespie's ideas, which are presented fully in his book. Yet they are not being used, mainly because the theory is so complicated. Einstein said: "A theory should be as simple as possible, but not simpler." Kimura's theory fits this criterion to a T. Scientists like simple theories, provided they are realistic. In addition, Kimura's theory has spawned a large number of statistical tests for studying polymorphism and evolutionary divergence.

There are many other arguments for neutrality, and Kimura was most inventive in finding them. At the same time geneticists have found a number of loci in which natural selection could be demonstrated. Nevertheless, these still make up a small fraction of the totality of molecular change. Of course, Kimura never believed that evolution of form and function happened by means other than natural selection. The issue was and is, what fraction of the genome is evolving neutrally? The jury is still out, but there is enough noncoding material, for which neutrality is a reasonable starting position, to lead to the standard assumption—often unstated—that neutral theory can be used for such purposes as inferring evolutionary rates and ancestral times in the evolutionary past.

It is common in science for a mathematical theory that has been worked out for its own sake to be useful later. Usually the mathematician is long dead by the time the theory is usefully applied. Kimura was lucky. The formulae that he worked out, many of which were produced during his graduate student days at the University of Wisconsin, turned out to be ready-made—that is, pre-adapted—for the neutral theory. The most obvious is the rate of evolutionary substitution. That this is equal to the mutation rate is easily seen, as I showed above. But it is clear that in order for this formula to be appropriate, the period of observation must exceed the time required for a substitution. But how long must this be? Kimura's later work supplied the necessary formula; it is $4N_e$ where N_e is Wright's effective population number, usually somewhat less than the adult census number. This factor immediately dictated the time scale over which the rate should be measured. The ready availability of such simple formulae—and formulae that could be simply interpreted—was a major reason for the widespread adoption of the neutral theory.

Kimura's theory also made possible the serious treatment of alleles that were very weakly selected. In particular, his colleague Tomoko Ohta, because of uncertainty as to whether the rate constancy is per year or per generation, believed that much of molecular evolution is due to fixation of very mildly deleterious mutations. This theory is more complicated, but she and Kimura worked out the necessary formulae.

There is much more to the theory than I have given here. The best sources are Kimura's own book and collected papers.[26]

KIMURA AS ADVOCATE

Although Kimura at first presented his theory somewhat tentatively, it did not take long for him to become a determined proponent. Every new molecular advance was examined from this viewpoint, and when possible used to bolster the case for the neutral theory. Kimura was particularly inventive in taking observations from many sources and bringing them to bear on his theory. He missed no chance to advocate the idea. Although he knew that most evolution of form and function is by natural selection, in his zeal for neutral changes he often underplayed this component. The result is that he was sometimes criticized as being anti-Darwinian.

Kimura was far from a passive participant in debates. He jumped in with both feet and became deeply involved emotionally. As a result, he became controversial not only for his theories but also for his take-no-prisoners advocacy. For the rest of his life, although he continued to produce new mathematics, his main thrust was to find support for his theory. Those who opposed his ideas were his enemies, in particular, people such as Gillespie, who provided the strongest arguments against his theory.[27]

His advocacy continued up to a short time before his death. In his late sixties, Kimura developed amyotrophic lateral sclerosis, and his health deteriorated very rapidly. He died of an accidental fall. Death is sometimes merciful.

KIMURA AS REBEL

Why was Kimura so vehement? Was this something that happened only after encountering criticisms of the neutral theory? My associations with Kimura as a student could not have been more pleasant for both of us. He had a wide intellectual interest, far beyond genetics, and we shared many happy discussions. Yet I was aware that he had strong personal antagonisms. I have already mentioned his dislike of Lush, and there were others whom Kimura detested. When I first visited him in Japan I noticed that he had friends and enemies. But these did not appear, to me at least, to have a major effect on his personality.

Why did his strongest antagonisms develop after he proposed the neutral theory? My guess is that basically Kimura did not change. His early mathematical work was widely accepted and highly praised. There was no reason for arguments. Only after presenting the neutral theory did he encounter widespread disagreement, and this led to animosity. His defensiveness, I believe, was already there; it took criticism to bring it out.

What effect did Kimura's belligerence have on acceptance of his theory? His polemical writings and the equally polemical responses from his opponents brought the theory wide attention. But did this hasten or retard its acceptance? I leave this question for future historians to mull over. But the usefulness of the theory for study of molecular evolution and population genetics was so clear that no advocacy would have been needed.

Kimura was highly honored. Early in his career he received the Weldon Prize from Oxford University. After his formulation of the neutral theory, new awards came in rapid succession, not only in Japan but also throughout the world. Altogether he had fourteen international prizes and awards, plus two honorary degrees at major American universities.[28] These brought him great satisfaction. He is particularly honored in his hometown of Okazaki, thanks largely to efforts of his brother. In addition to the museum mentioned above, there is a statue of him in the city.

THE PRESENT STATUS OF THE NEUTRAL THEORY

Haldane once said that the highest honor a scientist can have is for his theory to be so taken for granted that his name is no longer associated with it. We no longer mention Mendel when we conduct breeding experiments or Sturtevant when we map chromosomes. Likewise, neutral assumptions permeate modern studies of molecular evolution and population genetics.[29] In most cases Kimura is not mentioned; in fact, his contribution may not even be known. For example, it is accepted practice to infer phylogeny by using noncoding regions to avoid the complications of selection. Whether the theory is strictly correct or not no longer matters. Vast regions of the genome are near enough to being neutral for neutrality to be assumed. Virtually every study of molecular variability, evolutionary rates, and coalescence is based on this assumption.

Examples abound. The evidence for the "Out of Africa" hypothesis, to name but one, came from studying genetic variance, assumed to be neutral. Another widely used phenomenon is selective sweeps. If a new mutation has a selective advantage and increases rapidly in the population, the chromosomal region around the mutation will have reduced genetic variability. Hence, a reduced neutral variance around a locus provides the initial evidence and often the strongest evidence for the action of selection on the locus. Increasingly, the neutral theory is not so much an end in itself but a way to study other evolutionary processes such as migration.

No longer a subject of controversy, the neutral theory is now the standard null model. Ironically, it is often used to provide evidence for selection. I don't know whether Kimura would be pleased.[30]

NOTES

1. McClintock was also awarded a MacArthur "genius" grant. In her case it was for a lifetime, but she was seventy-nine at the time—a safe gamble for the foundation.

2. Much of this chapter is based on my "Motoo Kimura," *Biographical Memoirs of Fellows of the Royal Society of London* 43 (1997): 253–265. Several references that I have omitted here are given there.

3. Kimura's given name is pronounced Mo-toe, not Mo-too. The second syllable is protracted, and Kimura naturally thought that the best way to designate this in English would be to repeat the last letter. Alas, this led to a lifetime of having his name mispronounced.

4. When Kimura and I published a book together in 1970, he used his royalties to construct a tiny greenhouse at his home. He was unusually successful as an orchid breeder and produced several prize-winning clones.

5. Kimura did solve one problem for Kihara. Kihara was interested in introducing a foreign chromosome into a standard wheat variety. With back-crossing, half the introduced chromosomes are lost with each generation. Kimura solved the problem elegantly and completely, and this led to his first published paper, "The Theory of Chromosome Substitution Between Two Different Species," *Cytologia* 15 (1950): 281–284.

6. I must have been one of only a handful of people in the United States who knew of Kimura—perhaps I was the only one—for the Annual Reports of the Genetics Institute were almost unknown in this country.

7. Motoo Kimura, "Process Leading to Quasi-Fixation of Genes in Natural Populations Due to Random Fluctuation of Selection Intensity," *Genetics* 39 (1953): 280–295.

8. The three classical general references are J. B. S. Haldane, *The Causes of Evolution* (London: Longmans, 1932); R. A. Fisher, *The Genetical Theory of Natural Selection* (Oxford: Oxford University Press, 1930); Sewall Wright, "Evolution in Mendelian Populations," *Genetics* 16 (1930): 97–159. Fisher's book was reissued in a variorum edition, with numerous notes and corrections, 1999. Wright's work, and that of many others, is summarized in his

four-volume series, *Evolution and the Genetics of Populations* (Chicago: University of Chicago Press, 1968–1978). This is only a small part of population genetics theory, since I have included only that which led to Kimura's neutral theory. With several loci, the analysis becomes much more complicated, especially with linkage. I have omitted multiple-gene inheritance, of which a classical instance is Fisher's "Fundamental Theorem of Natural Selection," in his *Genetical Theory of Natural Selection,* 34, in which he shows that selection in nature follows least squares rules.

9. This is the process: a population starts with an arbitrary proportion of allele A, say, 60 percent. In the next generation, because of random sampling, the population frequency will usually deviate from 60 percent. If we observed many populations there would be a distribution of values, centering around 60 percent. In the following generation the process is repeated; each population generates a new distribution, and so on. Obviously, if the population is made up of more than a few individuals this becomes impossibly difficult. Kimura was able to approximate this enormously complicated process by a single partial differential equation, known as a diffusion equation. Wright, although much impressed by Kimura's mathematical skills, thought that starting the process at a fixed point was mainly of academic interest because most natural populations fluctuate somewhere near an equilibrium distribution. Wright greatly respected Kimura and had no intention of denigrating his result. Nevertheless, Kimura was extremely sensitive about his work, and Wright's casual remark was deeply distressing. Of course, Wright never knew this.

10. Motoo Kimura, "Stochastic Processes and Distribution of Gene Frequencies Under Natural Selection," *Cold Spring Harbor Symposia on Quantitative Biology* 20 (1955): 33–53.

11. Motoo Kimura, "Diffusion Models in Population Genetics," *Journal of Applied Probability* 1 (1964): 177–232.

12. A complete list of Kimura's publications, along with reprints of the most important papers, can be found in *Population Genetics, Molecular Evolution, and the Neutral Theory,* ed. N. Takahata (Chicago: University of Chicago Press, 1994).

13. It must have given Kimura considerable wry satisfaction, given his earlier low opinion of subdivision of epistatic variance when he was in Iowa, to discover the nonimportance of epistasis in most directional selection.

14. Motoo Kimura and James F. Crow, "On the Maximum Avoidance of Inbreeding," *Genetical Research* 4 (1963): 399–415.

15. James F. Crow and Motoo Kimura, "The Efficiency of Truncation Selection," *Proceedings of the National Academy of Sciences* 76 (1979): 396–399.

16. Emil Zuckerkandl and Linus Pauling, "Evolutionary Divergence and Convergence," in *Evolving Genes and Proteins,* ed. V. Bryson and H. J. Vogel (New York: Academic, 1965), 97–166.

17. Although Wright emphasized random allele frequency drift in all his writing, his viewpoint and Kimura's were entirely different. Wright thought random drift aided natural selection by permitting the population to go from one level of fitness to a higher one when intermediates were lower than either. Kimura was interested in random change as a phenomenon in itself.

18. Motoo Kimura, "Evolutionary Rate at the Molecular Level," *Nature* 217 (1968): 624–626.

19. Jack L. King and Thomas H. Jukes, "Non-Darwinian Evolution," *Science* 164 (1969): 788–798.

20. James F. Crow, "Molecular Genetics and Population Genetics," *Proceedings of the XII International Congress on Genetics* 3 (1969): 105–113.

21. Bryan Clarke said: "King and Jukes argue that random genetic drift has been primarily responsible for the majority of amino acid substitutions, but the weight of evidence does not support them." Bryan Clarke, "Evolution of Proteins," *Science* 168 (1970): 1009–1011. Rollin C. Richmond says: "The view that most evolutionary changes in DNA need not be attributed to natural selection rests on several lines of evidence. This article argues that the evidence leads more simply to a neo-Darwinian interpretation." Rollin C. Richmond, "Non-Darwinian Evolution: A Critique," *Nature* 225 (1970): 1025–1028. For an article more mathematical and more sympathetic to the neutral view, see Norman Arnheim and Charles E. Taylor, "Non-Darwinian Evolution: Consequences for Neutral Alleles," *Nature* 223 (1969): 900–903.

22. Motoo Kimura, *The Neutral Theory of Molecular Evolution* (Cambridge: Cambridge University Press, 1983).

23. The neutral theory was foreshadowed several years earlier by Noboru Sueoka, "On the Genetic Basis of Variation and Heterogeneity of DNA Base Composition," *Proceedings of the National Academy of Sciences* 48 (1962): 582–592, and Ernst Freese, "On the Evolution of Base Composition of DNA," *Journal of Theoretical Biology* 3 (1962): 82–101. Both were studying bacteria and invoked neutral mutation pressure to account for irregularities that were hard to explain by selection. In particular, Freese was concerned with the fact that there were wide discrepancies in amino acid content in different bacterial species despite only very minor changes in DNA and invoked neutrality as the explanation. Neutral alleles had been postulated before, for example for human blood groups, and enjoyed brief popularity based on little evidence. This ceased abruptly when selective factors were demonstrated. Likewise there was a spirited debate between Wright and Fisher, for example, over whether banding patterns in some snails were neutral or selected. None of these, however, led to Kimura's neutral theory, which grew from observations on molecular evolution.

24. Naoyuki Takahata, Kazushige Ishii, and Hirotsugu Matsuda, "Effect of Temporal Fluctuation of Selection Coefficient on Gene Frequency in a Population," *Proceedings of the National Academy of Sciences* 72 (1975):

4541–4545; John Gillespie, "The Molecular Clock May Be an Episodic Clock," *Proceedings of the National Academy of Sciences* 81 (1984): 8009–8013.

25. The argument is simple. In a diploid population of N individuals there are 2Nu new mutations per generation, where u is the per locus mutation rate. If we wait a long time, only one of the genes will have survived and become fixed. The probability that the new mutation is the lucky one is 1/2N. By multiplying, the 2N's cancel out and the evolution rate is simply the mutation rate.

26. Kimura, *Neutral Theory of Molecular Evolution.*

27. For a witty and nuanced account of the controversy and the personalities involved, see Deborah Blum, "Scientists in Open War Over 'Neutral Theory' of Evolution," *Sacramento Bee,* March 16, 1992.

28. For a list, see Crow, "Motoo Kimura."

29. For a recent account of the theory and its reception, see Michael R. Dietrich and John Beatty, "Molecular evolution," in *A Companion to Philosophy of Biology,* ed. Sahotra Sarkar and Anya Plutinski (Malden, Mass.: Blackwell, 2007).

30. Bret Payseur, Burt Singer, Naoyuki Takahata, Oren Harman, and Michael Dietrich all read the manuscript and provided helpful suggestions.

FURTHER READING

Kimura, M. *The Neutral Theory of Molecular Evolution* (Cambridge: Cambridge University Press, 1983).
Takahata, N. *Population Genetics, Molecular Evolution, and the Neutral Theory* (Chicago: University of Chicago Press, 1994).

CHAPTER SIXTEEN

Against the Grain:
The Science and Life of William D. Hamilton

ULLICA SEGERSTRALE

A paradigm shift in the evolutionary explanation of social behavior took place during the last quarter of the twentieth century. Concepts such as kin selection, inclusive fitness, and the gene's-eye view entered science, crowding out earlier notions of individual selection and individual fitness. Part of the novelty was the very idea that social behavior can evolve. This insight became the foundation for the emerging fields of behavior ecology and sociobiology.

The cornerstone of the new line of reasoning came in the 1960s with William Donald (Bill) Hamilton's crucial insight into the evolution of altruism. His 1964 paper "The Genetical Evolution of Social Behaviour" is one of the most frequently cited papers in the whole of science.[1] Darwin had regarded animal altruism as a major problem for his whole theory. Why would an animal ever behave so as to reduce its biological fitness by putting itself in danger or by forgoing reproduction? But Hamilton was able to show in the mathematical language of population genetics how such a counterintuitive trait could actually spread. What was required was simply that the benefits of an altruistic act did not fall simply on random members of a population, but on individuals genetically related (kin) to the donor.

Solving the puzzle of altruism was only the beginning. Hamilton went on to open up whole new fields of inquiry. Among these were parasite-host coevolution, the evolution of sex, mate choice, cooperation between non-relatives, sex ratios, dispersal, senescence, the evolution of insect sociality, and conflicts within the genome. For his contributions he received a great number of international prizes and honors, including the Crafoord Prize (the would-be Nobel for biology), the Kyoto Prize, and the Darwin Medal. Hamilton has been called the "primary theoretical innovator in modern Darwinian bi-

Figure 16.1. William Hamilton in 1986.
Photo by Hiroshi Ohira. Used by
permission.

ology, responsible for the shape of the subject today," and "Darwin's
heir."[2] Indeed, many have observed his uncanny similarity to Darwin
in taste, approach, and general reasoning style. In a sense, Hamilton
is Darwin plus mathematics.

But this is all in retrospect. Bill Hamilton's career represents a
classic case of misunderstood genius, or rather perhaps "resistance
of scientists to scientific discovery."[3] His were ideas whose time had
definitely *not* come. Every step of the way was a struggle with editors
and referees who could not follow his way of thinking. This battle
to get published and lack of timely acknowledgment for important
ideas led to bouts of loneliness. Sometimes he wondered if he was a
crank.[4]

Throughout his career Hamilton, largely self-taught and acting as
his own adviser and critic, followed his own research program. He
set very high standards for his papers. They often remained obscure
to his naturalist colleagues, however, because of their mathematical
language. But Hamilton was no abstract mathematician—rather, he

was a brilliant naturalist with deep empathy for all living things. It was simply that he, unlike many of his colleagues, was able to present his deep intuitions about nature in mathematical form. (Today, of course, mastery of the basics of population genetics and of the language of neo-Darwinism is a required part of biological education.)[5]

Hamilton was a man against the grain, questioning established belief, bursting with ideas, always looking for universal principles across observations, defending underdogs and underdog theories, and relishing evolutionary oddities and paradoxes. (In fact, the latter is the source of some of his major ideas.) For Hamilton, nature was a big puzzle that needed solving, and he was typically unsatisfied with prevailing orthodoxy. A physical and intellectual risk taker, he loved challenges; he was never happier than on an expedition hacking a new path through the jungle or faced with an unexpected problem. He was also extremely stubborn. He retained a boyish dislike for authorities throughout his life, had little patience with formalities, and escaped people for the natural world whenever he could.

Hamilton was a man against the grain also in another sense: the "grain" or new orthodoxy created as his pioneering work opened up new academic industries. Fiercely independent and prone to thinking "outside the box," Hamilton usually was a few steps ahead of his colleagues. This meant that as his colleagues finally started catching up with his ideas and engaging with them, he had already moved on to something else—a new unexplored area where he was again the pioneer.

Hamilton died in 2000 at the age of sixty-three after an expedition to collect evidence for an unpopular theory of the origin of AIDS. This act epitomizes Hamilton. It was not his own theory, but it had all the right attributes: it was interesting, it was plausible, and it was unpopular with the establishment. Hamilton felt moral and scientific outrage that truth was being suppressed. For him, science was a pursuit of truth, no matter what.

FINDING HIS PASSION

Hamilton's mother Bettina first introduced him to the beauty and puzzles of nature and encouraged his early interest in butterfly collecting. The Hamilton family lived in Kent on Badgers Mount in the greenbelt outside London, not far from the Thames estuary, and Bill and his sisters and brothers were allowed to roam freely. Bill used to go on overnight butterfly-catching expeditions, in part to avoid household chores.[6] His father, the engineer Archibald Hamilton, was an avid inventor who invited all his children to use his various tools. Archie was famous for a road through Kurdistan whose building he had single-handedly supervised and for the book about this adventure.[7] Archie encouraged his eldest son's mathematical and puzzle-solving interests, hoping for Bill to follow in his footsteps.[8]

Both parents were originally from New Zealand, which was very much a presence in the Hamilton house. Bettina was trained as a medical doctor but decided to devote herself fully to her family. She raised six children, of whom Bill was the second, after his older sister Mary. In the family there was much discussion of books, art, and ideas. There was also an emphasis on frugality and good habits (bedtime at 10 P.M.) and a premium on creativity and self-sufficiency. Anything fancy was usually frowned on. Bill's parents encouraged risk-taking behavior in their children. His younger brother Alexander lost his life in a climbing accident at age eighteen.[9]

World War II had a big effect on the Hamilton home. Archie was in charge of training the Home Guard and making explosives. The family learned to take cover in homemade bomb shelters during the raids. The war clearly affected Bill Hamilton's attitudes toward many things, such as his later "allergy" to anything associated with totalitarian-sounding ideas, including group selection.[10] Hamilton early on decided that he did not want to become an engineer. He had discovered his true passion: nature and evolution. As a teenager, he developed a total rational and scientific worldview with no place for such things as polite small talk or religious ritual.[11]

Hamilton attended the prestigious Tonbridge School as a "day boy." Just before the start of the school year, when he was twelve years old, an accident caused by his experiments with explosives at home cut off the top of some digits on one hand and filled his body with shrapnel. Hamilton's life was saved by quick action by his mother and hospital personnel. This incident earned him considerable prestige among the boys at school, as did his fearless performance in rugby, which he played with a protective metal plate sown under his shirt.[12] School-mates describe him as stocky and friendly with a naughty streak. Lacking a musical ear, he disliked being forced to "mouth" hymns in music class, and so once put he holly on the seat of the teacher.[13] The punishment he most enjoyed at school was having to run through the countryside. Hamilton's fortes were biology and chemistry. He also showed literary gifts; winning a couple of school prizes for essays, he chose *The Origin of Species* as a prize. After this followed two difficult years of mandatory military service, during which Hamilton found himself grappling with authority and nonsensical-seeming army rules. (He took his own revenge, including serving roadkill pheasant to army officers.)[14]

At Cambridge (1957–60), Hamilton studied botany, zoology, physiology and mathematics; during his final year he read genetics. He found the official curriculum rather boring and devoted himself instead to R. A. Fisher's book *The Genetical Theory of Natural Selection*—especially after being told by his teachers that Fisher, a geneticist and statistician, had nothing to offer biologists.[15] At the time animal behavior was assumed to be group selectionist ("for the good of the species"), and evolutionary questions were hardly addressed. For Hamilton, reading Fisher was like a personal revelation, and Fisher became his hero. A burning question for Hamilton had now become the evolution of altruism. Accepting the neo-Darwinian interpretation of evolution as a change in gene frequencies, Hamilton wanted to produce an exact mathematical formulation for the evolution of altruism in a Fisherian individual selectionist spirit. (He had already found and dismissed a mathematical attempt of Haldane's to explain altruism based on group selection.)[16]

THE STRUGGLE FOR ALTRUISM

In a postwar climate sensitive to any discussion of humans and genes, however, Hamilton's peculiar-seeming quest for the genetics of altruism was hopelessly outside genetic orthodoxy. He was finally accepted for graduate study in demographics at the London School of Economics and later was assigned an additional supervisor in genetics at University College. Now followed a long and lonely struggle with the mathematics for what was to become Hamilton's famous 1964 paper on inclusive fitness.[17] He soon realized that altruism could indeed develop if the benefit of an altruistic act (such as self-sacrifice or forgoing reproduction) fell on the donor's relatives. The problem was to find a general expression for the genetic relationship of donor and beneficiary. Finally, Hamilton found an answer in Sewall Wright's coefficient of relatedness, r. What Hamilton ultimately wanted to achieve was a generalization of Fisher's Fundamental Theorem of Natural Selection to include also the fitness effects of social behavior, or as he put it, "a quantity, inclusive fitness, which under the conditions of the model tends to maximize in much the same way as fitness tends to maximize in the simpler classical model."[18]

But Hamilton's supervisors were growing increasingly skeptical about his efforts. They doubted that this kind of research merited a Ph.D. It was too mathematical, and did not have examples. Did this kind of phenomenon really exist in nature? Knowing that he had something important, Hamilton made two decisions. One was to ignore his supervisors' myopic verdict and turn directly to the scientific community. The other was to collect overwhelming evidence to prove he was right. He needed to publish a journal article.

The best chance of having his article accepted was to submit it to the *Journal of Theoretical Biology.* He also prepared a short, readable overview, which he sent to *Nature,* which immediately rejected it. (This marked the beginning of his lifelong tug-of-war with that journal.) *American Naturalist* ended up publishing the short paper, which contained his two seminal ideas: "Hamilton's Rule" (altruism can evolve if the cost to the donor is less than the benefit to the recipient, taking into account the coefficient of relationship between donor and recipient) and "the gene's eye view" (whereby one anthropomorphizes the gene as an intentional actor and figures out what would be in its "interest" to do to ensure its spread).[19]

But enough with modeling and uncomprehending advisers! He wanted to learn from the true professionals and from nature. Hamilton arranged to go to Brazil to work on insects with the legendary Warwick Kerr. This trip was the beginning of the young naturalist's lifelong love affair with Brazil and the Amazon, to which he was to return again and again. This is also where he elaborated many of the examples for what was to become his famous 1964 paper, including the explanation of sterility in social insects. A referee had told him that the paper had to be made more convincing by restructuring and inclusion of worked-out examples.[20] This paper, with its theory of inclusive fitness (or kin selection, or "kinship theory"), was the principal basis for the many national and international honors and recognitions bestowed on Hamilton. For the first decade or so of its existence, however, "The Genetical Evolution of Social Behaviour" was seldom quoted.

DOUBLE REBELLIONS

In 1964, with two publications but no Ph.D., Hamilton landed a job as a junior lecturer at Imperial College, working at its field station, Silwood Park, near Ascot. Pioneering the tool of computer simulation combined with studies of actual insect mating behavior, he found that Fisher's famous 1:1 sex ratio was not universally valid after all. In cases of inbreeding (as with many insects), nature favors an often strongly female-biased sex ratio—sometimes skewed as much as 50:1. (This was later to make him wonder why males were needed at all and what the reason was for sexual reproduction.)[21] This paper became an exemplar for the possibility of prediction in evolutionary biology. The following year Hamilton finally obtained his Ph.D. on the basis of his published papers. His strategy of bypassing his advisers had worked. Meanwhile Hamilton had married Christine Friess, a student of dentistry and the daughter of a Lutheran pastor (a refugee from Nazi Germany).

Hamilton's career was clearly on the upswing. A short 1971 paper, "Geometry for the Selfish Herd," became one of *Current Contents'* most often-cited papers of all time.[22] There was even more potential mileage in his friendship and collaboration with the self-taught eccentric American genius George Price. Price had suggested a way to rederive the idea of inclusive fitness, with the help of his own newly developed covariance formula, which was applicable to every kind of natural selection, including group selection. Group selection was, of course, something that Hamilton early on had declared himself "allergic" to (and he had declared Haldane's formulae unworkable), but Price was undoubtedly correct, and his odd new formulae were strokes of genius. These absolutely needed to be published, along with Hamilton's new paper building on the work of Price! Why not rebel together against *Nature's* prickly publication policy? The coup succeeded. Hamilton and Price, by referring to each other, found a way to get *Nature* to publish both of their papers.[23]

The important point about the rederivation was that it showed that inclusive fitness was broader than kin selection because it could extend also to unrelated individuals. The simple criterion was that the benefit of altruism should fall on individuals that were altruists themselves (that is, carried genes for altruism), but these could exist anywhere in a population. (This was later developed by Dawkins as "the green beard hypothesis.") Hamilton was now able to clearly show that kin selection was not opposed to group selection, as some seemed to believe, but that there was a continuum between kin and group, and inclusive fitness covered both.[24] This constituted something of a "double rebellion" as far as Hamilton was concerned. Hamilton II, with his rederived inclusive fitness, was found rebelling against Hamilton I, with his dislike of group selection, at the same time he was rebelling against those who saw kin selection as an alternative to group selection.

It is remarkable how little attention was originally paid to Hamilton's rederivation (or to the contribution of Price, who died by his own hand in 1974). (Most recently David Sloan Wilson and Elliott Sober have pressed this point; its fiercest opponents are probably John Maynard Smith and Richard Dawkins.)[25] One reason may be that the rederivation was hidden away in a lengthy chapter of Hamilton's in an anthropology conference book. Moreover, his chapter makes for unusual reading. Titled "Innate Social Aptitudes of Man," it suggests among other things that in world history, invasion from barbarians may have served to increase the amount of altruism and self-sacrificial daring in the population and in this way brought creative impulses to established societies. It also discusses the important concept of multilevel or hierarchical selection, but Hamilton chooses human tribal organization to illustrate the point that a trait that is beneficial at one level may be detrimental at the next. For anthropologists this was clearly anathema. One convenient way to resolve the problem was to label his contribution "racist" and then ignore it. This is largely what seems to have happened.[26] A rebel can rebel, but he cannot force people to listen.

THE PARASITE PARADIGM

In the mid-1970s Hamilton's citation rate was growing fast, especially after the publication of E. O. Wilson's *Sociobiology: The New Synthesis* and Richard Dawkins's *Selfish Gene* and the new interest in sociobiology triggered by the sociobiology controversy.[27] But Hamilton was still a junior lecturer at Imperial College, having been passed over for promotion. This was the last straw. He knew there were complaints about his teaching, but he also knew the worth of his research. Hamilton made inquiries in the United States, ending up as Museum Professor of Evolutionary Biology at the University of Michigan at Ann Arbor, where he had intelligent and challenging graduate students.

He now threw himself into a new obsession, as intense as his pre-occupation with kin selection. His focus now was the evolution of sexual reproduction, Darwin's second unsolved problem. Why was there sex? Because sexual reproduction is more costly than asexual reproduction, it must confer some advantage. But what was it? The topic of sex had again come to the fore in academic circles, but none of the existing explanations satisfied Hamilton. As usual, he had to find his own answer. Using computer simulation to discover what types of environmental changes would favor sexual more than asexual reproduction, he realized that a good reason for sex could be the need to escape parasites (see figure 16.2). Sexual recombination would be capable of creating ever new types of DNA, fooling rapidly reproducing invaders that tried to mimic the current DNA.[28] Parasite-host coevolution could also answer the question why there was such diversity in human health.

The theme of parasites was further extended to sexual selection, Darwin's second mechanism for natural selection. What did females want? Hamilton and his student Marlene Zuk argued that females wanted healthy males. Biologically costly ornaments, colorful plumes, and complicated songs in birds served as honest signals of health. If this was true, colorful species ought to be particularly parasite-prone. Hamilton and Zuk's 1982 paper showed this relationship to hold up, fueling wide enthusiasm for the parasite paradigm as a new framework for research.[29]

OXFORD DON

The move to America may have been made in sheer protest, but it worked well as a career move, too. In the early 1980s, Hamilton received an offer to return to Britain, this time as a Royal Society Research Professor. Behind this offer was his former head from Imperial College, Richard Southwood, now in charge at Oxford University. He had already judiciously overseen the election of Hamilton to the Royal Society in 1980. Southwood, a naturalist himself, was a staunch believer in Hamilton. This new position would enable Hamilton to do exactly what he wanted: pursue his research with hardly any teaching and with a generous travel budget. The Hamilton family—now augmented by one more daughter—moved to Wytham village, near Oxford.[30] By way of explanation, Hamilton said that he was getting tired of political correctness.[31]

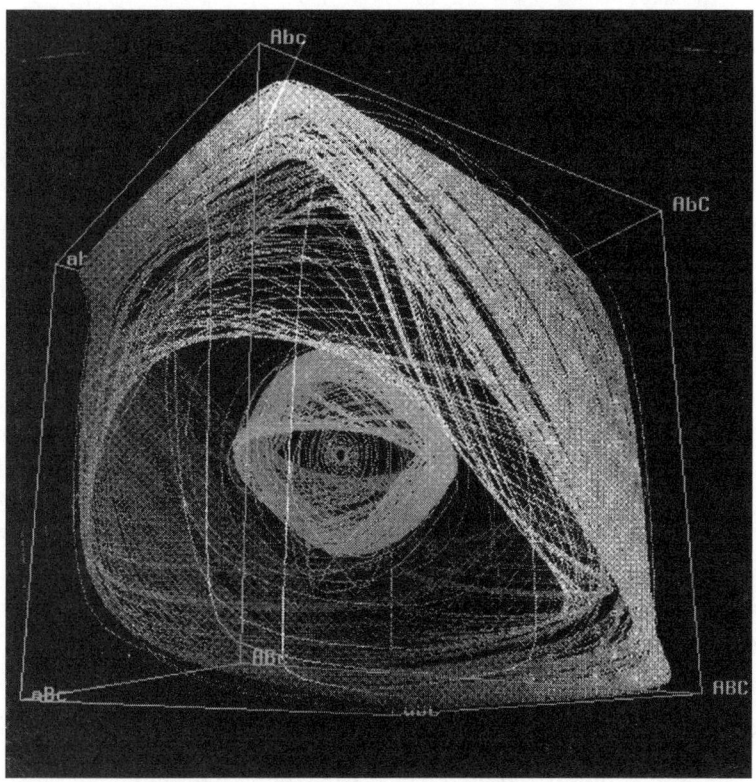

Figure 16.2. The "computer artwork" from Hamilton's simulations of host-parasite cycling. (The original computer graphic shows pink surrounded by green on a bright blue background.) Hamilton described this picture in the following terms: "Here's an example of the sort of cycle and sort of picture I produce, but not a nice one. I am always hoping to discover the face of God as I generate these cycles, but here I seem to have discovered a rather nasty supercilious eye that watches me. The green trajectory is of the parasite population: it swings wider because of having several generations to each generation of hosts. The pink is the host population, nicely herded to near the centre by the sheep-dog-like parasite population racing around, and consequently with no danger of any allele going extinct, which of course I like" (William Hamilton to Naomi Pierce, October 21, 1986).

Hamilton spent much of the 1980s refining his parasite-avoidance theory of sex, later termed the Parasite Red Queen theory.[32] He had now developed his simulation models under increasingly realistic conditions with the help of game theorist Robert Axelrod and a computer programmer. Their evolving simulated populations involved a good number of hosts and parasites and also several genetic loci. Hamilton and Axelrod believed that either *Nature* or *Science* would be happy to publish their final model "with its realism and its dramatic success under conditions others had deemed impossible for it."[33] How wrong they were! Publishing would become the greatest difficulty of Hamilton's whole career. The situation was an unpleasant instance of déjà vu: why did no one seem to understand? Why were models that were obviously much less carefully worked out being published in prominent places? After a three-year struggle with the referees of *Nature* and *Science* (the paper was rejected twice by *Nature* and once by *Science*) and a note of disinterest from the editor of the Royal Society journals, it was finally published in the *Proceedings of the National Academy of Sciences.*[34]

If we take into account the characteristic Hamiltonian pattern, however, this rejection was perhaps not very surprising. As soon as he embarked on something new (in this case host-parasite coevolution), he almost automatically left his faithful followers from his earlier stage (sociobiology, behavioral ecology) behind. At the same time, by moving to a new field, he was obviously putting himself willy-nilly at the mercy of new referees (in the case of parasites, his paper might get into the hands of baffled medical parasitologists). In other words, the story of uncomprehending referees was almost bound to repeat itself. Moreover, unlike many scientists, Hamilton did not like to keep things simple. He naturally operated, as it were, in three-dimensional space. This sometimes made his reasoning difficult to follow, because too many things were being varied at once. Or he might be introducing too many novel things in the same paper—new theory, new methodology, and new natural history evidence—making readers feel like rock climbers desperately searching for footholds.

Hamilton believed that the problem lay in the paper's attempt to include humans in the models: "Somehow the choice and mention of Homo appeared to annoy several of our reviewers almost as if they had been creationists," he noted, whereupon he advised any young reader: "Never expect honour for realism about humans except from the fewest of the few—even among evolutionists."[35] Today, Hamilton's

Parasite Red Queen theory of sex is an important research paradigm. The major rival theory suggests that sex evolved to eliminate danger-ous mutants. Hamilton himself concluded that both theories might be right in part.[36]

If Hamilton was surprised by the resistance of *Science* and *Nature,* he was totally unprepared for the sudden friendly fire coming from his own Zoology Department. In 1989 his colleagues at Oxford took the course of formally questioning the methodology and objectivity of Hamilton and Zuk's much-cited mate choice and parasite study.[37] For Hamilton, this was ungentlemanly behavior, if not high treason. He seriously considered an offer to move to Princeton and traveled there for a visit with his whole family.[38]

UNTRIVIAL PURSUITS

Hamilton was generating new ideas and theories all the time. Some interesting theories of Hamilton's had to do with why the leaves turn red in the autumn, the resistance of plants to herbivores, why wind-borne fruits color as they mature, why we like plants and animals, and why males prefer females with wide hips.[39] His credo was that ideas, however wild, should be given a chance before being discarded. This is why he took it as a challenge to look into James Lovelock's unortho-dox Gaia theory, certainly not a favorite among biologists in the mid-1990s. He was delighted when in 1996 he and Tim Lenton were able to develop the cloud formation hypothesis, which argued that Gaia could be the product of microorganisms' wish to propagate themselves through the air.[40] Those around Hamilton soon realized that he had theories about practically everything, including everyday life. He was not always right. (One of his dangerously mistaken theories was that he was immune to malaria.)[41] Hamilton did live in a world of models and explanations and had a perhaps obsessive need to theorize.

Bill Hamilton was a deeply moral person. He was aware of the dark side of human nature and felt that it was important to know the truth about it. This was also one reason why he worked hard to find the evolutionary rationale for positive human traits: altruism and cooperation. He sought the solutions to these problems in genet-ics and evolution, not in social or political organization. Later he worried about the degeneration of the human genome. He foresaw accumulations of mutations because of increased medical treatment; in his view the world was becoming one big hospital, a technological superorganism providing quick fixes. This would have serious long-term consequences for the human race and its ability to withstand a potential future crisis. He looked at mankind quite dispassionately as a biological population.[42]

THE FINAL DEFIANCE

In the 1990s Hamilton increasingly spoke out against what he saw as suppression of free thought and serious inquiry. He felt, not without reason, that he had the right to be listened to, having by then received Sweden's Crafoord Prize and the prestigious Kyoto Prize, as well as several other honors and medals. But it is also a matter of who the audience is and how one presents things. Hamilton was never very good with this, judging from reports about his lecture style and time management. Someone could have advised him in 1998 that the right way to convince the Pontifical Academy in the Vatican about population control and the importance of natural selection was certainly not by using shock therapy. (He would not have taken the advice.) The result was a total fiasco.[43]

In the early 1990s Hamilton tried in vain to advocate the need for testing the controversial polio vaccine (OPV) theory of the origin of AIDS. According to this theory, HIV originated from SIV, a chimpanzee virus found in contaminated polio vaccine in Africa in the early 1950s. Hamilton had been alerted to this by the journalist Edward Hooper, then working on *The River,* a book detailing the way this contamination and proliferation might have actually happened, and found the theory increasingly plausible.[44] For Hamilton, science was about truth and rationality, and testing all the alternatives, whatever the consequences. The notion of truth being deliberately suppressed enraged him. Using his influence, Hamilton tried to have the original vaccine samples released. He also wrote several letters to *Science* and *Nature,* asking for a general discussion about the origin of AIDS and its wider implications. His letters were rejected.[45] Hamilton recognized the situation only too well. (In this case, however, the issue was as much about science and truth as it was about the possibility of a major class-action law suit.)[46]

Undeterred, Hamilton decided to test the theory himself. His idea was to go to Congo to collect chimp feces in the suspected area around the original vaccination camp. An expedition with Ed Hooper did not yield enough data for analysis, so Hamilton decided to go back, assisted by two young companions. The second trip was to prove fatal. Taken ill with malaria, Hamilton was quickly given antidotes in Africa and rushed to England. It soon turned out that he had massive internal bleeding, causing coma and eventual death. In characteristic fashion, Bill Hamilton had entertained iconoclastic thoughts even about his own death, including a scenario in which he would be laid out in the Amazon jungle to be eaten by burying beetles.[47] He probably enjoyed the thought as much as its perceived shock value. He ended up being buried in ordinary fashion in Wytham village.

Hamilton was physically fit and proud of it. Visibly beating the process of aging, he hiked, climbed trees, and jumped over fences at any opportunity. He was also taking what his friends thought were unnecessary risks. One case in point was his skating on thin sea ice in Finland in 1994, causing a whole conference crowd to hold its breath.[48] He was already known for reckless bicycling, having once sailed through the back window of a car in Oxford, and for swimming in piranha-infested Amazon waters.[49] As a personality he loved challenges and finding clever solutions. But he disliked going to doctors and refused to recognize any debility in himself.[50] At the time of his death Hamilton was as busy as ever. He had found a new species of parasitized wasps. He was looking forward to going to a conference in the Bahamas about science and religion (for which he had written an outrageous paper), he had bought the nets for a trip to Brazil to study parasites in birds, and he was already planning volume 3 of his collected works.[51]

Before going to Congo, Hamilton had convinced the Royal Society to call a special urgent discussion meeting about the origin of AIDS.[52] When the meeting was finally convened, the vaccine theory was declared incorrect based on the available evidence. Had Hamilton been there, he would have fought. Most likely he would have stressed that the important thing was the ease with which the virus passed species boundaries—from SIV to HIV, whatever the mode of transmission—and the warning this implied for mankind in regard to such things as xeno-transplantation and future pandemics.[53] He would also most likely have panned modern medicine for selling out to the pharmaceutical industry.[54]

WHAT KIND OF REBEL WAS HAMILTON?

We know that he often thought of himself as a rebel. When misunder-stood or rejected, he liked to compare himself to Giordano Bruno and Galileo, speaking up for the Truth. On a more prosaic note, Hamilton also very much enjoyed reading Frank Sulloway's *Born to Rebel.*[55] He fitted that book's thesis—he was indeed the second-born, the one in a family who is most likely to rebel.[56] Let us do a quick inventory. Bill Hamilton disliked authority, hierarchy, taboos, organized piety, and the growing dependence of science and medicine on profit-seeking industry. He wanted open discussion and disliked suppression of truth. He disliked political correctness and the general sensitivity about applying evolutionary theory to humans. He also liked breaking rules, at least in small ways, and liked shocking people's beliefs. And he did act on each of these beliefs—by his pen or in person.

But this is not what Hamilton is best known for. He was foremost a scientific rebel, a deliberate challenger of prevailing paradigms. First there was the postwar good-of-the-species group selection idea, which he supplanted with kin selection (kinship theory); later there was the mutation elimination theory of the origin of sex, for which he offered the alternative Red Queen parasite avoidance explanation. In his role as scientific trailblazer, he introduced new ways of thinking, provid-ing the necessary theories and tools for others to follow. Throughout his academic life, Hamilton kept moving from field to field, preferring to be a path breaker, the thing that he did best. In this way, he kept going against the grain—including the grain that he created himself by opening up ever-new scientific industries. There is something of the Red Queen about Hamilton's whole approach. But in many re-spects, Hamilton simply behaved as many scientists would—if they only could—relentlessly pursuing his goals, no matter what.

Bill Hamilton has had an enormous influence on our understand-ing of how evolution works. He contributed to a number of scientific fields—sociobiology, ethology, ecology, animal behavior, and genetics —and inspired new fields such as evolutionary medicine and evo-lutionary psychology. His courage in coming up with odd, even unpopular-seeming ideas emboldened younger colleagues to follow in his footsteps and to dare to take risks themselves, especially as they saw how their hero, after meeting with resistance, eventually found the recognition that he deserved.

Hamilton had unusual gifts and idiosyncrasies, which he was able to put to good use, and was able to overcome obstacles to achieve what he wanted. But we have to look at the larger context, too. All his life he was fortunate to find sponsors for his often iconoclastic quests. Meanwhile we know that he was not a charismatic self-promoter; far from it. This does say something for the older academic system, which allowed space for unusual brilliance and for students (and faculty) who did not fit the mould. Would someone like Bill Hamilton have succeeded in today's academic system? Had Hamilton become a carpenter or a teacher (his Plan B), he would certainly have continued on his self-taught course, pursuing his exploration of nature and theorizing about evolution. But would he have had his papers published, would he have been taken seriously, and would we know about him today? Even rebels need a framework within which their rebellion will make sense.

NOTES

1. William D. Hamilton, "The Genetical Theory of Social Behavior," pts. 1 and 2, *Journal of Theoretical Biology* 7 (1964): 1–16, 17–32.

2. Alan Grafen, "Obituary of William D. Hamilton," *Guardian,* March 9, 2000; Olivia Judson, "Darwin's Heir," *Economist,* March 9, 2000.

3. Bernard Barber, "Resistance of Scientists to Scientific Discovery," *Science* 34 (1961): 596–602.

4. Descriptions of his feelings at different times of his career can be found in the autobiographical essays accompanying his collected works. See William D. Hamilton, *Narrow Roads of Gene Land: The Collected Papers of W. D. Hamilton,* vol. 1, *Evolution of Social Behaviour* (Oxford: Freeman, 1996, reissued by Oxford University Press, 1997) and William D. Hamilton, *Narrow Roads of Gene Land: The Collected Papers of W. D. Hamilton,* vol. 2, *Evolution of Sex* (Oxford: Oxford University Press, 2001). See also Ullica Segerstrale, *Defenders of the Truth: The Battle for Science in the Sociobiology Debate and Beyond* (Oxford: Oxford University Press, 2000), chap. 4.

5. For the neo-Darwinian revolution, see, e.g., Ullica Segerstrale, "Neo-Darwinism," in *Encyclopedia of Evolution,* ed. Mark Pagel, 107–110 (Oxford: Oxford University Press, 2002).

6. William D. Hamilton, "Between Shoreham and Downe: Seeking the Key to Natural Beauty," Inamori Foundation Kyoto Prize Commemorative Lecture (1996). (Republished in *Narrow Roads of Gene Land: The Collected Papers of W. D. Hamilton,* vol. 3, *Last Words,* ed. Mark Ridley [Oxford: Oxford University Press, 2005].) See also Mary Bliss, pers. comm.

7. Archibald Hamilton, *The Road Through Kurdistan: The Narrative of an Engineer in Iraq* (London: Faber and Faber, 1945).

8. Hamilton, "Between Shoreham and Downe."

9. Mary Bliss, pers. comm.

10. Hamilton, *Narrow Roads of Gene Land,* 1:354.

11. Hamilton, "Between Shoreham and Downe"; Martin Jacoby, "Bill Hamilton at School" (manuscript based on interviews with contemporaries), available at the William D. Hamilton Archive, British Library, London.

12. Mary Bliss, pers. comm.; Jacoby, "Bill Hamilton at School."

13. Helen Hamilton, pers. comm.

14. Mary Bliss, pers. comm.

15. Hamilton, *Narrow Roads of Gene Land,* 1:21.

16. Ibid., 1:22.

17. Hamilton, "Genetical Theory of Social Behavior."

18. Hamilton, *Narrow Roads of Gene Land,* 1:38. For more on Hamilton and Fisher, see Alan Grafen, "William Donald Hamilton," *Biographical Memoirs of Fellows, Royal Society of London* 50 (2004): 109–132.

19. William D. Hamilton, "The Evolution of Altruistic Behavior," *American Naturalist* 97 (1963): 354–356.

20. Hamilton, *Narrow Roads of Gene Land,* vol. 1, chap. 2.

21. William D. Hamilton, "Extraordinary Sex Ratios," *Science* 156 (1967): 477–488. Hamilton's idea of an "unbeatable strategy" in this paper inspired the idea of an "evolutionarily stable strategy," or ESS, first presented in John Maynard Smith and George Price, "The Logic of Animal Conflict," *Nature* 246 (1973): 15–18. This sex ratio paper also introduced the important idea of intragenomic conflict. For later work, see David Haig, *Genomic Imprinting and Kinship* (Brunswick, N.J.: Rutgers University Press, 2002).

22. William D. Hamilton, "Geometry for the Selfish Herd," *Journal of Theoretical Biology* 31 (1971): 295–311.

23. For Hamilton's "allergy" to group selection, see *Narrow Roads of Gene Land,* 1:354. For Hamilton and Price's collusion in regard to *Nature,* see *Narrow Roads of Gene Land,* vol. 1, chap. 5. Price's covariance paper is G. R. Price, "Selection and Covariance," *Nature* 227 (1970): 520–521.

24. It was John Maynard Smith who at an early point presented kin selection as an alternative to group selection: John Maynard Smith, "Group Selection and Kin Selection," *Nature* 201 (1964): 1145–1147. For a discussion about the relationship between Hamilton and Maynard Smith, see Segerstrale, *Defenders of the Truth,* chap. 4, and *Nature's Oracle: A Life of W. D. Hamilton* (Oxford: Oxford University Press, 2008).

25. Elliott Sober and David Sloan Wilson, *Unto Others* (Cambridge: Harvard University Press, 1998). For criticism see, e.g., John Maynard Smith, "The Origin of Altruism" (review of E. Sober and D. S. Wilson, *Unto Others*), *Nature* 393 (1998): 639–640.

26. William D. Hamilton, "Innate Social Aptitudes of Man: An Approach from Evolutionary Genetics," in *ASA Studies 4: Biosocial Anthropology,* ed.

Robin Fox, 133–153 (London: Malaby, 1975); Segerstrale, *Defenders of the Truth,* chap. 7.

27. For a discussion about the citations to Hamilton's paper see Seger-strale, *Defenders of the Truth,* 87–89, and Richard Dawkins, *The Selfish Gene,* 2d ed. (Oxford: Oxford University Press, 1989), 326.

28. Hamilton, *Narrow Roads of Gene Land,* vol. 1, chap. 2. The paper that introduced the parasite paradigm to a larger scientific audience was William D. Hamilton, "Sex Versus Non-Sex Versus Parasites," *Oikos* 35 (1980): 282–290.

29. William D. Hamilton and Marlene Zuk, "Heritable True Fitness and Bright Birds—A Role for Parasites," *Science* 218 (1982): 384–387.

30. Richard Southwood, pers. comm.

31. Hamilton, *Narrow Roads of Gene Land,* vol. 2, chap. 9.

32. Matt Ridley, *The Red Queen* (Harmondsworth: Penguin), 1993.

33. Hamilton, *Narrow Roads of Gene Land,* 2:605.

34. William D. Hamilton, Robert Axelrod, and Reiko Tanese, "Sexual Reproduction as an Adaptation to Resist Parasites (A Review)," *Proceedings of the National Academy of Sciences USA* 87 (1990): 3566–3573; Hamilton, *Narrow Roads of Gene Land* 2:609. No doubt one of the reasons for the original optimism was the success they had had with an earlier collaboration: Robert Axelrod and William D. Hamilton, "The Evolution of Cooperation," *Science* 211 (1981): 1390–1396, which had won the Newcomb-Cleveland prize for the year's best paper in science.

35. Hamilton, *Narrow Roads of Gene Land,* 2:608.

36. Ibid., vol. 2, appendix to chap. 16.

37. A. F. Read and Paul Harvey, "Reassessment of the Comparative Evidence for the Hamilton and Zuk Theory on the Evolution of Secondary Sexual Characters," *Nature* 339 (1989): 618–620; William D. Hamilton and Marlene Zuk, "Parasites and Sexual Selection, Hamilton and Zuk Reply," *Nature* 341 (1989): 289–290; Hamilton, *Narrow Roads of Gene Land II,* vol. 2, appendix to chap. 6.

38. Peter Grant, "William D. Hamilton," Biographical Memoirs, *Proceedings of the American Philosophical Society* 146, no. 4 (2002): 388–394.

39. William D. Hamilton and S. P. Brown, "Autumn Tree Colors as a Handicap Signal," *Proceedings of the Royal Society B* 268 (2001): 1489–1493; William D. Hamilton and Nancy Moran, "Low Nutritive Quality as Defense Against Herbivores," *Journal of Theoretical Biology* 86 (1980): 247–254; Hamilton, *Narrow Roads of Gene Land,* 2:321, 486–487, 304, 467.

40. William D. Hamilton and Tim Lenton, "Spora and Gaia: How Microbes Fly with Their Clouds," *Ethology, Ecology and Evolution* 10 (1998): 1–16.

41. Michael Worobey, pers. comm.

42. See, e.g., Hamilton, *Narrow Roads of Gene Land,* vol. 2, chap. 12.

43. Ibid., vol. 2, preface and chap. 12.

44. William D. Hamilton, preface to Edward Hooper, *The River: A Journey Back to the Source of HIV and AIDS* (Harmondsworth: Penguin, 1999).

45. Edward Hooper, *The River: A Journey Back to the Source of HIV and AIDS* (Harmondsworth: Penguin), chap. 37; William D. Hamilton, letter to *Science,* January 7, 1994 (rejected); William D. Hamilton to Daniel Koshland, editor of *Science,* January 7, 1994; William D. Hamilton to Daniel Koshland, editor of *Science,* February 23, 1994; William D. Hamilton, "AIDS Theory vs. Law Suit," letter to *Nature,* March 11, 1994 (rejected). The letters are available on the web site of Brian Martin: http://www.uow.edu.au/arts/sts/bmartin/dissent/documents/AIDS/Hamilton94/index/html.

46. Ed Hooper, pers. comm.

47. William D. Hamilton, "My Intended Burial and Why," *Ethology, Ecology and Evolution* 12 (2000): 111–122. (Originally published in *The Insectarium* 28, 238–247, in Japanese.)

48. David C. Queller, "W. D. Hamilton and the Evolution of Sociality," Hamilton Symposium, *Behavioral Ecology* 12, no. 3 (2001): 261–263; Lotta Sundstrom, pers. comm.

49. See, e.g., Richard Dawkins, foreword to William D. Hamilton, *Narrow Roads of Gene Land: The Collected Papers of W. D. Hamilton,* vol. 2, *Evolution of Sex* (Oxford and New York: Oxford University Press, 2001), xi–xix.

50. Luisa Bozzi, pers. comm.

51. Jeyaraney Kathirithamby, pers. comm. about wasps; William D. Hamilton, "Technosuperlife: Its Forerunners and Lies," paper prepared for the Evolution, Purpose and Meaning Symposium, the John Templeton Foundation, Nassau, The Bahamas, February 4–6, 2000, available at the William D. Hamilton Archive, British Library, London; and Luisa Bozzi, pers. comm.

52. The proceedings of the meeting, which was delayed after Hamilton's death, are published as a volume of *Philosophical Transactions of the Royal Society of London B* 356 (2001).

53. See Hamilton, preface to *The River.*

54. Mary Bliss, "In Memory of Bill Hamilton: Hazards of Modern Medicine." Presented at the meeting "Origin of HIV and Emerging Persistent Viruses," Accademia Nazionale dei Lincei, Rome, September 2001. Available online at William D. Hamilton Memorial Web site, maintained by Dieter Ebert. www.unifr.ch/biol/ecology/hamilton/hamilton.html.

55. Frank Sulloway, *Born to Rebel: Birth Order, Family Dynamics, and Creative Lives* (New York: Pantheon, 1996).

56. Even if one accepts the book's thesis, however, it may be unclear what counts as first- and second-born in this case. The first-born was Hamilton's sister Mary, but he was the first-born male, a fact that was of consequence in this family.

FURTHER READING

Dawkins, Richard. *The Selfish Gene* (Oxford: Oxford University Press, 1989).

Hamilton, William D. *Narrow Roads of Gene Land: The Collected Papers of W. D. Hamilton,* vol. 1, *Evolution of Social Behaviour* (Oxford: Oxford University Press, 1997).

———. *Narrow Roads of Gene Land: The Collected Papers of W. D. Hamilton,* vol. 2, *Evolution of Sex* (Oxford: Oxford University Press, 2001).

Matt Ridley, *The Red Queen: Sex and the Evolution of Human Nature* (Harmondsworth: Penguin, 1990).

———. *The Origins of Virtue: Human Instincts and the Evolution of Cooperation* (Harmondsworth: Penguin, 1996).

Segerstrale, Ullica. *Defenders of the Truth: The Battle for Science in the Sociobiology Debate and Beyond* (Oxford: Oxford University Press, 2000).

———. *Nature's Oracle: A Life of W. D. Hamilton* (Oxford: Oxford University Press, 2008).

CHAPTER SEVENTEEN

The Iconoclastic Research Program
of Carl Woese

JAN SAPP

Carl Woese has challenged doctrines and dichotomies at the core of twentieth-century biology. At a time when microbiologists declared that a phylogenetic classification of bacteria was impossible, Woese began such a research program based on comparisons of the ribosomal RNA molecule. His methods and concepts revitalized the study of microbial evolution and taxonomy. In so doing, he proposed a fundamentally new conception of the evolution of life on Earth in direct opposition to the canonical eukaryote-prokaryote dichotomy.

Articulated in 1962 and taught in textbooks from high school to university, the transition from prokaryotes (bacteria) to eukaryotes (cells like our own, which possess a nucleus and divide by mitosis) is taken to represent the greatest discontinuity in life. For Woese, the prokaryote-eukaryote dichotomy only served to mask unsolved problems about the origin and early evolution of life. He denied that the prokaryote represented a genealogical group that gave rise to eukaryotes. Instead he argued that there are three fundamental forms of life, each representing a separate lineage that evolved from a simpler, more primitive ancestor: the progenote, ancient life forms in the process of evolving the modern molecular translation machinery. In contrast to the conventional paradigm about the origin of life, which holds that the first modern organisms were heterotrophic organisms feeding in a rich primordial soup of organic compounds, Woese suggested that the first organisms were autotrophs, which synthesized their own organic compounds.

In tracing the origins of Woese's heretical concepts, we shall see that they are rooted not in microbiology per se but rather in his further unorthodox concepts about the evolution of the genetic code. A brief description of the state of microbial classification before his

Figure 17.1. Carl Woese. Courtesy of Carl Woese.

program will set the stage for what has been called "the Woesian Revolution."

A WORLD WITHOUT EVOLUTION

Microbial taxonomy of the first half of the twentieth century was steeped in methodological debates. Could one have a classification of bacteria based on genealogies, that is, one that reflected evolutionary relationships? Or should the classification of bacteria be determinative and based solely on utility, like the organization of library books? In the *Origin,* Darwin had argued, "All true classification is genealogical; that community of descent is the hidden bond which naturalists have been unconsciously seeking, and not some unknown plan of creation, or the enunciation of general propositions, and the mere putting together and separating objects more or less alike."[1]

When constructing genealogical trees, as Darwin argued, comparisons of adaptive characters (those that were most closely related to the habits of the organisms) were the least useful because they would be relatively recent developments particular to the species or variety. More fully encompassing genealogical trees required highly conserved ancient traits, those that were far removed from everyday life. Such phylogenetic classification among plants and animals could be based on comparative anatomy and embryology and on the fossil record, but bacteria lacked complex morphological traits and developmental histories and a fossil record. Bacteria did show enormous physiological diversity, but it was difficult to discern which physiological traits were old and which were recent adaptations. Constructing a genealogical classification of them seemed to be impossible, and by the early 1920s, many bacteriologists had given up on phylogeny.[2]

Throughout most of the twentieth century, biologists knew no more about the genealogical relations of bacteria than they had in the days of Pasteur. Nor was it clear what bacteria's relations were to other organisms. Since the nineteenth century there had been biologists who recognized a fundamental difference between bacteria that lacked a true nucleus and all other organisms whose cells contained a nucleus. In 1866, Ernst Haeckel had placed the bacteria in the order Moneres (later Monera) at "the lowest stage of the Protist kingdom."[3] Bacteria, as he later argued, were as different from nucleated cells as "a hydra was from a vertebrate" or "a simple alga from a palm."[4]

Despite Haeckel's scheme, biologists generally adhered to two kingdoms, plants and animals, and bacteria were usually considered to be plants. They were often called *schizomycetes* or *schizophyta* (fission fungi, or fission plants). Edwin Copeland asserted in 1927 that a plant kingdom that included the bacteria was "no more natural than a kingdom of the stones."[5] His son Herbert proposed in 1938 that Haeckel's subdivision Monera be granted its own kingdom on the grounds that bacteria and blue-green algae were "the comparatively little modified descendants of whatever single form of life appeared on earth, and that they were sharply distinguished from protists by the absence of nuclei."[6] He therefore proposed four kingdoms: Monera, Protista, Planta, and Animalia.

Still, bacteria remained ill-defined; it was not clear what organisms were members of the group or whether they really lacked a nucleus.[7] It was also not clear how viruses differed from small bacteria. Since the nineteenth century bacteria had been isolated using filters that were permeable to toxins but impermeable to bacteria. But some infectious agents were so small as to pass through a bacterial filter. They were called "filterable viruses." Other small obligate-parasitic bacteria of the rickettsial type, barely resolvable by the light microscope, were often thought to be transitional between the filterable virus and the typical bacterium. Thus, bacteria were thought to range in size from those of the size of some algae to those of the size of filterable viruses. In 1948 the editors of *Bergey's Manual of Determinative Bacteriology* suggested that both bacteria and viruses might belong to one kingdom: the "protophytes," or first plants.[8]

The molecular biology of the gene and the deployment of the electron microscope after World War II permitted refinement in the concept of the bacterium in terms of genetics, biochemistry, and morphology. In 1957 André Lwoff at the Pasteur Institute in Paris articulated the differences between the virus and the bacterium based on molecular structure and biochemistry. The virus contained either RNA or DNA enclosed in a coat of protein, and it possessed few if any enzymes, except those concerned with attachment to and penetration into the host cell. The virus was not a cell, and it did not reproduce by division like a cell. "Viruses should be treated as viruses," Lwoff concluded, "because viruses are viruses."[9] There were no known biological entities that could be described as transitional between a virus and a bacterium.

The terms *prokaryote* and *eukaryote* were introduced to the English-reading world by Roger Stanier and C. B. van Niel in their landmark 1962 paper titled "The Concept of a Bacterium."[10] They introduced their paper by noting that "the abiding intellectual scandal of bacteriology has been the absence of a clear concept of a bacterium."[11] Because of progress in knowledge since the Second World War, they proclaimed, "It is now clear that among organisms there are two different organizational patterns of cells, which Chatton (1937) [*sic*] called, with singular prescience, the eukaryotic and prokaryotic type. The distinctive property of bacteria and blue-green algae is the prokaryotic nature of their cells. It is on this basis that they can be clearly segregated from all other protists (namely, other algae protozoa and fungi), which have eukaryotic cells."[12]

Still, Stanier and van Niel could do little more than define the prokaryote negatively in relation to the eukaryote. Eukaryotes possessed a membrane-bound nucleus, a cytoskeleton, an intricate system of internal membranes, mitochondria that perform respiration, and in the case of plants, chloroplasts. Bacteria (prokaryotes) were smaller and lacked all of these structures; they lacked mitosis: "The principal distinguishing features of the procaryotic cell are: 1[.] absence of internal membranes which separate the resting nucleus from the cytoplasm, and isolate the enzymatic machinery of photosynthesis and of respiration in specific organelles; 2[.] nuclear division by fission, not by mitosis, a character possibly related to the presence of a single structure which carries all the genetic information of the cell; and 3[.] the presence of a cell wall which contains a specific mucopeptide as its strengthening element."[13]

Just as there would be no transitional forms between viruses and bacteria, there would be no transitional forms between bacteria and all other organisms. Stanier, Michael Douderoff, and Edward Adelberg declared in the second edition of *The Microbial World* (1963), "In fact, this basic divergence in cellular structure, which separates the bacteria and blue-green algae from all other cellular organisms, represents the greatest single evolutionary discontinuity to be found in the present-day world."[14]

In the revitalization of the prokaryote-eukaryote dualism, Stanier and his colleagues fully admitted that bacterial phylogeny concerning morphology and physiology was impossible. But they *were* confident that bacteria were derived from one stock, that they had a common ancestry. In *The Microbial World* they wrote: "All these organisms share the distinctive structural properties associated with the pro-karyotic cell (Chapter 4), and we can therefore safely infer a common origin for the whole group in the remote evolutionary past; we can also discern four principal sub-groups, blue-green algae, myxobacteria, spirochetes, and eubacteria, which seem to be distinct from one another. . . . Beyond this point, however, any systematic attempt to construct a detailed scheme of natural relationships becomes the purest speculation, completely unsupported by any sort of evidence."[15] Thus they concluded that "the ultimate scientific goal of biological classification cannot be achieved in the case of bacteria."[16]

It might seem paradoxical to maintain that one could not have a genealogical classification of bacteria based on structure and yet assert on the same basis that prokaryotes were a natural phylogenetic group. But no microbiologists of the 1960s questioned it. If bacteria were polyphyletic, then the category prokaryote would have no evolutionary or phylogenetic meaning. It would be an illusion, as was the nineteenth-century assumption that rhinoceroses, hippopotami, and elephants descended from a common large ancestor. It is now known that each of these animals evolved from a separate small ancestor, and the common ancestor of all of them was small and slightly built, with presumably thin skin and fur. Those animals share convergent characters that originated several times. When defined negatively, the taxon prokaryote might well be similar to the grouping "invertebrate," which includes such diverse creatures as insects and worms. Nonetheless, the articulation of the prokaryote-eukaryote dichotomy in the 1960s had a major impact: leading microbiologists immediately assigned to them the rank of superkingdoms: Prokaryotae and Eukaryotae.[17]

THE GENETIC CODE IN EVOLUTION

New paradigms often emerge from "outsiders," scientists who enter a field from a different discipline. Carl Woese's development of a new molecular-based microbial phylogenetics is a telling example. Woese was not immersed in the doctrines of microbiology about the impossibility of a genealogical classification of bacteria, nor was he among the main groups of molecular biologists of the 1960s. His initial interest was in the evolution of the genetic code: the correspondence between nucleic acid structure and protein structure.

He was born into a middle-class family in Syracuse, New York, in 1928, the elder of two children. He graduated from Amherst College in 1950, and three years later he completed a Ph.D. in biophysics at Yale University, working under the supervision of the distinguished physicist Ernest Pollard (1906–1997), founder of the biophysics program at Yale. Pollard had a famed intellectual pedigree: the physicists at Cambridge University. His mentor James Chadwick (1891–1974) had discovered the neutron and was awarded a Nobel Prize in physics in 1935. Chadwick's mentor, Ernest Rutherford (1871–1937), who postulated the existence of the atomic nucleus, was awarded the Nobel Prize in chemistry in 1908. Pollard made significant contributions to the development of radar at the radiation labs in Cambridge, Massachusetts, during the Second World War. After the war he began to apply physical principles to biology, using various types of ionizing radiation to determine the size and shape of macromolecules. Pollard was an inspiration to his students, always flowing with ideas. His scientific tradition was one of tremendous individualism, a quality that was apparent in Woese's own maverick career.

Woese completed his doctoral research in 1953, the year James Watson and Francis Crick published the double helical structure of DNA, for which they (with Maurice Wilson) were awarded the Nobel

Prize. Woese's research had concerned the inactivation by ionizing radiation and heat of viruses such as those responsible for influenza and New Castle disease. He subsequently entered medical school at the University of Rochester—for "two years and two days."[18] After those two days on the pediatrics ward, he quit. He returned to Pollard's lab, where he spent the next five years doing radiation studies, this time on bacterial spores.

In the fall of 1960 Woese signed on at General Electric. He worked in their main laboratory, the Knowles Lab in Schenectady. The Knowles Lab was equivalent to the Bell Labs at AT&T, and it took pride in its luminaries such as the famed physical chemist William David Coolidge (1873–1975), the inventor of the x-ray tube, and Irving Langmuir (1881–1957), the surface chemist and cloud seeder who was awarded the Nobel Prize in 1932. Woese was given a large laboratory on the top of a building where Langmuir once worked on cloud seeding.

While setting up his lab and waiting for an ultracentrifuge and other equipment to arrive, Woese turned to the most fundamental outstanding problem in molecular biology: cracking the genetic code. Deoxyribonucleic acid (DNA) was a long molecule comprised of thousands of nucleotide subunits made up of four bases: guanine, cytosine adenine, and thymine (G, C, A, and T). It was understood that there was a triplet code (a codon) for each of the twenty amino acids, but it was not known how a specific codon, say, CCC, corresponded to a specific amino acid (proline).[19]

This cryptographic problem was essentially solved by 1964. Codon assignments for amino acids were deciphered. For the famed founders of the field, the cracking of the code signaled the beginning of the end of the great conceptual era of molecular biology, but for Woese the genetic code raised fundamental evolutionary questions: How did such codon assignments come to be? What was the reason for the correspondence between nucleotide triplet (codon) in nucleic acid and its amino acid? Why did UUU (uracil) code for phenylalanine? In posing such questions, Woese's thinking was already outside what was rapidly becoming dogma in molecular biology.

There were essentially two ways of conceptualizing the correspondence between codon and amino acid. It might have emerged by pure chance; the historical relation between a codon sequence and an amino acid would in that case be meaningless. The code was simply a historical accident. This view was called "the frozen accident theory." Francis Crick was its chief advocate.[20] The genetic code was a language, and even if it was due to chance, once the relation between sign and meaning was established, Crick argued, it would be difficult to change.

Woese disagreed.[21] Certainly it was difficult to see why nature adopted one particular correspondence rather than another, why one nucleic acid triplet "means" a certain protein subunit and not another. Though one could not see any reason for the correspondence today, that did not mean it did not exist at all. Perhaps primitive organisms had some constraints of structure that molecular biologists knew nothing about. The precise univocal correspondence between each group of three nucleic acid subunits and each protein subunit could not have arisen in a single stroke, of course. The translation machinery was complex, involving different types of RNA and many protein-enzymes. In this process, DNA was transcribed to RNA, which carried the message to the ribosomes (complex structures comprised of RNA and protein), the sites of protein synthesis. There transfer RNAs, each carrying a specific amino acid, lined up amino acids like beads on a string in accordance with the RNA codon sequence. That convoluted process, Woese argued in 1965, had to have evolved in steps, with improvements to its speed and accuracy accruing over time.[22]

His big career break occurred in 1962, when he spent four months at the Pasteur Institute, where Jacques Monod and François Jacob had just completed a synthesis of their Nobel Prize–winning work concerning gene regulation in *E. coli*. Sol Spiegelman, one of the leading American molecular geneticists, passed through, and he was looking for a molecular biologist to fill a vacant position at the University of Illinois at Champaign-Urbana. Working with bacteria and bacteriophage, Spiegelman and his collaborators had just published their famed techniques of RNA-DNA hybridization.[23]

Spiegelman found Woese's ideas about the evolution of the genetic code intriguing and invited him to Illinois for a visit. Woese arrived in Urbana in the fall of 1963, and while he was chatting with Spiegelman in his office, the department head, Kim Atwood, arrived and offered him a position. He was granted tenure immediately. He moved to the University of Illinois in February 1964. That position gave him the continued freedom to pursue his evolutionary questions far from the main streams of molecular biology, evolutionary biology, and microbiology.

During his early years at Illinois, Woese wrote *The Genetic Code* (1967).[24] It appeared two years after James Watson's *Molecular Biology of the Gene* (1965).[25] Watson's book became standard reading. It helped form the conceptual framework for molecular biology. There was another striking difference between the two. Unlike Woese's book, Watson's made virtually no reference to evolution. Woese's aim was to create an experimental system for investigating how the complex mechanism for translating nucleic acids into the amino acid sequences of proteins had evolved. Given a bacterial phylogeny, he reasoned, perhaps one might be able to track life back to its universal ancestor, which might not possess fully evolved, modern translation machinery.

THE RIGHT MOLECULE FOR THE JOB

The field of molecular evolution was emerging, and molecular approaches to taxonomy were under way in the 1960s.[26] Instead of comparative anatomy and physiology one could construct family trees on differences in the order of amino acids of proteins and nucleotides of genes. The molecular approach to phylogeny was extremely logical. Genetic mutations that either have no effect or that improve protein function would accumulate over time. As two species diverge from an ancestor, the sequences of the genes they share also diverge, and as time advances, the genetic divergence will increase. One could therefore construct genealogies and make phylogenetic trees by comparing the sequence divergence of genes or of proteins isolated from those organisms.[27]

The key to a phylogenetic classification of bacteria lay in choosing the right molecule. In 1955 the British chemist Frederick Sanger (b. 1918) had succeeded in developing techniques for amino acid sequencing and first used them to deduce the complete sequence of insulin. For this work he was awarded the Nobel Prize in chemistry in 1958.[28] Emile Zuckerkandl and Linus Pauling subsequently pioneered the use of amino acid sequence comparisons to infer primate phylogeny with data from hemoglobin sequences.[29] Walter Fitch and Emmanuel Margoliash compared amino acid sequences of cytochrome c to infer phylogenetic relationships among eukaryotes from such diverse species as horses, humans, pigs, rabbits, chickens, tuna, and baker's yeast.[30] Cytochrome c, the terminal enzyme in the respiratory chain, is located in the inner membrane of mitochondria, the respiratory organelles of eukaryotes. Cytochrome c is also present in bacteria that breathe oxygen (aerobes), but many bacteria live in the absence of oxygen. It would therefore be less useful for the deep phylogeny that Woese sought.

Woese looked to the translation machinery and considered not proteins but the RNAs that, together with proteins, comprised ribosomes, the organelles in which translation from nucleic acid to protein occurs. This great insight, the choice of ribosomal RNA (rRNA) for phylogenetic purposes, was obvious from his perspective; he sought the origins of the genetic code. From the point of view of microbial phylogenetics, rRNA also had all the right attributes. The cells of all organisms from bacteria to elephants need rRNAs to construct proteins. Therefore their similarities and differences could be used to track every lineage of life. Ribosomes are also abundant in cells; there are thousands of them in each bacterium. The ribosomes, and the RNAs which they contain, were also far removed from the usual vicissitudes of phenotypic change. Woese suspected that they would be among the best "conserved" elements in all organisms (evolving far more slowly than most proteins) and therefore would make the best recorders of life's long evolutionary descent of roughly 3.5 to 4 billion years.

The technique for his project was announced in 1965, when Sanger published a method for sequencing and cataloguing short RNA nucleotide sequences.[31] Two years later Spiegelman acquired a postdoctoral student from Sanger's lab, David Bishop, who set up the Sanger system in his lab to analyze viral nucleic acids.[32] Woese asked his own doctoral student, Mitchell Sogin, to learn the procedure from Bishop. Then, in 1969, Spiegelman moved to Columbia University to be the director of Columbia's Institute of Cancer Research and professor of human genetics and development in the College of Physicians and Surgeons. Bishop left as well. Woese acquired Spiegelman's setup and moved into one of Spiegelman's labs.

Developing the Sanger technique, Woese's laboratory separated small fragments (oligonucleotides) of ribosomal RNA and catalogued them. The long string of nucleotides (adenine, cytosine, uracil, and guanine, ACUG) was first broken into small fragments several nucleotides long ("oligo-" denotes few) by cutting at every G residue with the enzyme T_1 ribonuclease. Thus the RNA "text" was cut into short segments ending with a single G, as in AACUCG. Such pieces were short enough to be sequenced by available techniques. Sequencing of the entire 16S rRNA gene was not feasible until the late 1970s and early 1980s.

Ribosomes comprise two subunits, a smaller one slightly nestled inside a larger one; both contain RNA and protein. Woese focused on the RNA of the smaller subunit: 16S rRNA (about fifteen hundred nucleotides long). He and his co-workers made catalogue collections of oligonucleotide sequences, "dictionaries" characteristic of organisms in various taxa. Matching oligonucleotides in different bacteria could be compared to one another to determine how closely the organisms were related. The more similar the catalogues were, the more closely related the corresponding organisms.

The work was tedious; it was experimentally demanding; it required considerable expertise; and it was expensive. Woese built up a strong team of researchers and technicians. His postdoctoral fellow George Fox produced the first 16S rRNA dendrograms (phylogenetic trees). Woese grouped bacteria together phylogenetically by "a seat of the pants" method using what he called "signatures." If he found an oligonucleotide sequence that was present in about 95 percent of the organisms in some cluster of bacteria, but found almost nowhere else, then that oligonucleotide would be a characteristic of the group; all oligonucleotides of this nature would then be a signature for that cluster.

This approach would provide data that would challenge the prokaryote-eukaryote dichotomy. But Woese had already laid the conceptual foundation for that challenge before his research program got under way. The dichotomy had been founded on differences discernable by the electron microscope, but in 1970 Woese began to conceptualize what separated them at the molecular level. Based on differences in their translation apparatus, he argued that eukaryotes and prokaryotic lineages had diverged much earlier than was generally assumed and that prokaryotes may not have preceded eukaryotes. Instead, both lineages may have emerged from a more primitive life form in the process of evolving its translation machinery.[33] Woese and Fox later called these more primitive life forms "progenotes."[34] This shift away from one of the fundamental canons of biology was the prelude to Woese's conception of the archaebacteria as a third form of present-day life.

A THIRD FORM OF LIFE

By 1980 the 16S rRNA of almost two hundred species of bacteria and eukaryotes was characterized. Those results often contradicted the standard classification based on morphological similarities of bacteria.

For example, *Bergey's Manual* distinguished the gliding bacteria, the sheathed bacteria, the appendaged bacteria, the spiral and curved bacteria, the rickettsias, *Flavobacterium,* and *Pseudomonas.* Woese argued that these groups had no biological or evolutionary meaning; they were really paraphyletic, or polyphyletic, that is, they are not genealogically coherent groups. But no studies caused more controversy and attracted more interest than the discovery of the archaebacteria.

The discovery of the "third form of life" had occurred in June 1976. Woese had requested advice from many microbiologists about what organisms to analyze; many would send microbes to classify with the new technique. One of his colleagues at the University of Illinois, Ralph Wolfe, suggested that Woese try his technique on an odd group of methane-producing bacteria. Methanogens were chemo-autotrophs; they derived their energy from carbon dioxide and hydrogen. They were initially found in low-oxygen environments such as marshes and the hot springs of Yellowstone National Park. By the time Woese tested the methanogens, his laboratory had made oligonucleotide catalogues for about thirty kinds of bacteria and a few representative eukaryotes. He had created a list of oligonucleotides characteristic of all prokaryotic 16S rRNA. Methanogens were missing almost all the 16S rRNA signature sequences that he knew to be characteristic of prokaryotes. The prokaryotes could not be a coherent genealogical group.

In 1977 Woese and Fox named the new form of life, or "urkingdom," *archaebacteria* to distinguish them from true bacteria, or *eubacteria.*[35] As they wrote: "The apparent antiquity of the methanogenic phenotype plus the fact that it seems well suited to the type of environment presumed to exist on earth 3–4 billion years ago lead us tentatively to name this urkingdom the archaebacteria."[36] Over the course of the next two years, Woese's group expanded the archaebacterial urkingdom to include other organisms that lived in extreme environments: halophiles, found in brines five times as salty as the oceans, and the thermophiles *Sulfolobus* and *Thermoplasma,* found in geothermal environments that would cook other organisms.

During the late 1970s the organisms that Woese's laboratory had grouped as archaebacteria were subsequently shown by others to have certain unusual phenotypic traits in common: their cell membranes possessed lipids with a structure that differed sharply from the lipids characteristic of eubacteria; the cell walls of archaebacteria lacked the complex molecule peptidoglycan (or murien) typical of prokaryotes; and the transcription enzymes of archaebacteria possessed a structure that is characteristically very different from their counterparts in typical bacteria.

RADICAL REVISIONS

Woese and his collaborators synthesized their views in a landmark 1980 paper in *Science* titled "The Phylogeny of Prokaryotes."[37] This paper was a synthesis of several interrelated concepts that had shaped Woese's research program in the previous decade and would shape much of the emerging field of microbial phylogeny in the next two decades. The paper was an exercise in iconoclasm.

First, the proposed urkingdoms directly confronted the prokaryote-eukaryote dichotomy. That dualism had entered all biological textbooks as fundamental to biological education, but for Woese and his co-workers it was a false dichotomy, a phylogenetic deception, and a fundamental obstacle to understanding the evolution of life. As he and Fox had put it in 1977, "Dividing the living world into *Prokaryotae* and *Eukaryotae* has served, if anything, to obscure the problem of what extant groupings represent the various primeval branches from the common line of descent. The reason is that eukaryote/prokaryote is not primarily a phylogenetic distinction, although it is generally treated so."[38]

There was another reason why the dichotomy was faulty. Eukaryotes were chimeric. Stanier and van Niel had defined the eukaryote in terms of a single genome in the nucleus. That definition proved incorrect the next year when DNA was discovered in mitochondria and chloroplasts. Those organelles also possessed their own transcription machinery distinct from that of the nucleus. The idea reemerged that these organelles may have arisen as engulfed symbionts in the remote past. That notion had floated on the margins of biology throughout the twentieth century, but like bacterial phylogeny itself, it had been dismissed as idle speculation.[39] In this instance, too, the rRNA technology offered a means for resolving the issue. In the mid-1970s the rRNA technology was exported from Urbana to the laboratories of Ford Doolittle and Michael Gray at Dalhousie University, Halifax, Canada. Doolittle's laboratory and Woese's laboratory focused on chloroplast rRNA;[40] Gray's laboratory focused on mitochondrial rRNA.[41] Collectively, their results showed that chloroplasts and mitochondria had origins that were independent from each other and from rRNA derived from the nucleus. The extent to which the nuclear genome itself was chimeric remained unclear.

In Woese's scheme, prokaryotes did not lead to the eukaryotes, as they did in previous models. Instead, all three lineages diverged early from protocells, "progenotes," which were in the throes of evolving their translation mechanisms in terms of both precision and speed. The characteristic differences between archaebacterial and eubacterial cell walls, membranes, and transcription enzymes suggested that these features might be in the process of development at the progenote stage of evolution.

Woese confronted the concept that the first organisms were heterotrophs.[42] This virtually unquestioned tenet was the cornerstone of the still-accepted Oparin-Haldane scenario, according to which life emerged in a bountiful ocean rich with prefabricated organic compounds. Woese speculated that the first eubacteria may have been photosynthetic autotrophs and that the first archaebacteria may have been chemo-autotrophic methanogens.[43]

In what microbial evolutionists called "the Woesian revolution" there followed an upheaval in bacterial systematics and a major revision of texts. Classical microbiologists were generally skeptical of his approach and of his claims, but as techniques for sequencing RNA and DNA dramatically improved in the 1980s, the field of microbial phylogeny expanded. For the new generation of researchers, Woese had led the way to showing how molecular techniques could resolve the problem of bacterial phylogeny. The 16S rRNA approach had dramatically transformed microbial taxonomy from a nonevolutionary descriptive pursuit to an exciting experimental field of evolutionary biology.

The discovery of the archaebacteria signaled the great depth and diversity to be explored in the microbial world. To further emphasize that prokaryotes do not have a common ancestry and to counter the notion that the archaebacteria are "just bacteria," in 1990 Woese, O. Kandler, and Mark Wheelis renamed them the Archaea.[44] In doing so they made a formal taxonomic proposal of three "domains" of life: the Bacteria, the Archaea and the Eucarya. Woese received microbiology's highest awards and honors, including the Bergey Award from the Bergey's Manual Trust (1983) and the Leeuwenhoek Medal (1990) from the Royal Netherlands Academy. For his discovery of the Archaea, in 2003 Woese was awarded the Crafoord Prize, awarded by Royal Swedish Academy of Sciences for accomplishments not covered by the Nobel Prizes.

Revolutions are rarely complete, however. Archaea's place in nature remained contentious. Some molecular evolutionists objected that the lineage was not as old as the name implied.[45] The Archaea had many traits in common with eukaryotes, and it was possible that they emerged after eubacteria, not before them. Traditional evolutionists whose work focused on eukaryotes protested on different grounds. For them, the three-domain proposal obscured the phenomenal morphological differences between prokaryotes and eukaryotes.[46] They therefore insisted on the superkingdoms of Eukaryotae and Prokaryotae and placed the Archaea within the latter as a kingdom. The three-domain proposal was based on phylogenetic distance and common ancestry, but, they argued, it failed to consider the amount and nature of evolutionary change. No one argued this more fiercely than Ernst Mayr. For Mayr, Woese was an outsider unaccustomed to (his own) established rules of taxonomy.[47] For Woese, Mayr's was an eye-of-the-beholder approach to order and complexity, one that was to be replaced in the twenty-first century.[48]

LATERAL THINKING

Outsiders make big changes. They bring new attitudes, new ways of seeing; they develop new techniques for solving old problems and open up fields to new ones. Small changes, refinements, occur within disciplines; large-scale changes may result from the sharing of innovations between them. The development of microbial phylogenetics led by Woese vividly illustrates the saltational effects of the lateral transfer of concepts and techniques between fields. Woese was not brought up within the tradition of microbiology, nor was he a member of the group of luminaries that are so highly celebrated as the founders of molecular biology. Yet his union of those fields led to a new era in microbiology, one that exposed unquestioned assumptions and revealed new possibilities.

Woese's iconoclasm, fostered by lustrous intellectual ancestry and institutional conditions, was rooted in his views of the evolution of the genetic code and his determination to bring the "spirit of evolution" to twentieth-century biology. His heretical research program arose from his search for an experimental system with which to investigate the evolution of translation. Theory intermingled indissolubly with experimentation. His theory that prokaryotes and eukaryotes were each derived from an ancestral form in the process of developing the translation machinery set the stage for the archaebacterial concept six years later. His great insight, the choice of rRNA as the basis for microbial phylogenetics, was derived from a mixture of theoretical preparation and technological opportunity. The 16S rRNA approach was a singular initiative. It was not "in the air"; no other labs developed it independently. Woese and his collaborators had the field virtually to themselves for a decade.

NOTES

1. Charles Darwin, *On the Origin of Species,* with an introduction by Ernst Mayr, facsimile edition of 1859 (Cambridge: Harvard University Press, 1964), 420.

2. Jan Sapp, "The Bacterium's Place in Nature," in *Microbial Phylogeny and Evolution,* ed. Jan Sapp (New York: Oxford University Press, 2005), 1–52.

3. Ernst Haeckel, *Generelle Morphologic der Organismen,* 2 vols. (Berlin: George Reimer, 1866).

4. Ernst Haeckel, *The Wonders of Life: A Popular Study of Biological Philosophy,* trans. Joseph McCabe (New York: Harper and Brothers, 1905), 205.

5. Edwin B. Copeland, "What Is a Plant?" *Science* 65 (1927): 388–390, 390.

6. Herbert F. Copeland, "The Kingdoms of Organisms," *Quarterly Review of Biology* 13 (1938): 383–420, 386.

7. Sapp, "Bacterium's Place in Nature."

8. R. S. Breed, E. G. D. Murray, and A. Parker Hitchens, *Bergey's Manual of Determinative Bacteriology,* 6th ed. (Baltimore: Williams and Wilkins, 1948), 9.

9. André Lwoff, "The Concept of Virus," *Journal of General Microbiology* 17 (1957): 239–253, 252.

10. For a detailed deconstruction of the prokaryote-eukaryote dichotomy, see Jan Sapp, "The Prokaryote-Eukaryote Dichotomy: Meanings and Mythology," *Microbiology and Molecular Biology Reviews* 69 (2005): 292–305.

11. Roger Y. Stanier and C. B. van Niel, "The Concept of a Bacterium," *Archiv für Mikrobiologie* 42 (1962): 17.

12. Ibid., 20–21.

13. Ibid., 21.

14. R. Stanier, M. Douderoff, and E. Adelberg, *The Microbial World,* 2d ed. (Engelwood Cliffs, N.J.: Prentice-Hall, 1963), 85.

15. Ibid., 409.

16. Ibid.

17. See Sapp, "Prokaryote-Eukaryote Dichotomy."

18. Interview with Carl Woese, November 18, 2001.

19. Carl Woese, "Nature of the Biological Code," *Nature* 194 (1962): 1114–1115.

20. Francis H. Crick, "The Recent Excitement in the Coding Problem," *Progress in Nucleic Acid Research and Molecular Biology* 1 (1963): 163–217; Francis H. Crick, "The Origin of the Genetic Code," *Journal of Molecular Biology* 38 (1968): 367–379, 370.

21. Carl Woese, "Universality in the Genetic Code," *Science* 144 (1964): 1030–1031; Carl Woese, "Order in the Genetic Code," *Proceedings of the National Academy of Sciences USA* 54 (1965): 72–75, 74–75.

22. Carl Woese, "On the Evolution of the Genetic Code," *Proceedings of the National Academy of Sciences USA* 54 (1965): 1546–1552, 1549.

23. Dario Giacomoni, "The Origin of DNA: RNA Hybridization," *Journal of the History of Biology* 26 (1993): 89–107.

24. Carl Woese, *The Genetic Code* (New York: Harper and Row, 1967).

25. James D. Watson, *Molecular Biology of the Gene* (New York: Benjamin, 1965).

26. Michael Dietrich, "The Origins of the Neutral Theory of Molecular Evolution," *Journal of the History of Biology* 27 (1994): 21–59; Michael Dietrich, "Paradox and Persuasion: Negotiating the Place of Molecular Evolution Within Evolutionary Biology," *Journal of the History of Biology* 31 (1998): 85–111. See also G. L. Morgan, "Emile Zuckerkandl, Linus Pauling, and the Molecular Evolutionary Clock, 1959–1965," *Journal of the History of Biology* 31 (1998): 155–178.

27. Emile Zuckerkandl and Linus Pauling, "Molecules as Documents of Evolutionary History," *Journal of Theoretical Biology* 8 (1965): 357–366.

28. F. Sanger, "Chemistry of Insulin," *Science* 129 (1959): 1340–1344.

29. Zuckerkandl and Pauling, "Molecules as Documents of Evolutionary History."

30. W. M. Fitch and E. Margoliash, "The Construction of Phylogenetic Trees—A Generally Applicable Method Utilizing Estimates of the Mutation Distance Obtained from Cytochrome c Sequences," *Science* 155 (1967): 279–284.

31. F. Sanger, G. G. Brownless, and B. G. Barrell, "A Two-Dimensional Fractionation Procedure for Radioactive Nucleotides," *Journal of Molecular Biology* 13 (1965): 373–398.

32. D. H. L. Bishop, J. R. Claybrook, and S. Spiegelman, "Electrophoretic Separation of Viral Nucleic Acids on Polyacrylamide Gels," *Journal of Molecular Biology* 26 (1967): 373–387.

33. Carl Woese, "The Genetic Code in Prokaryotes and Eukaryotes," in *Organization and Control in Prokaryotic and Eukaryotic Cells,* ed. H. P. Charles and B. C. J. G. Knight, *Symposium Society for General Microbiology* (Cambridge: Cambridge University Press, 1970), 20:39–54.

34. Carl Woese and G. E. Fox, "The Concept of Cellular Evolution," *Journal*

of Molecular Evolution 10 (1977): 1–6; 2. See also Carl Woese, "The Universal Ancestor," *Proceedings of the National Academy of Sciences USA* 95 (1998): 6854–6859.

35. Carl Woese and G. E. Fox, "Phylogenetic Structure of the Prokaryotic Domain: The Primary Kingdoms," *Proceedings of the National Academy of Sciences USA* 74 (1977): 5088–5090, 5089.

36. Ibid.

37. G. E. Fox, E. Stackebrandt, R. B. Hespell, J. Gibson, J. Maniloff, T. A. Dyer, R. S. Wolfe, W. E. Balch, R. S. Tanner, L. J. Magrum, L. B. Zablen, R. Blakemore, R. Gupta, L. Bonen, B. J. Lewis, D. A. Stahl, K. R. Luerhsen, K. N. Chen, and C. R. Woese, "The Phylogeny of Prokaryotes," *Science* 209 (1980): 457–463, 457.

38. Woese and Fox, "Phylogenetic Structure," 5088.

39. Jan Sapp, *Evolution by Association: A History of Symbiosis* (New York: Oxford University Press, 1994).

40. L. Bonen and W. F. Doolittle, "On the Prokaryotic Nature of Red Algal Chloroplasts," *Proceedings of the National Academy of Sciences USA* 72 (1975): 2310–2314; L. B. Zablen, M. S. Kissil, C. R. Woese, and D. E. Buetow, "Phylogenetic Origin of the Chloroplast and Prokaryotic Nature of Its Ribosomal RNA," *Proceedings of the National Academy of Sciences USA* 72 (1975): 2418–2422; W. F. Doolittle, C. R. Woese, M. L. Sogin, L. Bonen, and D. Stahl, "Sequence Studies on 16S ribosomal RNA from a Blue-Green Alga," *Journal of Molecular Evolution* 4 (1975): 307–315; L. Bonen and W. F. Doolittle, "Partial Sequence of 16S rRNA and the Phylogeny of Blue-Green Algae and Chloroplasts," *Nature* 261 (1976): 669–673.

41. L. Bonen, R. S. Cunningham, M. W. Gray, and W. F. Doolittle, "Wheat Embryo Mitochondrial 18S Ribosomal RNA: Evidence for Its Prokaryotic Nature," *Nucleic Acids Research* 4 (1977): 663–671; M. W. Gray and W. F. Doolittle, "Has the Endosymbiont Hypothesis Been Proven?" *Microbiology Reviews* 46 (1982): 1–42.

42. Carl Woese, "An Alternative to the Oparin View of the Primeval Sequence," in *The Origins of Life and Evolution,* ed. H. O. Halvorson and K. E. Van Holde, 65–76 (New York: Liss, 1980).

43. Fox et al., "Phylogeny of Prokaryotes."

44. C. R. Woese, O. Kandler, and M. L. Wheelis, "Towards a Natural System of Organisms: Proposal for the Domains Archaea, Bacteria, and Eucarya," *Proceedings of the National Academy of Sciences USA* 87 (1990): 4576–4579.

45. Patrick Forterre, "Neutral Terms," *Nature* 355 (1992): 305; T. Cavalier-Smith, "Bacteria and Eukaryotes," *Nature* 356 (1992): 570.

46. Lynn Margulis and Ricardo Guerrero, "Kingdoms in Turmoil," *New Scientist* (March 23, 1991): 46–50; Ernst Mayr, "A Natural System of Organisms," *Nature* 348 (1990): 491.

47. Ernst Mayr, "Two Empires or Three?" *Proceedings of the National Academy of Sciences USA* 95 (1998): 9720–9723.

48. Carl Woese, "Default Taxonomy: Ernst Mayr's View of the Microbial World," *Proceedings of the National Academy of Sciences USA* 95 (1998): 11043–11046.

FURTHER READING

Sapp, Jan. "The Bacterium's Place in Nature. In *Microbial Phylogeny and Evolution,* ed. Jan Sapp (New York: Oxford University Press, 2005), 1–52.

———. "The Prokaryote-Eukaryote Dichotomy: Meanings and Mythology," *Microbiology and Molecular Biology Reviews* 69 (2005): 292–305.

———. "Two Faces of the Prokaryote Concept," *International Microbiology* 9 (2006): 163–172.

———. "The Structure of Microbial Evolutionary Thought," *History and Philosophy of Science,* in press (2007).

Woese, Carl. "Default Taxonomy: Ernst Mayr's View of the Microbial World," *Proceedings of the National Academy of Sciences USA* 95 (1998): 11043–11046.

———. "The universal ancestor," *Proceedings of the National Academy of Sciences USA* 95 (1998): 6854–6859.

———. "Interpreting the Universal Phylogenetic Tree," *Proceedings of the National Academy of Sciences USA* 97 (2000): 8392–8396.

———. "A New Biology for a New Century," *Microbiology and Molecular Biology Reviews* 68 (2004): 173–186.

CHAPTER EIGHTEEN

Stephen Jay Gould, Darwinian Iconoclast?

DAVID SEPKOSKI

To the scientifically literate public, the proposition that Stephen Jay Gould was some kind of scientific iconoclast might seem odd. Gould was, after all, one of the twentieth century's most ardent defenders of evolution and the scientific method, and he devoted countless books, essays, and reviews to fostering wider public appreciation of the truth and robustness of Darwinian evolutionary theory. A professional pale-ontologist or biologist, however, might react quite differently: Gould's scientific oeuvre was regarded by many of his colleagues with sus-picion and often outright hostility during his lifetime, so much so that the English geneticist John Maynard Smith once wrote that "the evolutionary biologists with whom I have discussed his work tend to see him as a man whose ideas are so confused as to be hardly worth bothering with," adding, "he is giving non-biologists a largely false picture of the state of evolutionary theory."[1] Depending on whom one asks, then, Gould was either a noble soldier on the side of the Darwin-ian angels or a self-promoter intent on inserting his name alongside Darwin's as the author of a major revision to evolutionary theory. Which picture is correct?

In any case, it can be fairly stated that Gould was probably the late twentieth century's most prominent iconoclast about evolution-ary theory. He himself often spoke of a "Darwinian orthodoxy" that he hoped to unseat (perhaps to create a foil for his own "pluralistic" vision), and he made no secret of his intention to revise several of the major pillars of received evolutionary theory. Gould's target was the version of Darwinism known as the Modern Synthesis, which empha-sized the gradual, uniform nature of evolution and focused on the ge-nome of living populations as the significant level of resolution. Gould objected to several major premises and corollaries of this approach, most significant among them (1) the view that phyletic evolution (the

Figure 18.1. Stephen Jay Gould. Courtesy
of Harvard University.

evolution of lineages) was always gradual in tempo and uniform in
mode; and (2) the belief that the functionality of organisms can always
be explained as a direct consequence of adaptation to environmental
pressures. This led Gould, in the first case, to reject the central Darwinian tenet (inherited from Charles Lyell) of "uniformitarianism,"
and in the second, to question the ubiquity of natural selection. As
much as he was driven, however, by a fierce and personal desire to
make a mark on the field of evolutionary biology by challenging many
of its orthodoxies, in more reflective moments he showed an almost
gravitational pull back toward convention. Despite spending most of
a career tilting with colleagues in evolutionary biology, paleontology,
and philosophy about such central Darwinian concepts as adaptation,
uniformity, and selection, to the end of his days he proclaimed his
allegiance to something he called "Darwinism."

     For most of his career, Gould was at odds with supporters of one or
more established evolutionary principles. His interpretation of Darwinism was, without doubt, idiosyncratic. Yet Gould himself was deeply
concerned with history and historical continuity: as much as he may
have hoped to reserve a special place for himself in the history of

evolutionary thought, he was also conscious of the pull of consensus, and he deeply respected tradition. He read and appreciated Thomas Kuhn's theory of revolutionary transformation in science but did not necessarily subscribe to its interpretation of scientific change.[2] He often argued that his major ideas had been developed with the expectation of supporting, not challenging, Darwinian orthodoxies. It is this "conservative" feature of Gould's thought that complicates his easy characterization as a rebel. Gould may have been iconoclastic, but he was a rebel with an uneasy conscience.

In this chapter I argue that although Gould preached an iconoclastic brand of evolutionary theory, he did not reject Darwinism, as some have contended. Gould's most iconoclastic stance was to promote the importance of *contingency* as a crucial element in the history of life: his approaches to evolutionary pattern, adaptive ubiquity, and even ideological, social, and political commitment shared a deep appreciation for the significance of nondeterministic and random factors in the causal structure of historical events. In contingency he saw a revision and expansion of Darwin's theory. In this chapter, therefore, Gould will not be presented as an anti-Darwinian rebel but rather as a "Darwinian iconoclast."

BRIEF BIOGRAPHY

Stephen Jay Gould was born in 1941 in Queens, New York to eastern European Jewish immigrants and raised in a secular household strongly flavored by his parents' working-class Marxism. In 1958 Gould left New York for Antioch College, a small, progressive liberal arts college in Yellow Springs, Ohio, whose founding president had been the liberal educator and abolitionist Horace Mann. Gould majored in geology with a minor in philosophy and spent a year in Leeds, England, where he first received formal training in paleontology. He also met the Columbia University paleontologist Norman Newell during a field course, which ultimately led him to enroll in the graduate program at Columbia following his graduation from Antioch in 1963.[3]

At Columbia, Gould's thesis topic was the biology and evolution of Caribbean snails, but he quickly developed an even deeper interest in asking large theoretical questions about evolution. Newell was one of the few American paleontologists at the time who was actively examining large-scale evolutionary patterns, and John Imbrie, who was the resident invertebrate specialist, had recently pioneered new approaches in the mathematical analysis and modeling of fossil data. But as influential as his teachers were, Gould was not content to leave his education in the hands of others. As fellow student Niles Eldredge recalls, "I always felt that 98 percent of my training in practical pale-ontology and how to think came from my fellow graduate students."[4] Gould, in fact, was the ringleader of a group of students interested in evolutionary theory, and according to Eldredge, Gould organized an unofficial seminar in macroevolution.[5] By the time he received his degree in 1967 he had developed something of a rebellious streak. Eldredge recalls Gould commenting to him at this time, "What are you supposed to do—wait until you're sixty before you can sit down and write something theoretical? We should do this now, when we're young!"[6]

Gould was prepared to follow his own advice: in 1965 he published an essay with the provocative title "Is Uniformitarianism Necessary?" in the *American Journal of Science*. The paper argued that the concept of uniformitarianism, or the belief that natural processes remain con-stant and consistent throughout Earth's history, is both anachronistic and false. In order to appreciate just how iconoclastic this statement could seem, recall that the uniformity of process, a major pillar of Charles Lyell's geology, was one of the major assumptions in Darwin's theory of evolution via gradual modification and descent.

In only five pages of text, Gould dismissed substantive uniformi-tarianism as untenable and argued that descriptive uniformitarian-ism (the belief that natural laws work constantly and inexorably) was simply an affirmation of the scientific principle of induction.[7] One needed no special name for this rule, he continued, unless one had doubts about the status of paleontology. Otherwise, "the term today is an anachronism; for we need no longer take special pains to affirm the scientific nature of our discipline."[8] When Chester Longwell, a former chair of the Geology Department at Yale University, published an arch response in the same journal (beginning "The question posed by S. J. Gould calls for brief comment in words of few syllables"), Gould's published reply breezily ignored the criticisms and happily concluded, "I am glad the apparent controversy seems to resolve itself into agreement."[9] Even as a second-year graduate student, Gould was clearly not a person cowed by position, age, or authority.

Gould's early reputation was based not on his dissertation but rather on a long review paper about allometry (the mathematical study of rates of growth) written during his third year in graduate school.[10] This led to his appointment as assistant professor at Harvard University in 1967, before his thesis had been completed. As Gould recalls, Bernhard Kummel "wanted to get a young guy who did the opposite of what he did . . . he wanted someone who could do quantitative evolutionary work," and he simply asked Newell to recommend a suitable candidate. Gould accepted on the condition that he could defer the position for a year to finish his thesis; he arrived in 1967, was granted tenure in 1972, and was still on the faculty, as Alexander Agassiz Professor of Zoology, when he died more than thirty-five years later. In Gould's own words, "I came by the job somewhat dishonorably by modern standards, or at least by my own. [But] I kept it totally honorably."[11]

TWO ICONOCLASMS

We now have a fairly clear if somewhat ambiguous picture of Gould as he began his professional career. On one hand we see an iconoclastic son of Marxist radicals, educated at one of the most liberal colleges in the United States, who refused to bow to tradition and authority and was unafraid to speak his mind. On the other, we have something of a prodigy who, though having demonstrated promise in a fairly traditional area of descriptive paleontology, nonetheless had not produced a major piece of original research (the allometry paper, thought brilliant, was a review paper). It would not be an exaggeration to say that at this point, Gould's career could have gone in a number of different directions. There was certainly no guarantee that he would become the person who, on being (posthumously) awarded the prestigious Paleontological Society Medal for lifetime achievement thirty years later, would be lauded for "the establishment of paleontology as a legitimate, important player in the high stakes game played out at the High Table of evolutionary theory."[12]

Punctuation and Orthodoxy

The turning point came in an innocuous invitation to participate in a small symposium titled "Models in Paleontology" at the 1971 meeting of the Geological Society of America. Tom Schopf, a young paleontologist at the University of Chicago (and quite an iconoclast in his own right), was deeply concerned that paleontology had lost its theoretical rigor and biological sensitivity, and he hoped to generate interest in a new, models-oriented approach to the field. Schopf asked Gould to contribute a paper about "models in speciation," and Gould's response was somewhat ambivalent: "A damned good idea, your symposium. I'm flattered by your invitation and will gladly accept. My only hesitation is that you have given me a topic that ranks only third on your list in terms of my competence [behind models in morphology and phylogeny]."[13] Gould was, however, familiar with Eldredge's dissertation and follow-up paper in the journal *Evolution,* which described patterns of stasis and speciation in Devonian trilobites, and he suggested a collaborative project exploring these patterns more generally. The eventual compromise was to co-author a paper with Eldredge that Gould presented at the conference but for which Eldredge was first author in the published proceedings. The result was the now-famous essay "Punctuated Equilibria: An Alternative to Phyletic Gradualism."[14]

The classic Darwinian evolutionary model assumes that species change very gradually over vast amounts of time (tens of millions of years or more), developing in response to equally slow and gradual changes in environment that produce adaptations which ultimately lead to the appearance of new species. A major assumption is that this is a constant, inexorable process and that the tempo of evolution is unchanging. The major driving force in Darwinian evolution is the mechanism of natural selection, which presses individual organisms to compete with one another and their environments, rewards beneficial adaptations, and punishes less successful species with extinction. That the fossil record was incomplete—with its intermittent gaps and jumps and the notorious absence of transitional "missing link" species—was of no concern to evolutionists, who contended that it was simply imperfectly preserved.

Beginning with the unorthodox intuition that the fossil record was in fact a much more accurate record of the history of life than had been previously assumed, Gould and Eldredge proposed a radical revision to this standard narrative. They argued that the pattern of evolutionary history really is composed of fits and starts, consisting of long periods of evolutionary stasis (or equilibria) punctuated by shorter periods of rapid speciation. This proposal presented some very significant revisions of Darwinian evolutionary theory: by suggesting that species can act as independent "units" in natural selection, Gould and Eldredge upset the orthodox Darwinian assumption that natural selection can only bring about adaptive advantages in single organisms. Punctuated equilibria instead proposed that entire lineages (or clades) have "life spans" and that selection at higher taxonomic levels acts independent of neo-Darwinian microevolution.

To what extent was this theory genuinely rebellious? As Gould and Eldredge protested frequently, punctuated equilibrium owes much to earlier biological theories. The most frequently cited touchstones are Ernst Mayr's work concerning peripheral isolation and allopatric speciation and G. G. Simpson's briefly held theory of "quantum evolution."[15] Indeed, Mayr had written in 1954 that "many puzzling phenomena, particularly those that concern paleontologists, are elucidated by a consideration" of peripherally isolated populations, including "unequal (and particularly very rapid) evolutionary rates, breaks in evolutionary sequences and apparent saltations, and finally the origin of new 'types.'"[16]

Generally, however, most critics of the theory have found reasons to implicate Gould in a more radical agenda; Mayr himself commented that since "the Eldredge and Gould theory was expressly based on my 1954 paper one might think that I would be completely behind the Gould theory." Mayr continued ominously, "this, however, is only partially the case. I strongly object to the Goldschmidtian interpretation of rapid speciation in founder populations and I likewise do not agree with the complete stasis of other species."[17] What Mayr is suggesting is that Gould and Eldredge had attempted to smuggle in the German geneticist Richard Goldschmidt's much-ridiculed theory of genetic saltations under the cover of a legitimate discussion of allopatric speciation. Mayr went on to specify in a 1982 essay that there are two possible readings of punctuated equilibrium: the first, a "moderate" or "Mayr version," involves only a "slight translation" of his 1954 theory "into vertical terms." The second, a "drastic or 'Goldschmidtian version'" of punctuated equilibrium, was presented in Gould and Eldredge's 1977 follow-up to the original paper, which suggests that speciation is based on drastic mutations.[18]

There is indeed an unmistakably radical edge to all of the formulations of the theory, including the 1972 version. What is genuinely iconoclastic, however, is not the notion that rates of speciation vary with respect to geographical separation and isolation, but rather the historical claim that the fossil record is much more complete than had been traditionally assumed. A better way of understanding Gould's iconoclasm is to consider the sociological ramifications of punctuated equilibrium. Eldredge calls Gould's primary contribution the instigation of a "behavioral revolution in fundamental approaches to the fossil record and its interpretation . . . the switch from paleontology's passive role as dutiful documenter of what-came-before-what in the history of life, to its much more exciting, active, even downright New York feisty, role in challenging the idea that all we need to know about evolution comes from Drosophila experiments and the mathematical formulations of population genetics."[19] In other words, by challenging the restrictive notion that the fossil record is an unreliable source of evolutionary data, Gould helped open the door for paleontologists to demand a greater role in the community of evolutionary biology.

The effects were profound and immediate. Within three years of the publication of the original essay, Schopf had founded a new journal—*Paleobiology*—devoted to theoretical, evolutionary paleontology; Steven Stanley had introduced the concept of "species selection" (largely an extrapolation from Gould and Eldredge's work); and an exciting new research program focusing on mathematical modeling and simulation of paleontological data had been established as a joint enterprise among Gould, Schopf, and several colleagues. All of these developments benefited intellectually and institutionally from the inspiration and notoriety of punctuated equilibrium, and each in its own way helped define the direction of evolutionary paleontology well beyond the decade of the 1970s. In particular, punctuated equilibrium's influence has been felt most strongly in the establishment of a hierarchical view of macroevolutionary patterns and processes (the proposal that evolution viewed at the higher taxonomic levels reflects causal factors acting independent of those at levels below the species), where an entirely new field of macroevolutionary theory emerged more or less as a direct result of Gould's encouragement.

So punctuated equilibrium was iconoclastic—particularly as a statement and a call to arms—if it was not radically anti-Darwinian. In considering whether Gould himself was an iconoclast, however, it is worth remembering that (a) he did not invent the theory—it was adapted from Eldredge's dissertation, (b) he was not looking to make a major splash in the theory of speciation—it had not occurred to him to do so before Schopf's unexpected invitation, and (c) all his life he maintained in unequivocal language that punctuated equilibrium is an interpretation of the *rates* at which Darwinian processes operate, not a rejection of those processes (see figure 18.2).

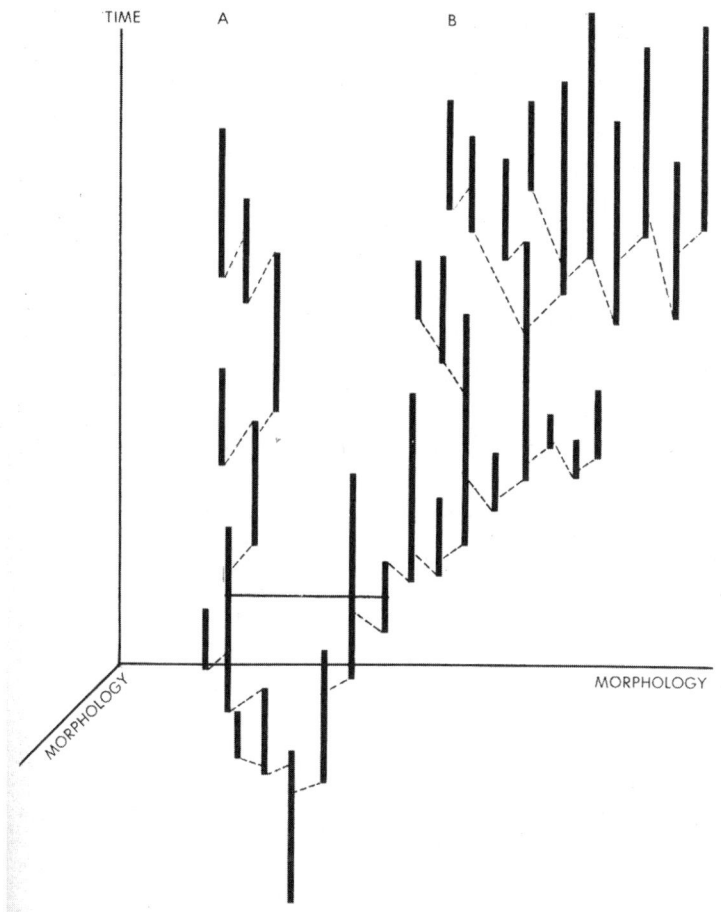

Figure 18.2. Punctuated equilibrium. Niles Eldredge and Stephen Jay Gould, "Punctuated Equilibria: An Alternative to Phyletic Gradualism." In *Models in Paleobiology*, ed. Thomas J. M. Schopf (San Francisco: Freeman, Cooper, 1972), 113. Courtesy of Niles Eldredge.

Critique of Adaptation

If much of Gould's popular reputation rests on his co-authorship of punctuated equilibrium, probably his most significant scientific contribution was his critique of the paradigm of "adaptationism" in evolutionary thinking. Briefly, in a number of works published around 1977 onward, Gould challenged the strictly functionalist principle of adaptation held by many evolutionary theorists. Orthodox Darwinian natural selection assumes that every adaptation is a response to environmental conditions which produce, in theory, random variations. Nature then "selects" the modifications that prove successful. Gould modified this notion by suggesting that internal constraints, such as developmental "pathways" encoded by genes or guided by behaviors, can prevent certain kinds of modifications in individual organisms—in other words, organisms can resist some of the force of selection. He also proposed, along with Richard Lewontin, that not all features of organisms are the result of adaptations. Using the architectural analogy of a kind of leftover space created in the construction of domed arches in medieval cathedrals, Gould argued that certain evolutionary features are "spandrels" (an architectural term), or accidental by-products of some other combination of adaptations. In a controversial turn of phrase, Gould cheekily labeled the adaptationist program the "Panglossian paradigm," after the character in Voltaire's *Candide* who resolutely maintains (despite mounting evidence to the contrary) that we live in the best of all possible worlds. The world we live in, Gould stressed repeatedly, is one shaped not by inevitability but by unpredictable contingency.

The full significance of Gould's iconoclastic notion of spandrels must be placed in context with work in which he participated between 1972 and 1975 as part of a collaboration with Schopf, David Raup, and Daniel Simberloff on stochastic (random) models of evolutionary development. This research program set out to test whether randomly generated phylogenetic lineages formed patterns similar to those found in the existing fossil record. The simulations, performed using a computer program written by Raup, showed a surprising result: the computer-generated diagrams depicting clade shape (called "spindle diagrams") were virtually indistinguishable from those modeled on real data. In other words, the patterns formed at random (the program ran through a series of steps, randomly determining the next evolutionary move for the lineage at each stage) showed the same patterns of speciation, stasis, and extinction found in the fossil record. This did not prove that clade evolution is *necessarily* random, but it certainly provided an incentive to reconsider whether phyletic evolution *must* be explained as the result of directional causes.[20]

Stochastic modeling had a profound effect on Gould's commitment to traditional Darwinian adaptationism. Many observers have questioned the motivation and timing of Gould's critique of the adaptationist paradigm, appearing as it did in the late 1970s after the publication of E. O. Wilson's controversial *Sociobiology: The New Synthesis.* The contention is that Gould was motivated not by empirical scientific evidence, but rather by political conviction, to attack all forms of biological determinism whose consequences for social and racial equity he found distasteful. Paul Gross maintains that Gould's scientific stance was inextricably linked to an "attempted scientific assassination of E. O. Wilson" on the ground that accepting biological determinism "means that we can never mitigate social and political horrors: it's all biology, the end of hope for change."[21] Similarly, Michael Ruse contends that "Gould was not always antiprogressionist. . . . [U]p to and including the writing of his *Ontogeny and Phylogeny* in 1977, he was in favor of progress, a process apparently triumphing with our own species. Then Gould swung round against the idea."[22]

Gould may have felt that biological determinism was morally troubling, but treating his opposition to adaptationism as a political crusade simply does not fit with the facts. Indeed, although Gould has written that he graduated from Columbia in the late 1960s "a philosophically committed adaptationist," it was not the introduction of sociobiology that caused him to reexamine his views.[23] Rather, he was moved to "reassess the methodological requirements for advancing a claim of adaptation" during his earlier participation in the stochastic modeling project with Raup and Schopf. In a 1974 follow-up, Gould and Raup introduced the variable of randomly changing morphology into their computer runs of simulated phylogenies. Their stated goal was to reexamine the role of random changes in morphology in the shape of evolutionary patterns and to interpret their significance. If, they reasoned, random processes can produce an ordered pattern, then there is no reason to suppose that unidirectional selection is a necessary inference from the shape of the tree of life.[24] The surprising result of the computer runs was a set of "well ordered arrays of morphology with few of the chaotic elements that might have been anticipated in a system employing random change," leading the authors to conclude that the evolution of morphological steps in an evolutionary sequence are at least as likely to be constrained by the "topological properties of the abstract form of the tree itself" as from "any special biological process."[25] As Gould recalled many years later, "I, at least, was surprised . . . at the range of apparent order that could be generated by random models but that, unhesitatingly, had always been viewed as *prima facie* evidence, sufficient all by itself, for adaptation."[26]

This was the context of Gould's 1977 *Ontogeny and Phylogeny* and his later work with Lewontin on spandrels. Gould's stance may have been iconoclastic, but it was *consistently* iconoclastic, and we have no reason to question Gould's contention that he "did not attack adaptationism because I disdained sociobiology; I disliked sociobiology because I regarded its central premise as fatally flawed."[27] Indeed, there is nothing particularly progressionist or adaptationist about *Ontogeny and Phylogeny*: the main argument of the book is that "heterochrony," or change in developmental timing during ontogeny and development (such as precocious sexual maturity or retention of juvenile characteristics), offers a significant revision of classic Darwinian understanding of phyletic evolution. He maintains that "Darwinian theory has been overly burdened by a rigid insistence upon very slow, continuous, adaptive transformations" and offers instead a set of heterochronic mechanisms that "frees morphology to experiment not only by releasing selective control altogether (and abandoning Darwinism), but by directing it elsewhere."[28] The result is the potential to "save Darwinism from an embarrassing situation usually swept under the rug of orthodoxy—the difficulty of explaining transitions between major groups if the transitions must be gradual and under the continual control of selection upon morphology."[29]

When Gould and Lewontin wrote "The Spandrels of San Marcos," then, Gould had already established a consistent interest in examining alternatives to selection as the sole directing force for phyletic evolution. The essay begins by pointing out the tendency for evolutionary biologists to ignore constraints on development and to look only for direct adaptations. This was not a new point for Gould; he had been interested in morphological constraints ever since he wrote his review of allometry as a graduate student, and his insight was honed by his work with Raup and Schopf. Although Gould and Lewontin castigated Panglossian adaptationism, they did not deny the importance of adaptation or selection. Their actual proposal is, relatively speaking, fairly moderate. They grant the traditional efficacy of both adaptation and selection but add the caveat that initial adaptation often arises not as a direct response to environment but instead as "a secondary utilization of parts present for reasons of architecture, development or history."[30] In other words, many features of organisms first appear as nonadaptive consequences of "architectural" constraints on size and body plan and only later develop adaptive significance. For this reason, Gould and Lewontin urge biologists to eschew "just-so stories" that attempt to provide causal explanations for the existence of every morphological feature of an organism.

Note that in offering their theory as a "pluralistic" compromise, Gould and Lewontin repeatedly invoke Darwin's own appreciation for pluralism (they present their critique "in Darwin's spirit"). The point, as they conclude, is not to establish an anti-Darwinian dogma but rather to mitigate a tendency in evolutionary biology to focus only on the level of the gene: "A pluralistic view could put organisms, with all their recalcitrant, yet intelligible, complexity, back into evolutionary theory."[31] Over the course of the following decade Gould expanded this program, eventually coining the term *exaptation* (with Elizabeth Vrba) for the mechanism that produces adaptive, nonselective features: characters "evolved for other usages (or no function at all), and later 'co-opted' for their current role."[32] The central message in Gould's anti-adaptationism is one of compromise: evolutionists, from Darwin's day to the present, have consistently felt bound to choose between two alternatives. On one hand, functionalists have maintained that adaptive pressures drive the evolution of specialized parts to meet the functional needs of organisms. On the other, formalists (or structuralists) have countered that internal forces or constraints determine the appearance of the most suitable (and, indeed, possible) parts. Gould's solution was quite simple: functionalism often obtains, but it is constrained by developmental prerequisites (formal requirements) and exaptation—and none of these features alone can explain evolution.

CONCLUSION

In his magnum opus, *The Structure of Evolutionary Theory,* published just before his death in 2002, Gould makes a lengthy argument for a pluralistic revision of orthodox Darwinian theory. According to Gould's account, what has gone badly wrong with neo-Darwinism is its insistence on producing a set of inviolable laws and principles of evolution that can be reduced to mathematical rules of genetics. In his mind, this misses the spirit of the original: Darwin himself, Gould argues, stressed that in most cases "answers must be sought in the particular and contingent prior histories of individual lineages, and not in general laws of nature that must affect all taxa in a coordinated and identical way."[33] Gould offers a plea for old-fashioned naturalism—for the power of narrative explanation in revealing the historically contingent path evolution has taken. Although it departs from the letter of neo-Darwinism, he claimed that his vision would return evolutionary theory to something more closely matching Darwin's intent: "This book attempts to expand and alter the premises of Darwinism, in order to build an enlarged and distinctive evolutionary theory that, while remaining in the tradition, and under the logic, of Darwinian argument, can also explain a wide range of macroevolutionary phenomena lying outside the explanatory power of extrapolated modes and mechanisms of microevolution, and that would therefore be assigned to contingent explanation if these microevolutionary principles necessarily build the complete corpus of general theory in principle."[34] In other words, there are a lot of things traditional Darwinism explains, but there are also many features of evolution that Darwin could not have conceived of in his day. Gould hopes that he can convince his reader that this expanded Darwinism is, if not what Darwin actually thought, then what he likely *would* have thought were he alive to comment today.

It is, however, fair to ask whether Gould's humility is genuine or whether it is a cover (as many critics appear to suspect) for an ambition to replace Darwinism with Gouldism. It is useful to put aside much of the published rhetoric—Gould's own and the responses of his critics—and examine his private sentiments. In an unusually candid letter to Tom Schopf, written during a meditative afternoon on the day after Thanksgiving, 1977, Gould offered the following reflection on his career to date: "For me, at least, ideas don't seem to be tapering off, but this may be a peculiarity of my frame of mind, so correctly identified by you at one of those dinners as 'conservative' scientifically, despite general social and political views. As I've told you, when I should have been most iconoclastically creative, in my early 20's, I saw my task in life as documenting some elegant, inductive, multivariate demonstrations of Simpsonian paleontology. I have been dragged literally (once by you and once by Niles) into the two creative things I have done. Once in, I think that I have some skill in drawing connections and seeing consequences, and in generating potential empirical tests—so I am a reasonably useful member of a team, if not an initial innovator."[35]

Of course, when Gould wrote this, spandrels and exaptation were yet to be presented, and his astonishing popularity as a public intellectual had only just begun (his first book of collected essays, *Ever Since Darwin,* had appeared earlier that year). But, following Gould's own assessment, one might draw the following conclusion: in advocating an expanded, modified Darwinism, he was as much as a *conservator* of the proper role of natural history and paleontology as he was a rebel against the established orthodoxy of Synthetic Darwinism. If he was a revolutionary, then the kind of revolution he preached was more consistent with the pre-Enlightenment sense of the word: a cyclical return to an original state, not a violent upheaval and annihilation of an old regime. Or, as Richard Lewontin and Richard Levins—two of his Harvard colleagues of long standing—observed in his obituary, he was a radical in the true meaning of the term, someone who "consider[ed] things from their very root [*radix*]" and who practiced "perfectly acceptable 'normal science' that adds richness to Darwin's original scheme."[36] Time will tell whether any of his major innovations will have lasting impact on the field of evolutionary biology, and indeed he may be remembered ultimately as a gadfly who provoked fruitful debate rather than as a theoretical heavyweight. But the final analysis, as with many aspects of his career, must be that Gould defies conventional labels: as he suggested in his letter to Schopf, even Stephen Jay Gould's iconoclasm was iconoclastic.

NOTES

1. John Maynard Smith, "Genes, Memes, and Minds," *New York Review of Books* 42, no. 19 (1995).

2. Thomas S. Kuhn, *The Structure of Scientific Revolutions* (Chicago: University of Chicago Press, 1970).

3. Stephen Jay Gould to Thomas J. M. Schopf, November 25, 1977. Thomas J. M. Schopf Papers, Smithsonian Institution Archives, RU 007429 (hereafter Schopf Papers).

4. Niles Eldredge, interview with David Sepkoski, January 19, 2006.

5. Niles Eldredge, "Presentation of the Paleontological Society Medal to Stephen Jay Gould," *Journal of Paleontology* 77 (2003): 812.

6. Eldredge interview.

7. Stephen Jay Gould, "Is Uniformitarianism Necessary?" *American Journal of Science* 263 (1965): 226.

8. Ibid., 227.

9. Chester R. Longwell, "Response to 'Is Uniformitarianism Necessary?'" and Stephen Jay Gould, "Reply," *American Journal of Science* 263 (1965): 918–21.

10. Stephen Jay Gould, "Allometry and Size in Ontogeny and Phylogeny," *Biological Reviews* 41 (1966): 587–640.

11. Gould, interview November 2, 1998, quoted in Patricia M. Princehouse, "Mutant Phoenix: Macroevolution in Twentieth-Century Debates over

Synthesis and Punctuated Evolution" (Ph.D. diss., Harvard University, 2003), 250–51.

12. Eldredge, "Presentation," 812.

13. Stephen Jay Gould to Thomas J. M. Schopf, March 13, 1970. Schopf Papers, RU 007429.

14. Niles Eldredge and Stephen Jay Gould, "Punctuated Equilibria: An Alternative to Phyletic Gradualism," in *Models in Paleobiology,* ed. Thomas J. M. Schopf, 82–115 (San Francisco: Freeman, Cooper, 1972).

15. See Ernst Mayr, "Change of Environment and Speciation," in *The Growth of Biological Thought: Diversity, Evolution and Inheritance* (Cambridge: Harvard University Press, 1982 [1954]), 188–210; and George Gaylord Simpson, *Tempo and Mode in Evolution* (New York: Columbia University Press, 1944), 206–7.

16. Mayr, "Change of Environment and Speciation," 206.

17. Ernst Mayr to Thomas J. M. Schopf, February 9, 1982. Schopf Papers, RU 007429.

18. Stephen Jay Gould and Niles Eldredge, "Punctuated Equilibria: The Tempo and Mode of Evolution Reconsidered," *Paleobiology* 3 (1977) 115–51; Ernst Mayr, "Speciation and Macroevolution," *Evolution* 36 (1982): 1127.

19. Eldredge, "Presentation," 812.

20. David M. Raup, Stephen Jay Gould, Thomas J. M. Schopf, and Daniel S. Simberloff, "Stochastic Models of Phylogeny and the Evolution of Diversity," *Journal of Geology* 81 (1973): 525–42.

21. Paul R. Gross, "The Apotheosis of Stephen Jay Gould," *New Criterion,* October 2002, 80.

22. Michael Ruse, *The Evolution Wars: A Guide to the Debates* (New Brunswick, N.J.: Rutgers University Press, 2001), 245.

23. Stephen Jay Gould, "Fulfilling the Spandrels of Word and Mind," in *Understanding Scientific Prose,* ed. Jack Selzer (Madison: University of Wisconsin Press, 1993), 318.

24. David M. Raup and Stephen Jay Gould, "Stochastic Simulation and Evolution of Morphology: Towards a Nomothetic Paleontology," *Systematic Zoology* 23 (1974): 306.

25. Ibid., 319–20.

26. Gould, "Fulfilling the Spandrels," 319.

27. Ibid., 320.

28. Stephen Jay Gould, *Ontogeny and Phylogeny* (Cambridge: Harvard University Press, 1977), 340.

29. Ibid., 341.

30. Stephen Jay Gould and Richard Lewontin, "The Spandrels of San Marco and the Panglossian Paradigm: A Critique of the Adaptationist Programme," *Proceedings of the Royal Society of London, Series B, Biological Sciences,* 205, no. 1161 (1979): 593.

31. Ibid., 597.

32. Stephen Jay Gould and Elizabeth Vrba, "Exaptation—A Missing Term in the Science of Form," *Paleobiology* 8 (1982): 6.

33. Stephen Jay Gould, *The Structure of Evolutionary Theory* (Cambridge: Harvard University Press, 2002), 1335.

34. Ibid., 1339.

35. Stephen Jay Gould to Thomas J. M. Schopf, November 25, 1977. Schopf Papers, RU 007429.

36. Richard Lewontin and Richard Levins, "Stephen Jay Gould—What Does It Mean to Be a Radical?" *Monthly Review* (November 2002), 18–20.

FURTHER READING

Gould, Stephen Jay. "Fulfilling the Spandrels of Word and Mind." In *Understanding Scientific Prose,* ed. Jack Selzer (Madison: University of Wisconsin Press, 1993).

———. "Is a New and General Theory of Evolution Emerging?" *Paleobiology* 6 (1980): 119–30.

———. *Ontogeny and Phylogeny* (Cambridge: Harvard University Press, 1977).

———. *The Structure of Evolutionary Theory* (Cambridge: Harvard University Press, 2002).

Gould, Stephen Jay, and Richard Lewontin. "The Spandrels of San Marco and the Panglossian Paradigm: A Critique of the Adaptationist Programme." *Proceedings of the Royal Society of London, Series B, Biological Sciences,* 205, no. 1161 (1979): 581–98.

Sepkoski, David. "Stephen Jay Gould, Jack Sepkoski, and the 'Quantitative Revolution' in American Paleobiology," *Journal of the History of Biology* 38 (2005): 209–37.

Michael Shermer. "The View of Science: Stephen Jay Gould as Historian of Science and Scientific Historian, Popular Scientist and Scientific Popularizer," *Social Studies of Science* 32 (2002): 489–524.

CHAPTER NINETEEN

Culture and Gender Do Not Dissolve into How Scientists "Read" Nature: Thelma Rowell's Heterodoxy

VINCIANE DESPRET

"The males of nearly every social primate play a special role in challenging predators, particularly if an infant is threatened," the primatologist Alison Jolly wrote in her 1972 book *The Evolution of Primate Behavior*. More precisely, "defense seems to be a male role throughout at least the monkeys and apes. Furthermore it may be concentrated among the dominant males, as in macaque troops, or even be the clearest sign of dominance, as in the cebus monkey. . . . When a savanna-living baboon troop encounters a big cat, it may retreat in battle formation, females and juveniles first, the big males with their formidable canines last, interposed between the troop and the danger." This beautiful pattern, however, has one exception: "Rowell's forest-edge baboons simply run away to the safety of the trees, each at his own speed, which means strongest males first and females and infants lumbering at the rear."[1] In the case of these baboons, as Thelma Rowell later stated, there was no heroism going on at all.[2]

Jolly also mentions that among Rowell's baboons, young male infants get more attention from the males than do young females, a fact that has never been described in any other baboons.[3] But the strangest fact appears when Jolly compares the social behavior of these baboons with that of the baboons observed in other studies. In Rowell's troop, males were extremely peaceful: they formed a cohort, "constantly aware of each other's movements, but with scarcely any aggressive interactions."[4]

These eccentric baboons had been observed from the beginning of the 1960s at the edge of the forest of Ishasha, in Uganda, by the primatologist Thelma Rowell. From her very first descriptions, Row-

Figure 19.1. Thelma Rowell. Courtesy of Vinciane Despret.

ell's observations contrasted sharply with those of colleagues working with similar animals.[5] Not only were Rowell's baboons peaceful,
but males, bizarrely, were not competitive. There was much positive
or friendly interaction. Aggression was rare, even at feeding places:
baboons almost never stole food from each other. The typical day of
an Ishasha adult baboon involves long periods of social interaction,
playing, and reciprocal grooming. Of course, there are some ten-
sions between males, but these are mainly expressed by an absence
of grooming between themselves and a high frequency of exchange of
gestures of politeness or conciliation. "The dominant impression of
interaction between males," Rowell concluded, "was that of active
cooperation."[6]

Baboons had hitherto been unanimously described as extremely
competitive, intensively aggressive toward each other, and involved
most of the time in fight over food or females. This picture originated
at the end of the 1920s with the observations that the zoologist Solly
Zuckerman[7] had made in the colony of hamadryas baboons in the
London Zoo. The story of this colony considerably influenced the

construction of this image and the theories that accounted for social organization of primates: of the one hundred baboons, mostly males, that founded this colony in 1925, twenty-seven adults died during the first six months, most of them showing wounds indicating recent participation in fights. In 1927, thirty adult females were added to the remaining population; things only got worse, and fights over females raged.

Today, we realize that the colony contained too many animals, strange to each other, in too small a space and combined in an inappropriate sex ratio. Zuckerman, however, believed he was watching normal behavior and extrapolated from his observations a general thesis of social organization in primates: sexual instincts formed the cement of the group, and sexual competition was the basis of primate society.[8]

From Zuckerman's observations grew a theory that was to become a hallmark of primatology: dominance hierarchy is the most important principle of all primate social organizations. Defined as priority of access to desirable objects, dominance was based on the ability to fight.[9] Ultimately, however, the function of dominance was to reduce the amount of aggressive behaviors in the group; once dominance was established, supplanting and avoidance interactions replaced the fights over desirable objects. Dominance had also selective advantages: because the desirable objects included oestrous females, dominance would imply increased numbers of descendants for males. According to this theory, females had no social role and had no rank in the hierarchy other than as subordinates. "Female baboons are always dominated by their males," Zuckerman wrote, "and in many situations the attitude of a female is of extreme passivity."[10]

With subsequent researches, in captivity and in the wild, scientists' interest in dominance theory continued to grow; they also retained their conviction regarding females' social insignificance. These views of primates' social organization still held true at the end of the 1950s in the writings of the most influential primatologists: the physical anthropologist Sherwood Washburn, observing baboons in southern Rhodesia, and his student Irven DeVore, working in Kenya, wrote that "the main characteristics of baboon social organization . . . are derived from a complex dominance pattern among adult males that usually ensures stability and comparative peacefulness within the group, maximum protection for mothers and infants, and the highest probability that offspring will be fathered by the most dominant males."[11]

As the philosopher of science Donna Haraway notes, dominance became to primatologists what kinship had been to anthropologists: "at once the most mythical, most technical, and discipline-grounding of a field's conceptual tools."[12] Dominance hierarchy was so commonly accepted, Rowell herself remarked, that where groups had been observed in which the usual criteria of rank were not obvious, the concept of "latent dominance" was used to explain an apparent lacuna in an otherwise universal phenomenon. If we reconsider the baboons that she observed, we may now understand their oddity. In all other studies the same pattern of organization was observed: baboon societies were male-centered, marked by competition over females and food, very aggressive, rigidly organized, and hierarchical. Social roles were sexually distributed: males were leaders, defenders, and policers whereas females were described as dedicated mothers to small infants and sexually available to males in order of the males' dominance rank, but otherwise of little social significance.

In Rowell's troop, not only did it seem as if the baboons were living in peace and harmony, with little or no competitive interaction, but there seemed to be no observable hierarchy as well. It could neither be detected among males, nor, more surprising, inferred in the relationships between males and females. In fact, females seemed to have what all other studies considered to be the dominant male's role. Whereas in other troops males were leaders, in Rowell's group the older females determined the daily route of the group to food and water. Everywhere else, males had been the center of the troop. In Rowell's group females acted as a focus of the group's social activity. They formed the nucleus of the troop.

CONTRARINESS

Are we to imagine, then, that these animals have been infected by some sort of "contrariness"—what has since been called, by some primatologists, the "Thelma effect"?[13] Of course, *contrariness* had actually been used to characterize a human heuristic device, the practice of thinking thoughts opposite to the currently accepted ones; but how are we to understand how a scientist's having such thoughts can lead to animals' behaving bizarrely or unexpectedly?

One could close this debate by referring to the arguments put forward by science studies experts who conclude that animals are "guided" by the expectations of those who study them. We are reminded of Bertrand Russell's astonishment at the high incidence of animals conforming to the behavior expected of them by observers.[14] Before Rousseau's time they were seen as ferocious beasts but they subsequently conformed to his cult of the noble savage. During the Victorian period, primates were virtuous monogamists; during the post-Freudian era of sexual liberation, one could have been appalled by the considerable deterioration of their moral standards.

If we adopt this research framework, the tracks are laid out for us: we could assume that baboon females may claim a social role when a female primatologist observes them or that the male baboons act considerably less heroically or more peaceably when their story is told by a woman. Thelma Rowell's "contrariness" can be put down to a matter of gender. Solly Zuckerman had affirmed in 1963 that "among field workers the observer's own temperament and sex might be an important filter in determining, for example, the amount of agonistic behavior observed and reported in groups of primates."[15]

In their introductory chapter to the book *Primate Encounters,* the primatologists Shirley Strum and Linda Fedigan note the considerable change in scholars' interpretation of primate behavior: "We have moved from a general vision that primate society revolves around males and is based on aggression, domination, and hierarchy to a more complex array of options based on phylogeny, ecology, demography, social history and chance events. The current image of primate society . . . would be a strong counterpoint to the earlier view. It would highlight the importance of females within society, emphasize tactics other than aggression (particularly those that rely on social finesse and the management of relationships), and argue that hierarchy may or may not have a place in primate society, but that males and females are equally capable of competition and rank ordering."[16] These changes, the authors add, have generated a great deal of interest among feminist historians of science and in the popular media because they have been linked to a provocative claim: that women scientists played a major role in the revision of the primate's image.

Numerous observers among primatologists and science studies scholars have suggested that women observe differently than do men. For some, women's patience makes them ideal observers.[17] The well-known paleoanthropologist Louis Leakey deliberately chose to send women—Jane Goodall in the early 1960s and later Diane Fossey and Biruté Galikas—into the field because he assumed that they were better observers of primates and would be more emotionally connected to their subjects.[18] This characteristic has been largely supported by what has been called the "*National Geographic* effect," which "had done much to create the myth that primatology is a type of mothering activity."[19] This conception gives rise to the idea that women are better observers of animals because they have a special relationship with nature.

Another line of argument is proposed by the feminist philosopher of science Donna Haraway, who suggests that "the unifying theme in the primatology done by women has been their high likelihood of being skeptical of generalizations and their strong preference for explanations full of specificity, diversity, complexity, and contextuality. In the 1960s, consider Jane Goodall, Thelma Rowell, Alison Jolly, Phyllis Jay, and Suzanne Ripley."[20] We can certainly find some similarities between Thelma Rowell and the few women who were working in the field of primatology at the beginning of the 1960s. For instance, Jolly is often compared to Rowell in that both are recognized[21] for having shown very early on that the generalizations about primate behavior based on very few field studies were, as Rowell stated, "too slender pillars to support the edifice built on them."[22] It is no coincidence that Jolly, when she reviewed the results of primate studies, enthusiastically underlined the baboon extravagances described by Rowell. Jeanne Altmann may also be compared to Rowell:[23] not only did Rowell suggest that females are the nucleus of the troop, but she also stressed the value of adopting the female monkey's point of view in that this tactic has the power to challenge accepted explanations,[24] and Altmann, for her part, proposed new sampling methods that encouraged the inclusion of females as research subjects. Goodall and Rowell had shared characteristics as well: notably, they were both credited with having succeeded in shifting primatology's focus on group dynamics to the individuality of monkeys, a very different approach to that taken by ethology.[25] Still, at the beginning of the decade, Phyllis Jay described a troop of hanuman langurs in central India whose peaceful and relaxed males lived in perfect harmony; dominance was not particularly visible or important in langur life. Like Rowell, she suggested that females form the core of the group's social life.[26] Shirley Strum, though she appeared more than ten years later, also challenged the dominance model in baboons—with equally iconoclastic observations.[27]

All these women had one feature in common: they have taken an original stance in their field. May these marginal positions be linked to the scientist's gender? I put this hypothesis to the test by reviewing the events that led Thelma Rowell to take up heterodox positions.

BEING A WOMAN PRIMATOLOGIST

Originally, Thelma Rowell had no ambition to become a primatologist.[28] Born in Bradford, Yorkshire, in 1935, she studied zoology at Cambridge University and planned to investigate rodents. At the beginning of the 1960s, after she finished her ethological dissertation on maternal behavior in golden hamsters, the zoologist Robert Hinde invited her to work as a research assistant in the Madingley Laboratory at Cambridge University. She started to work with captive monkeys and undertook with Hinde a number of mother-infant studies.

In 1961, inconspicuously signing herself T. E. Rowell, she submitted a paper about her findings to the journal of the Zoological Society of London. The society was impressed and invited T. E. Rowell to come down from Cambridge and give a talk to the fellows, but when it was discovered that T. E. Rowell was a woman, there was some embarrassment. She was able to give the lecture but not to sit with the fellows for dinner because of her sex. The solution was to ask her to sit behind a curtain, out of sight, and eat her meal. She declined.

This anecdote provides a good illustration of the situation faced by women in the field of ethology, particularly those wishing to pursue an academic career. There was no question of employment for women at a university at that time. Women could, however, obtain grants for research. That is what Rowell did.

In 1962 she followed her husband, a neuroethologist, to Uganda, where he held a post at Makere University. She decided to continue her work with primates and embarked on a study of the olive baboon in a forested area of the Queen Elizabeth Park—the forest of Ishasha. Rowell habituated the Ugandan forest baboons to her presence and began her research, spending two weeks with the baboons and two weeks in Kampala working with captive monkeys. The main focus of her research was to compare the social behavior of wild and caged baboons and to try to find a way of assessing the differences between them. This research broke new ground in two ways. First, for the baboons in captivity, Rowell invented a sophisticated cage design that considerably reduced stress and competition among the animals. She built large cages in which cover was provided by a series of solid partitions so arranged that each animal could choose to be out of sight of his companions; the group could thus subdivide in more complex ways. She also took care to give more than one source of food and water so as to reduce competition. Second, she would stay five years with the baboons of Ishasha, an exceptionally long period for this era.

Another common feature among women working in primatology at that time is that all remained in the field longer than the majority of men. The hypothesis according to which women were better observers because they were more patient or more emotionally connected to their animals therefore deserves a slight revision: women had observed other things because they had stayed longer with their animals. The hypothesis that they stayed longer because they were women is, of course, correct, provided we give historical and social meaning to the word *woman*—one having no access to an academic position.

Staying in the field for a longer period had major consequences. Rowell was able to achieve an unexpected proximity; habituation allowed for closer observation and for a greater ability to recognize the individuals of the troop. Take, for example, the iconoclastic proposition according to which females are at the core of social organization. One could suggest that it was because Rowell was a female that she took an interest in females. The situation, however, was slightly more complicated. The hypothesis of the female's central role actually came from a totally unexpected observation: in Rowell's troop, no adult male had stayed for the entire study period; males constantly moved from one troop to another. Until that time primatologists had held the firm belief that individuals stay in the same group for their entire life. No scientist had remained long enough in the field—or been sufficiently familiar with his animals to be able to recognize individuals—so as to detect the males' nomadic movements. This finding could radically challenge all the accepted ideas about social organization in baboons: indeed, Rowell concluded, if males never remain more than a few weeks or a few months, they surely cannot act as the nucleus of the troop; who can, therefore? Rowell's observations answered the question: all the animals of the troop repeatedly solicited grooming from the females, who as a result acted as a focus of the group's social activity. This role was accentuated by the interest all baboons had in the young infants, who are, of course, with their mothers. In the same vein, one may understand why the older females played the role of leader in the daily foraging trips. After all, who, in a given territory, knows where to find the figs at a given moment or water in periods of drought? Those who have lived their entire life in that environment and who have learned all the secrets of this environment from their mothers and grandmothers. These are the females.

ISN'T DOMINANCE A MALE'S PROBLEM?

We are still left with an unresolved oddity of the baboons of Ishasha: they were not aggressively competitive and, above all, did not demonstrate the rigid dominance hierarchy that was the characteristic of all primate species. Here again, the hypothesis of the influence of the observer's gender had been suggested—I have already referred to Phyllis Jay's description of peaceful and relaxed langur males for which dominance was not particularly important. It is also worth mentioning the early accounts that Goodall gave of her chimpanzees' associating fluidly and forming a harmonious, open society.[29] Should we, then, accept Zuckerman's argument that there was a difference between the ways men and women detect or construe competition and aggressive behaviors?[30]

Rowell, in fact, chose a radically different path. Of course, whether or not the animals are competitive, whether or not they are organized in a rigid hierarchy, actually depends on the observer. It is not, however, a matter of how the observer interprets, construes, or subjectively perceives the situation, as earlier theories assumed. Rowell's explanation is much more provocative: hierarchy is really nothing but an effect of observation. In other words, dominance hierarchy only exists where the observer creates it. It is an artifact.

In captivity, Rowell explained, hierarchies are known to flourish under two conditions: those in which monkeys are total strangers to each other and those in which they lack the facilities ordinarily available in the wild. For example, hierarchical behavior might be induced by reducing the available space or by making animals compete for food. This is precisely how dominance experiments are usually carried out. Rowell wrote, "The experimenter will report that his trials have demonstrated a dominance relationship between the monkeys, while in fact they (the trials) have actually caused it."[31] In the wild, the process is very similar. Note that before the practice of habituation was introduced, the covert method used by primatologists to achieve relative proximity to their animals (in a very short time) was provisioning.[32] Unknown to the scientists, this changed the behavior of the animals. For instance, the baboons that were described as aggressive toward each other and involved in frequent dominance interactions were, in fact, brought into competition by tidbits thrown from the observation vehicle. Rowell stresses that in the film made in Nairobi by the American primatologist Irven DeVore for his students, "the commentary points out very clearly the central position of the dominant male. If you turn off the sound, the students are more likely to spot the peanuts being thrown during filming. The center, in this case, was defined by the trajectory of the peanuts, which were mostly intercepted by the adult males."[33]

It seems quite possible that many of the characteristics which had been thought to belong to the normal repertoire of baboons, Rowell concluded, explicitly mentioning DeVore's work of 1964, "might in fact be related to artificial feeding. One such character is a high degree of aggressiveness and obvious hierarchy among adult males, which were described for macaques and baboons which were fed, but not seen in the Ishasha baboons which were not."[34] Ishasha baboons, therefore, were not so eccentric: they were simply observed by a cautious primatologist. And other baboons testify only about one thing: rigid hierarchy is nothing but the animal's answer to the social disorder created by the setting; it is the answer to the competition-inducing stress imposed by the observers.

Until then, no researcher had even questioned the existence of dominance. This was understandable, because this concept was the pillar of all the models of organization of primates. Remember that the concept of latent dominance was used to describe groups in which the usual criterion of rank was not obvious. Rowell, in this respect, by challenging this assumption, had been an iconoclast. We cannot, however, neglect the fact that other women would also criticize the dominance-hierarchy model. Alison Jolly did not seem to deny the concept itself but challenged the view that aggressive dominance is universal in primates.[35] In the 1970s Shirley Strum would adopt a more radical stance by claiming that hierarchy is a myth. Other feminist scientists, such as Fedigan, would likewise criticize the model. Should we not assume, then, that being a woman might have made certain scientists more suspicious concerning this type of theory? A hypothesis taken from the feminist "standpoint theories" could be put forward: because women's lives and roles are different from men's, women hold a different type of knowledge. In this case, the fact of having suffered the effects of sexism—the effects of being subordinate in a rigid system of male dominance—may have given women scientists an "epistemic privilege."[36] This standpoint brings with it an acute sensitivity and creates a critical consciousness concerning the way male scientists give preference to models based on competitiveness and dominance. Without a doubt this critical awareness led Thelma Rowell to ask whether our own species is not "more than usually bound by hierarchical relationships, at least among the males, who have written most about this subject."[37]

An alternative hypothesis, arising from the same standpoint theories insofar as it is close to the peculiarity of Rowell's practice, may be more convincing. Remember that Rowell's mistrust of the dominance-hierarchy theory was rooted in a critical analysis of the empirical conditions of research. According to feminist theorists, this characteristic is linked to the gender of the scientist: women are more attentive to the empirical conditions of knowledge. It is true that the scientists could respond that the detection of the artifacts belongs to the know-how of good scientists rather than being a matter of gender. But not only did Rowell insist on the fact that hierarchy and competition are the consequences of situations in which animals are subjected to stressful conditions; she also took care to offer the best conditions to the animals she studied. Some feminists have also claimed that women's ways of practicing science could be related to "caring labor": a knowledge that is at the same time completely "immersed in the practical world" and attentive to the demands of those we study.[38] This is a knowledge that weaves affects, sensitivity, concrete and material conditions, and bodies, a practice that, above all, involves the one who questions.[39] This characteristic is not, however, confined to women scientists. The contrast one could draw between genders is traced along lines similar to those between scientific knowledge and that which we call "informal knowledge": the knowledge of animal keepers, trainers, and breeders.[40] For them, too, knowing and caring are inseparable. We only really know what we care for; we only take real care of that which we know well.[41]

BEYOND PRIMATES

This is the main constant of Rowell's work: she has always paid intense attention to the conditions that allow the animal to deploy a full, flexible, and varied repertoire. This preoccupation had been foremost when she compared the behavior of monkeys in captivity and in the wild; it was at the core of each of her iconoclastic stances and, most especially, in her latest research—her latest heterodoxy.

In 1968 she left Uganda and, in 1970, went to California. After conducting research in the San Francisco Zoo she moved to the Zoology Department of the University of California at Berkeley. She stayed in Berkeley until her retirement in 1994, dividing her time between teaching and field work with diverse monkeys. Although Rowell's study in the late 1980s was not done with primates, the scenario described at the beginning of this chapter was reproduced: like the baboons of Ishasha, these animals appeared to contradict entirely what was expected of them. But this time the rebellious animals were . . . sheep.

According to ethologists, sheep were organized in a rigid hierarchy: the dominant male led the flock, followed by the other males and then the females.[42] Relationships between individuals were very simple: they were competitive and were based on the dominance hierarchy pattern. Moreover, extensive studies had shown why sheep lack an essential skill for us to be able to give them the title "socially sophisticated"—they do not form long-term relationships.[43] In Rowell's flock, the sheep are led by the oldest female rather than the male. There is no hierarchy in the current sense of the word; a ram may invite others to follow him, getting up and pointing his nose in a given direction; at times they will oblige, at other times they will not.[44] Still, in Rowell's flock the sheep form lasting bonds; males weave individualized friendship networks—noticeable, particularly, in their choice of certain "friends"—whereas conflicts are rare and limited in duration. Actually, they do work hard at maintaining bonds and group cohesion, especially in the pre-mating period, when the tension mounts. Moreover, it seems that when conflicts do occur, sheep demonstrate increasingly friendly behaviors. They frequently stop fighting to rub their heads and cheeks together. These gestures should, according to Rowell, be interpreted as reconciliatory, similar in their function to those discovered by the primatologist Frans De Waal in chimpanzees.[45]

The comparison with chimps' behavior is not due to hazard. It is the consequence of the research device itself. This time the heterodoxy is not confined to the behavior described and to the challenge to the theories; the heterodoxy lies within the methodology. Thelma Rowell had decided to treat sheep as chimps, to ask sheep questions hitherto only addressed to primates. We can evaluate how iconoclastic her proposition was by considering the reaction of the sheep experts who had to review her paper when she tried to publish the first results of her studies in the beginning of the 1990s: they were all appalled by what they saw as anthropomorphy and had difficulty understanding why she was interested in questions of social organization.[46] The prejudice of these ethologists deeply reflects the well-accepted ideas about sheep: they are socially stupid, whereas primates are socially sophisticated. But are they really? This prejudice might simply be a consequence of the way in which the researches are organized. According to Rowell, "We have given primates multiple chances: the more research advances, the more interesting the questions about apes become, and the more these animals turn out to be endowed with elaborate social and cognitive competences."[47] By contrast, sheep have been victims of questions of little relevance compared to their ability to organize themselves socially. Moreover, on closer examination we immediately see that their conditions made it very unlikely that sheep could prove to have sophisticated social behaviors: most of the research was carried out on groups formed for the experiment, consisting of animals bought for that purpose and that had never met before. Only a miracle could have allowed lasting bonds to be established.

In other words, Rowell proposed to give sheep a better chance of being socially complex,[48] creating a troop with respect to equilibrium in sex and age, allowing them to form links, and ensuring the absence of all stress—notably when she would give them extra food by distributing it in such a way as to avoid competition.[49] Rowell understood the extent to which an impoverished, stereotyped, and oversimplified behavioral repertoire might be nothing but the result of a bad setting. Baboons had been testifying for years.

Indeed, the baboons described nowadays are much more similar to those of Ishasha than to those that dominated their companions—and the theories of the 1960s. One could return to the beginning of this chapter and reread the lines by Jolly describing the changes in baboons. Of course, these changes are also due to the work of numerous scientists: they no longer perceive the baboons in the same manner since they have learned to ask them other questions; more radically, baboons do not behave in the same fashion because those who study them have learned to question them differently. We may, however, assume that both the change of questions and the transformation of methodology are in part the results of Rowell's work. Rowell's intense attention to the conditions of settings deeply influenced her colleagues; her savvy in-cage design allowing captive primate behavior to take place in unusual depth and complexity is a recurrent theme of the interviews carried out by Haraway.[50]

The effect of her work goes deeper. What has been called the "Thelma effect," the practice of thinking thoughts opposite to the currently accepted ones, led Rowell to challenge the primatologists' most authoritative orthodoxy: the belief that baboons everywhere are hierarchically organized and aggressively competitive. She boldly claimed that these observations were no more than the results of the research conditions. Moreover, the emphasis on the variability of the behavior across and among species eventually succeeded in causing reexamination of "the" model that had so much weight in this domain—but we must surely take into account the role of Phyllis Jay, Alison Jolly, and many others.

Of course, if we are reminded that, as Haraway stated it, the unifying theme in primatology practiced by women researchers has been their high likelihood of being skeptical of generalizations and their strong preference for specific explanations, we could reduce the Thelma effect to a matter of gender. Indeed, Haraway's well-accepted assumption among primatologists could fairly account for Rowell's heterodoxies. One could even presume that Rowell would accept it. But that would be leaving aside the Thelma effect: Rowell might as well construe Haraway's assumption as another currently accepted theory, if not a generalization, that should also be challenged. So she does when, she replies to that hypothesis with disarming simplicity: "That was my Cambridge training. We were always taught to question authority: the more authoritarian it is, the more you question it."[51]

NOTES

1. Alison Jolly, *The Evolution of Primate Behavior* (New York: Macmillan, 1972), 73.

2. Thelma Rowell, interview with the author, June 2003. This interview was carried out June 1–3, 2003 during the making of a documentary (Vinciane Despret and Didier Demorcy, *Non Sheepish Sheep*, 2005) for the exhibit "Making Things Public: Atmospheres of Democracy," Zentrum für Kunst und Medientechnologie, Karlsruhe, Germany, Spring 2005. See also Vinciane Despret, "Sheep Do Have Opinions," in "Making Things Public: Atmospheres of Democracy," ed. Bruno Latour and Peter Weibel (Cambridge: Zentrum für Kunst und Medientechnologie/MIT Press, 2005), the catalog of this exhibition.

3. Jolly, *Evolution of Primate Behavior*, 250.

4. Ibid., 181.

5. Thelma Rowell, "The Habit of Baboons in Uganda," *Proceedings of the*

East African Academia 2 (1964): 121–127; "Forest-living baboons in Uganda," *Journal of Zoology of London* 149 (1966): 344–364.

6. Thelma Rowell, *Social Behaviour of Monkeys* (Harmondsworth: Penguin, 1972), 44.

7. Born in South Africa, Solly Zuckerman carried out his studies at the London Zoo, where he was a research anatomist in the early 1930s.

8. Solly Zuckerman, *The Social Life of Monkeys and Apes* (1932; reprint, London: Routledge, 1981), 218–219.

9. Sherwood Washburn and Irven DeVore, "Social Behavior of Baboons and Early Man," in *Social Life of Early Man*, ed. Sherwood Washburn (London: Methuen, 1962), 91–105, 95.

10. Zuckerman, *Social Life of Monkeys and Apes*, 237.

11. Quoted by Allison Jolly, "The Bad Old Days of Primatology?" in *Primate Encounters: Models of Science, Gender and Society*, ed. S. Strum and L. Fedigan (Chicago: University of Chicago Press, 2000), 71–84, 78.

12. Donna Haraway, *Primate Visions: Gender, Race, and Nature in the World of Modern Science* (London: Verso, 1992), 164.

13. Shirley Strum, "Science Encounters," in *Primate Encounters: Models of Science, Gender and Society*, ed. Shirley Strum and Linda Fedigan (Chicago: University of Chicago Press, 2000), 475–497, 484.

14. Bertrand Russell, *An Outline of Philosophy* (London: Allen & Unwin, 1927).

15. Quoted by Thelma Rowell, "Variability in the Social Organization of Primates" in *Primate Ethology*, ed. Desmond Morris (London: Weidenfield & Nicolson, 1967), 219–235, 222.

16. Shirley Strum and Linda Fedigan, "Introduction," in *Primate Encounters: Models of Science, Gender and Society*, ed. Shirley Strum and Linda Fedigan (Chicago: University of Chicago Press, 2000), 3–49, 5.

17. For a good account of these issues, see Londa Schiebinger, *Has Feminism Changed Science?* (Cambridge: Harvard University Press, 1999).

18. This hypothesis, I should note, discloses some of the privileged links that our cultural tradition makes among women, nature, and emotions. See Vinciane Despret, *Our Emotional Makeup* (New York: Other Press, 2004), Donna Haraway, *Primate Visions*.

19. Linda Fedigan, "Science and the Successful Female: Why There Are so Many Women Primatologists," *American Anthropologist* 96 (1994): 529–540.

20. Haraway, *Primate Visions*, 397, n. 13.

21. Ibid., 124.

22. Rowell, "Variability in the Social Organization of Primates," 220.

23. Jeanne Altmann was a mathematician who worked in primatology with her primatologist husband, Stuart Altmann.

24. Rowell, "Introduction: Mothers, Infants and Adolescents," in *Female*

Primates: Studies by Women Primatologists, ed. Meredith Small (New York: Allan and Liss, 1984), 13–16, 16.

25. Haraway, *Primate Visions,* 301.

26. See Jolly, *Evolution of Primate Behavior,* 181; Strum and Fedigan, "Introduction," 12.

27. Strum showed that males' investment in "special relationships" with females had greater payoff than did a male's rank in a dominance hierarchy.

28. Considering the increasing number of women primatologists in the 1980s, some commentators have suggested that the choice of studying primates was linked to gender (e.g., Fedigan, "Science and the Successful Female").

29. I say "early" accounts because her descriptions of the chimpanzees changed dramatically at the end of the 1960s. See below for a plausible hypothesis.

30. Note that, in the beginning of the 1930s, Clarence Ray Carpenter observed howler monkeys in the island of Barro Colorado (Panama) and reported that howlers rarely threatened each other at all. Chacma baboons, observed by Ronald Hall in the beginning of the 1960s in South Africa, behaved, according to Rowell, much more like "her" baboons than the ones Washburn and DeVore observed, "so it was assumed that they belong to a different species." Rowell, "Forest Baboons—A Recantation," paper prepared for the Seminar of the Wenner-Green Foundation, "Baboon Field Research: Myths and Models," June 25–July 4, 1978.

31. Thelma Rowell, "The Concept of Social Dominance," *Behavioral Biology* 11 (1974): 136.

32. "This," according to Rowell, "may also explain the historical change in perspective about chimpanzee society observed by Goodall's team, with the original stories from each of several study sites being amazed that the peaceful nature of chimp society, and then increasingly stories about aggression." Provisioning had progressively accentuated competition among chimpanzees and produced social disruption. (Rowell interview; Rowell is referring to Margaret Power's *The Egalitarian: Human and Chimpanzee* [Cambridge: Cambridge University Press, 1991].)

33. Thelma Rowell, "A Few Peculiar Primates," in *Primate Encounters: Models of Science, Gender and Society,* ed. Shirley Strum and Linda Fedigan (Chicago: University of Chicago Press, 2000), 61.

34. Rowell, "Social Behaviour of Monkeys," 72.

35. Jolly, *Evolution of Primate Behavior,* 172.

36. On Sandra Harding's work, see Sarah Brake and Maria Puig de la Bellacasa, "Building Standpoints," 309–316, and Alison Wylie, "Why Standpoint Matters," 339–352, both in *The Feminist Standpoint Theory Reader,* ed. Sandra Harding (New York: Routledge, 2004).

37. Rowell, "Concept of Social Dominance," 132.

38. Sarah Ruddick, "Maternal Thinking as Feminist Standpoint," in *The*

Feminist Standpoint Theory Reader, ed. Sandra Harding (New York: Routledge, 2004), 161–168, 163.

39. Isabelle Stengers, *Power and Invention: Situating Science* (Minneapolis: University of Minnesota Press, 1997).

40. Concerning this issue, read the work of circus and zoo animals expert Henri Hediger, *Les animaux sauvages en captivité: Introduction à la biologie des jardins zoologiques* (Paris: Payot, 1953). See also Donna Haraway, *The Companion Species Manifesto: Dogs, People, and Significant Otherness* (Chicago: Prickly Paradigm, 2003).

41. Vinciane Despret, "The Body We Care For: Figures of Anthropo-Zoo-Genesis," in *Body & Society,* special issue, *"Bodies on Trial,"* ed. M. Berg and M. Akrich, 10 (2004): 111–134. Let me simply stress that the comparison with informal ways of knowing may avoid an essentialist definition of woman's way of thinking and considerably enlarge the definition of gender.

42. Valerius Geist, *Mountain Sheep: A Study in Behavior and Evolution* (Chicago: University of Chicago Press, 1971).

43. See, e.g., Alistair Lawrence, "Mother-Daughter and Peer Relationships of Scottish Hill Sheep," *Animal Behavior* 39 (1990): 481–486.

44. Thelma Rowell and C. A. Rowell, "The Organization of Feral Ovies Aries Ram Groups in the Pre-Rut Period," *Ethology* 95 (1993): 213–232. See also Thelma Rowell, "Till Death Do Us Part: Long-lasting Bonds Between Ewes and Their Daughters," *Animal Behavior* 42 (1991): 681–682.

45. Thelma Rowell, "The Ethological Approach Precluded Recognition and Reconciliation," in *Natural Conflict Resolution,* ed. Filippo Aureli and F. De Waal, 227–229 (Berkeley: University of California Press, 2000).

46. Rowell, "A Few Peculiar Primates," 69.

47. Rowell interview; see also Thelma Rowell, "The Myth of Peculiar Primates" in *Mammalian Social Learning: Comparative and Ecological Perspectives,* ed. Hilary O. Box and Kathleen R. Gibson, 6–16 (Cambridge: Cambridge University Press, 1999), 9.

48. Bruno Latour, "A Well Articulated Primatology: Reflections of a Fellow Traveler," in *Primate Encounters: Models of Science, Gender and Society,* ed. Shirley Strum and Linda Fedigan, 358–382 (Chicago: University of Chicago Press, 2000) 367.

49. "What I do is to give as many bowls as there are sheep, plus one, so that every sheep can always find a bowl for itself without having to compete with another sheep. And the bowls are far enough apart that you can't reach one from another. And my hope is that it would reduce their need for fighting between animals over food, which I don't want to happen." Rowell interview.

50. Haraway, *Primate Visions,* 398, n. 16.

51. Rowell interview.

FURTHER READING

Despret, Vinciane. "Sheep Do Have Opinions," in *Making Things Public: At-mospheres of Democracy,* ed. Bruno Latour and Peter Weibel (Cambridge: Zentrum für Kunst und Medientechnologie/MIT Press, 2005), 360–368.

Haraway, Donna. *Primate Visions: Gender, Race, and Nature in the World of Modern Science* London: Verso, 1992.

Rowell, Thelma. *Social Behaviour of Monkeys* (Harmondsworth: Penguin, 1972).

———. "The Concept of Social Dominance," *Behavioral Biology* 11 (1974): 131–154.

———. "The Myth of Peculiar Primates," in *Mammalian Social Learning: Comparative and Ecological Perspectives,* ed. Hilary O. Box and Kathleen R. Gibson (Cambridge: Cambridge University Press, 1999), 6–16.

Schiebinger, Londa. *Has Feminism Changed Science?* (Cambridge: Harvard University Press, 1999).

Strum, Shirley, and Linda Fedigan, eds. *Primate Encounters: Models of Science, Gender and Society* (Chicago: University of Chicago Press, 2000).

CHAPTER TWENTY

Bringing Statistical Methods to Community and Evolutionary Ecology: Daniel S. Simberloff

WILLIAM DRITSCHILO

When Daniel S. Simberloff began his call for more rigorous scientific methods in ecology, it was already under pressure for change. This infused arguments that should have been quietly technical with something of the *sturm und drang* of melodrama, both enriching and obscuring those issues. The neutral observer T. F. H. Allen, remarking that the dialogue being raised was something the discipline needed, nonetheless found "images of hand-to-hand combat or a bar-room brawl" coming to mind on reading certain exchanges.[1] One between the philosopher Marjorie Grene and Simberloff that Allen quoted in full shows both the combativeness and the humor of Simberloff's personality. Those two traits served him well in continuing what at times must have seemed like a losing struggle.

"Many biologists," Grene wrote, "when they turn to philosophical (epistemological or ontological) questions, abandon the standards of accuracy that, at least in the layman's view, ought to govern discourse as scientists. Simberloff's argument forms an unusually flagrant example of this practice."[2]

"Many philosophers," Simberloff responded, "when they turn to biological questions, abandon in favor of captious logomachy the quest for epistemological or ontological enlightenment that, at least in the layman's view, ought to govern their discourse. Grene's argument forms an unusually flagrant example of this practice."[3]

BACKGROUND: ECOLOGY IN REVOLUTION

It was the 1960s. There was a war in Vietnam, and burglars seemingly were operating out of the White House. There was riot and there was rebellion. And there was Earth Day at the end of the decade. From all this a single word, learned for the first time by many from Rachel Carson's 1962 book *Silent Spring,* summed up both what seemed to be wrong and how things could be set right. It was *ecology.* Then a quiet, arcane subject, ecology had been born essentially simultaneously with Darwin's theory of natural selection. Too often seen as a soft science, it was in the middle of redefining itself into a rigorous one that could hold its head high in the twentieth century. Although the ecological point of view, like natural selection, had seemingly permeated all of biology, ecology still had to fight for a separate identity in order to gain scientific legitimacy. For example, "What is ecology?" was asked in journals and books in almost every decade of the twentieth century. The question seldom had a simple answer, except that it should not be seen as the pursuit of nineteenth-century gentlemen "butterfly collectors."[4]

One useful way to make sense of the situation that Simberloff came upon is by looking at it as the aftermath of two major traditions in collision. One was the study of ecological communities based on their material and energy flows; the other was through mathematical description of interactions between species. One, then associated with Eugene P. Odum, came from a Midwestern tradition of classification, experimentation, and field manipulation; the other, associated with Robert H. MacArthur, took the observations of naturalists (those butterfly collectors) and added mathematical and Ivy League refinement to it. At stake were the normal rewards of prestige that come to the leaders of a field, ratcheted up at this time by the financial support that was being funneled into "Big Science" as an outgrowth of the Manhattan Project and by the National Environmental Policy Act. All of this was going on as notoriety in the public arena was linking the science with a philosophy of life, increasing the need for scientific legitimacy. Both camps, of course, claimed to be able to best deliver that legitimacy. In the early 1970s, it was the MacArthur forces that seemed to hold the upper hand to the point where dissenters complained in print of suffering censorship at their hands.[5]

Into this contentious arena stepped the newly minted Ph.D. Dan Simberloff. He hardly had the appearance of a revolutionary or even a manner that could be called rebellious, just a dry sense of humor

Figure 20.1. Dan Simberloff. Courtesy of Dan Simberloff.

and a tenacious opposition to illogic. Yet without question he did fashion a revolution that was very much an example of a T. S. Kuhn paradigm shift.[6]

CONFLICT: SHOULD ECOLOGY FOLLOW SCIENTIFIC METHOD?

The orthodoxy against which Simberloff rebelled, known as community ecology, had at its heart the fruits of a successful collaboration between Edward O. Wilson, an unprepossessing self-described naturalist but a weak mathematician, and the perfect partner for him: Robert MacArthur, who has been called "the James Dean of ecology" and who happened to know his math. By the time their book, *The Theory of Island Biogeography,* was published in 1967, MacArthur had arguably become the head of "establishment" ecology. His influence continued after his death in 1972. It was proclaimed as the "New Ecology" in a news article in *Science,* signifying a victory of sorts over the ecosystem ecology of Odum.[7]

Ecology is the science of working out how organisms are and become adapted to their environments. Defined most simply, ecological communities are groups of organisms interacting in the well-known roles found in predation, competition, and mutualism. Community ecologists codified the idea that Darwinian natural selection shapes species' characteristics into the Competitive Exclusion Principle, which states that no two species could occupy the same niche (the jargon of the field for identical ecological characteristics). Based on this principle, ecologists in the 1940s and 1950s began to build a body of theory to explain the biogeographic patterns that so influenced the ideas of Darwin and Wallace. This grew into the mathematical community ecology that came to an apex of sorts in the MacArthur and Wilson book. It combined evolutionary theory (natural selection through competition), mathematics (Lotka-Volterra–type population growth equations), and observations of community structure (which species were found where with respect to other species, something for which an answer was promised in the mathematics). Sometimes also known as species-packing or niche theory, it was an impressive recipe with which to build up a science, and it quickly attracted adherents, especially among young scientists, the best and brightest of their generation, being drawn into ecology by the social forces of the time.[8]

Simberloff, inspired to switch from math to biology while an undergraduate at Harvard University not much more than a year after the publication of *Silent Spring,* stayed on to get his Ph.D. there with encouragement from Frank Carpenter, E. O. Wilson's mentor. An introductory biology course Simberloff had taken from George Wald had brought about the switch, continuing an interest in the natural world that hitherto he had not considered as the source of a possible career. (He claims to have helped found a "Society of Entymologists [*sic*]" at the age of four while spending time in the woods and fields near Easton, Pennsylvania.)[9]

Carpenter told him, "Many areas of ecology and evolution were exciting fields where people with math backgrounds could do interesting work and were badly needed." It had been at a time when many ecologists were eagerly striving to add physico-chemical and mathematical tools to their methodology.[10]

For his dissertation Simberloff did something unusual in community ecology. He tested island biogeography theory experimentally by totally defaunating a number of mangrove islands off the Florida Keys. Some of the islands were also cut in two. He found data to support the theory's prediction that the number of species of arthropods represented on each island was a result of an equilibrium between the rate at which the species colonized it and the rate at which they went extinct on it. The rates also increased (or decreased) with island size, as predicted.

The work became an instant classic, winning Simberloff and Wilson the Ecological Society of America's prestigious Mercer Award in 1971.[11] It also represented one of the few experiments done in community ecology and one of the few of those that lent support to an important theory. Simberloff was therefore poised to receive the mantle of leadership of the MacArthur School as it settled into being establishment ecology, except . . . Except that there were rumblings to be heard. Nonmathematical types, those supposedly anachronistic butterfly collectors, began to grumble that mathematicians, some of whom were totally free of any mud on their boots (or had no field boots to muddy) were taking over their science from the cozy confines of their offices and computer labs. Presumably in defense of such accusations, one of ecology's mathematicians is known to have proudly shown off a gash he obtained in a fall while scrambling through the rocky intertidal zone at which he had aimed his mathematics.

Simberloff chose a more difficult path, however. He became one of the grumbling voices, not against the mathematicians—his thinking is structured by its formalism, after all—but against how the math was being used. As he read the ecological literature early in his career, it struck him that many of the MacArthur-inspired mathematical models "more or less fit some data" but had no reason to be preferred to others. "Being probabilistically inclined," in his own words, he began to ponder how one might choose one model, island equilibrium, for example, instead of another, given a particular community pattern. To Simberloff, dogma needed not only mathematical but also experimental justification. Experiments in ecology were hard to come by, however. Simberloff saw another approach: use the data available, but treat them statistically as one should in any other experiment. Soon after publishing his dissertation he made his first attempt at doing so. He started by examining the logic behind explanations for why the number of species of a particular genus found on islands is less than it is in mainland areas. As a baseline for comparison, he tried a simulation model in which species were randomly placed on islands. Its results were little different than the actual geographic data. The 1970 paper had little impact, but it was the start of his thinking about null models in ecology. He also at around this time collaborated with another group of iconoclasts (one of whom, Stephen Jay Gould, has his own place in this book) who were trying to bring mathematical techniques, in particular statistical ones that could curb unwarranted speculation, into the seemingly nonmathematical fields of paleontology and cladistics.[12]

Simberloff's rebellion became very much a collaborative effort when he took a post at Florida State University and recruited the like-minded but very independent Donald R. Strong to its faculty. An iconoclast Simberloff was, but he was no Darwin or Mendel, toiling alone in his garden. He was not Bateson or Goldschmidt, producing monographs and books by himself. Simberloff led a small army (if that term can apply to academics) that came to be known as the Tallahassee Mafia. For narrative convenience, I have focused on Simberloff.

Reading Amyen MacFadyen's 1973 Presidential Address to the British Ecological Society, though, was what motivated Simberloff to continue to develop his skepticism and helped him refine his ideas and find the right language to express them. He began to read works about the philosophy of science. He also recognized about then that "people desperately wanted to publish just to publish."[13]

Simberloff's campaign to reform ecology came at about the same time that the Canadian ecologist Robert Henry Peters was also questioning the logical basis for much of ecology and evolutionary biology, starting with the circularity of its most basic tenet: natural selection. Both Simberloff and Peters had read the European philosopher Karl Popper, as had MacFadyen in warning against the "bright ideas" that might tyrannize ecology. Being essentially an outsider and a single voice who went on to make himself even more of one by arguing that most of ecology (the MacArthur-style kind, in particular) should be abandoned in favor of studies that look in comparison like basic environmental engineering, Peters was easy for the field to dismiss. He was mostly ignored. Simberloff's response to the New Ecology was a different matter, however: instead of making a general attack, Simberloff looked at specific community studies. One of the first was his own.[14]

SLOSS: ISLANDS AND REFUGES

In order to understand the controversy about methodology that follows, one must first understand a related controversy: SLOSS (single *large* or several *small* refuges). More than merely related, they are intricately bound together both in logical foundation and in the personalities in contention.

MacArthur and Wilson's mathematical model for species' numbers on islands, $S = CA^z$, the species-area relationship, on first view appears to be every bit as elegant as $F = ma$ and $E = mc^2$. Conservation scientists called it the "most seminal branch of ecological theory" and "one of community ecology's few genuine laws." But in order to prove that the number of species on an island was the result of the equilibrium postulated by the theory, according to Simberloff, one had to prove that species were actually going extinct on the islands while others were colonizing them, which his work with Wilson had done. The same could not be said for any of the oceanic island chains to which the theory was being applied. The theory, wrote Simberloff, rather than being proved by his work, remained a hypothesis still.[15]

Meanwhile, investigators found all manner of applications of island biogeography theory merely by finding data that fit the species-area graph. Particular attention was drawn by confirmation of the species-area relationship for bird species in forest "islands," bearing out a suggestion made by MacArthur and Wilson. Imagine the landscape that is the eastern megalopolis of the United States and the significance of such forest "islands" becomes obvious. Others found evidence of historical "faunal collapse," a loss of species, when areas were cut off from similar habitats to become islands. Reduce area, and reduce the number of species at the new equilibrium in consequence.

Given that the theory of island biogeography applies to all sorts of situations that are not at all islands, there is no limit to its useful applications. Habitat islands in urban and suburban ecosystems? Why not? How about nature reserves? How about the entire earth?

To apply island biogeography theory to the whole world was not only tempting but was urged on by reputable ecologists. So was its application to parks and refuges. One of the most urgent and eloquent voices for it was that of the bird biogeographer and kidney physiologist (and a now best-selling anthropologist) Jared Diamond. Reasoning from species-area data with which he was most familiar, that of the birds of the New Hebrides Islands, Diamond argued that some "species would be doomed by a system of many small reserves, even if the aggregate area of the system were large." Conservation scientists immediately began to urge adoption of his ideas.[16]

Diamond had jumped into the conservation arena a bit too fast with his ideas, however, for Simberloff. He and Florida State colleague Lawrence Abele quickly pointed out that there were neither theoretical grounds nor evidence from data to justify the way that island biogeography theory was being applied to conservation. They showed mathematically that, based on MacArthur and Wilson's theory, two smaller areas would support more species than a single reserve of equal area. What is more, they had data: Simberloff's mangrove islands. Fragmented islands held more species in combination than any individual fragment of similar size.[17]

Simberloff and Abele could not slow the rush of wildlife conservationists to embrace Diamond's ideas, however. The conservation community came gunning for the two. It led to a very visible airing out of differences that could not have been good for the status of theoretical ecology as a useful scientific tool, and it helped add venom to the controversies about the methodology of science also being raised by Simberloff. "Simberloff and Abele pass over lightly" the need to minimize extinctions, hold a position that "depends on biologically unrealistic assumptions and should not be applied to any practical problem of conservation," and ignore "the fact that it is much easier to convert a natural area into a housing development than vice versa," according to some of the published rebukes.[18]

Jared Diamond, in his dissent to Simberloff and Abele's paper, pleaded from the outset that "*biologists have felt intuitively that most wildlife refuges are too small* to avert extinction of numerous species. However, because there has been no firm basis for even approximately predicting extinctions in refuges, *biologists have had difficulty convincing government planners* faced with conflicting land-use pressures of the need for large refuges" (emphasis added). To Diamond, the theory of island biogeography gave scientific credence to an intuitive belief. He went on to caution that "because those indifferent to biological conservation may seize on Simberloff and Abele's report as scientific evidence that large refuges are not needed, it is important to understand the flaws in their reasoning." The flaws, according to Diamond, were that Simberloff and Abele had not taken into consideration that some species have poor dispersal ability and others are especially susceptible to extinction at the low population sizes on small islands. Diamond did offer as a compromise "one refuge as large as possible plus some smaller ones."[19]

Simberloff and Abele replied that island biogeography theory is absolutely silent as to the identity of species. If there are species that need special attention, they should get it by way of models fitted to those species' particular ecology, rather than the untested models of community theory. The two, every bit as conservationist as the others, lamented being "cast as the *bêtes noire* of conservation," but they stuck to their principles.[20]

Eventually, although island biogeography theory remains in many

current textbooks, conservation-minded ecologists found other ratio-
nales for large preserves. Simberloff has remained active in conservation biology, particularly invasion biology. He continues to redress
logical and evidentiary fallacies that he finds there, but he has abandoned refuge design as "more political than scientific."[21]

The animosity that built up between Diamond and Simberloff starting with the SLOSS argument soon became personal. Simberloff's
reserved personality, which made him seem aloof and difficult to
those who did not know him, when combined with his sense of justice,
may have fueled the rancor that eventually colored their disagreement. ("He would gladly cut the throat of an ill-conceived idea and
swat down an innocent misstatement like a fly," David Quammen has
written of Simberloff in his popular book.)[22] There may be reason to
speculate that Simberloff, having grown up in a working-class Jewish
home, may not have been as attuned as others to the social nuances
of the gentlemanly club that had grown up around MacArthur in a
science not notable up to that time for its diversity.

Within a few years, Diamond and Simberloff were no longer speaking to each other. Diamond would not accept invitations to meetings
at which Simberloff was present. Communication was through inter-
mediaries or journal pages only.

Aspects of island biogeography theory unrelated to conservation
were also in dispute between junior members of the Tallahassee Mafia and acolytes of Robert May at Princeton, but these did not directly involve Simberloff and need not be touched on further, except
to emphasize the polarization into camps and the spilling over into
another arena—that of the role of competition in structuring communities, in particular island communities. The two camps began to
look like Ivy-League-and-West-Coast establishment versus southern
school upstarts.[23]

THE DARK HEART OF THE MATTER: NULL MODELS

Science normally progresses through a Cartesian process of deduction and testing that has evolved into hypothesis and experiment,
the scientific method. Obviously, not all science fits that statement,
nor can it. Geologists, for example, are limited in the kind of experimentation they can do. They most often use patterns, present and
historical, that can be explained or not by competing theories. Some
of the theories, most often based on an inferred history, have much

explanatory power but few actual data to support them. They have been called "geopoetry." Biogeographers and community ecologists have similarly made explanations based on observed patterns, rather than experimentation, but on islands they have had available to them "natural experiments" resulting in those patterns. It is not an easy matter to take a chunk of mainland and add or remove species to it, then follow changes for a hundred—or ten—years and try to make some sense of them. On islands, though, exactly that has apparently happened naturally. Often, as in the Galapagos, there are even similar islands present that can serve as replicates. Islands are natural experiments waiting for interpretation. The rewards of doing so can be great, as so many, starting with Darwin, have found.

Problems arise, however, when patterns on islands are used to explain, well, the patterns on islands—a tautology. Simberloff accused Diamond of doing just that in Diamond's highly applauded theory of community assembly in explanation of the occurrence of bird species on the various islands of the New Hebrides chain. Simberloff was perhaps accusing Diamond—and others of the MacArthur School—of "ecopoetry."

The paper in which Diamond presented his community assembly rules was part of a symposium of Robert MacArthur's "friends, colleagues, and relatives" who gathered together at Princeton in a sort of a wake after his death. Like MacArthur (of whom it has been said by an admirer that "having bypassed the normal review process, he was well on his way to becoming famous") and MacArthur's mentor, G. E. Hutchinson, Diamond also took to publishing his most important work in unrefereed or scarcely refereed publications. Self-refereed it was, but Hutchinson, by then the Grand Old Man of Ecology, called Diamond's chapter "a very impressive work" in his influential textbook. Ed Wilson and a co-author implied in the same volume as Diamond's paper that his assembly rules put it "within the power of science not merely to hold down the rate of species extinction but to reverse it." It was the "species packing problem," a basic idea in community ecology related to competitive exclusion, on which hopes could be based, with Diamond's assembly rules being one of its "more sophisticated" developments. Simberloff, of course, was unimpressed. He and his student Edward F. Connor, in a paper having to do with studies of Darwin's iconic finches, dismissed Diamond's ideas as "based on weak inference."[24]

The matter wound up totally unsettled, with disputants invoking homilies from Uncle Remus on one side for sympathy, the insult of mathematical degeneracy on the other, and parodies, musical and otherwise, on both sides. However, the central question throughout was quite clear: should ecology follow scientific method?[25]

There were those who argued that it should not. Diamond claimed scientific affinity for ecology with geology and astronomy, among other fields, in which there is reasonably little experimental tradition. Jonathan (now Joan) Roughgarden, standing in for the ghost of Robert MacArthur, perhaps, claimed common sense as method enough, justifying it by the need to maintain the rising status of ecology. Most, however, conceded that ecology should follow scientific method, but began to take issue with the specifics of Simberloff's methodology. One pair of noted biologists offered to clear up Simberloff's confusion using a "Marxist approach."[26]

But so much for the sociology of the events; they cannot be fully analyzed within the confines of this chapter. The methodological issues are more tractable. Experiments in biology, even those with good controls, rely on statistics to separate real results from the nonsense produced by uncontrolled random variation. For natural experiments, Simberloff and his Tallahassee Mafia were suggesting that a pattern observed in nature that was thought to result from some particular process should be tested for the likelihood that it was created simply by chance. In a series of papers seemingly churned out furiously at Florida State University, they tested a number of old (and new) chestnuts in ecology against null models of their creation, Charles Elton's species-to-genus ratio, W. L. Brown and E. O. Wilson's character displacement, Hutchinson's 1.3 size ratio, and Diamond's assembly rules for communities foremost among them. What they shared was competition and competitive exclusion, a major prop of almost all evolutionary theory. Not even Darwin's finches were immune to their scrutiny. They could find no difference between the patterns for which competition was inferred and patterns that they generated at random. Competition, they claimed, should be rejected on elementary statistical principles.

They were not denying that competition existed but said that it should not be assumed in explanation without direct proof. Often, based on the vituperativeness of reaction against them, that disclaimer may have gone unheard or unbelieved. What they were saying was that, absent controlled experimentation for confirmation, here was their null model with which to see if at least the pattern being explained required any biology in explanation.

Like any test against a null hypothesis, null models require a certain sophistication of design. The simplest null model for island studies was one in which species were placed on islands at random. Every species has an equal chance of being on any island. This is unrealistic, but it can be modified to allow for islands to have only a certain number of species, differences in dispersal abilities for species, and so forth. These changes, of course, introduce biology into what ostensibly is a random model.[27]

For simplicity, the simplest null model has been called "random colonization." Unfortunately, the term has introduced an element of semantic confusion that has clouded the importance of the issues, especially to those looking in from outside the science. The two sides have been falsely characterized as competition (Diamond) and random colonization (Simberloff). Instead, the Tallahassee Mafia's null models of random colonization were meant to be *nonexplanatory,* null hypotheses in every sense. One could as easily have set up a null model by which all the species occupied all the islands, then went extinct randomly, putting cards back into the deck, in a sense. The important idea behind the null models was that they represented statistical tests of data. Nothing more, nothing less. But that was plenty.[28]

Simberloff and his colleagues admitted the difficulty in developing appropriate null models, but for some of their detractors, the difficulty was enough to bar their use. Others, all very closely tied to the MacArthur School (Diamond and Roughgarden, for example), fought their use on higher grounds. These can be summed up (in Roughgarden's numbering) as follows: (1) science has no formal rules such as a specific scientific method, (2) Wittgenstein's philosophy has overshadowed Popper's, (3) null models are hard work, and (4) the success of models based on Lotka-Volterra–type equations should not be underestimated. After all, these were not only the basis of island biogeography theory, but they led to Robert May's discovery of chaos in ecology.

Funding for ecology also was at issue in the dispute. This came about in two ways. Simberloff and colleagues were admonished by Roughgarden not to jeopardize the success of the MacArthur School in obtaining more funding for ecology. It was not acknowledged, however, that in a zero-sum process, MacArthur's followers were taking funds away from other avenues of research. This was brought home in no uncertain terms to those at Florida State. Not only were promising studies denied funding, but grant reviewers' comments left their antipathies toward null models undisguised, based on their level of vitriol. Requests to the National Science Foundation for funding "got vicious, vicious grant reviews back," Nicholas Gotelli remembers. "I mean some really nasty, unbelievable stuff."[29]

EPILOGUE

Ecology came around, however. The unrelenting logic of Simberloff's arguments wore down that of the opposition, moving ecology from a mainly descriptive science, mathematical or otherwise, in which deduction was often supported by the observation generating it, to its current state. The constitution of a proper null model for a particular observation is still unresolved, but that ecology is essentially an experimental science requiring statistical rigor—null models, if applicable—is no longer in doubt.

NOTES

1. T. F. H. Allen, "The Noble Art of Philosophical Ecology," *Ecology* 62 (1981): 870.

2. Marjorie Grene, "A Note on Simberloff's 'Succession of Paradigms in Ecology,'" *Synthese* 43 (1980): 41.

3. Daniel Simberloff, "Reply," *Synthese* 43 (1980): 79.

4. See, e.g., letters by Charles E. Bessey, W. F. Ganong, and Theo Gill, "The Word 'Ecology,'" *Science* 15 (1902): 593–94; A. G. Tansley, "Presidential Address," *Journal of Ecology* 2 (1914): 194–202; V. E. Shelford, "Guide to the Study of Animal Ecology," *Science* 39 (1914): 581; Walter P. Taylor, "What Is Ecology and What Good Is It?" *Ecology* 17 (1936): 33–46; Lee R. Dice, "What Is Ecology?" *Scientific Monthly* 80 (1955): 346–51; Paul B. Sears, "The Place of Ecology in Science," *American Naturalist* 94 (1960): 193–200; Edward S. Deevey Jr., "What an Ecologist Isn't," *Bulletin of the Ecological Society of America* 53, no. 2 (1972): 5–6; Robert Platt, "Who Speaks for Ecology?" *Bulletin of the Ecological Society of America* 55, no. 4 (1974): 3–8; P. F. Owen, *What Is Ecology?* 2d ed. (Oxford: Oxford University Press, 1980); Mark Westoby, "What Does 'Ecology' Mean?" *Trends in Ecology and Evolution* 12 (1997): 166, and William Dritschilo, *Earth Days: Ecology Comes of Age as a Science* (iUniverse, 2004), http://www.ebookmall.com/ebooks/earth-days-dritschilo-ebooks.htm.

5. Leigh Van Valen and Frank Pitelka, "Commentary: Intellectual Censorship in Ecology," *Ecology* 55 (1974): 925–26.

6. Ecologists have readily accepted that they were undergoing a Kuhnian paradigm shift at this time. See Michael H. Graham, Paul K. Dayton, and Mark A. Hixon, "Paradigms in Ecology: Past, Present, and Future," *Ecology*

83 (2002): 1479–80, and papers that follow; but see Robert T. Paine, "Advances in Ecological Understanding: By Kuhnian Revolution or Conceptual Evolution?" 1553–59, for a dissenting view.

7. "James Dean" is from David Quammen, *The Song of the Dodo: Island Biogeography in an Age of Extinction* (New York: Simon & Schuster, 1996), 177. The news articles are Gina Bari Kolata, "Theoretical Ecology: Beginnings of a Predictive Science," *Science* 183 (1974): 400–1, 450; Roger Lewin, "Santa Rosalia Was a Goat," *Science* 22 (1983): 636–39.

8. Note that throughout the history of ecology there were plant and animal physiologists who thought of themselves as ecologists and still do. See G. Richard Tracy and J. Scott Turner, "Commentary: What Is Physiological Ecology? Introduction," *Bulletin of the Ecological Society of America* 63 (1982): 340–31, and letters that follow. In addition, many ethologists still align themselves with ecology.

9. Daniel S. Simberloff, emails to author, October 23, 2002, and September 6, 2005.

10. Simberloff email, October 23, 2002.

11. Daniel S. Simberloff and Edward O. Wilson, "Experimental Zoogeography of Islands: The Colonization of Empty Islands," *Ecology* 50 (1969): 278–96, the paper that won the Mercer Award, is included as one of forty most important papers of ecology in Leslie A. Real and James S. Brown, eds., *Foundations of Ecology* (Chicago: University of Chicago Press, 1991).

12. Simberloff email, October 23, 2002. Daniel S. Simberloff, "Taxonomic Diversity of Island Biotas," *Evolution* 24 (1970): 23–47; David M. Raup, Stephen Jay Gould, Thomas J. M. Schopf, and Daniel S. Simberloff, "Stochastic Models of Phylogeny and the Evolution of Diversity," *Journal of Geology* 81 (1974): 525–42.

13. Daniel S. Simberloff, email to author, August 29, 2005.

14. Amyan MacFadyen, "Some Thoughts on the Behaviour of Ecologists," *Journal of Animal Ecology* 44 (1975): 351–63. See Robert Henry Peters, "Tautology in Evolution and Ecology," *American Naturalist* 110 (1976): 1–12; Robert Henry Peters, *A Critique for Ecology* (Cambridge: Cambridge University Press, 1991); Daniel Simberloff, "Species Turnover and Equilibrium Island Biogeography," *Science* 194 (1976): 572–78.

15. O. H. Frankel and Michael E. Soulé, *Conservation and Evolution* (Cambridge: Cambridge University Press, 1981), 101; Thomas W. Schoener, "The Species-Area Relationship within Archipelagos: Models and Evidence from Island Land Birds," *Proceedings of the International Ornithological Congress* 16 (1974), 629, cited in Stephen J. Gould, "An Allometric Interpretation of Species-Area Curves: The Meaning of the Coefficient," *The American Naturalist* 114 (1979), 335–343; Simberloff, "Species Turnover."

16. Jared M. Diamond, "The Island Dilemma: Lessons of Modern Biogeographic Studies for the Design of Nature Reserves," *Biological Conservation* 7 (1975): 143.

17. Daniel S. Simberloff and Lawrence G. Abele, "Island Biogeography Theory and Conservation Practice," *Science* 191 (1976): 285–86.

18. John Terborgh, Robert F. Whitcomb, James F. Lynch, Paul A. Opler, and Chandler S. Robbins, "Island Biogeography and Conservation: Strategy and Limitations," *Science* 193 (1976): 1029–32.

19. Jared M. Diamond, "Island Biogeography and Conservation: Strategy and Limitations," *Science* 193 (1976): 1027–29.

20. Daniel S. Simberloff and Lawrence G. Abele, "Island Biogeography and Conservation: Strategy and Limitations," *Science* 193 (1976): 1032.

21. Simberloff email, October 23, 2002.

22. Quammen, *Song of the Dodo,* 483.

23. Nicholas Gotelli, interview with author, January 5, 2004.

24. G. Evelyn Hutchinson, *An Introduction to Population Ecology* (New Haven: Yale University Press, 1978), 165, n. 34; Edward O. Wilson and Edwin O. Willis, "Applied Biogeography," in *Ecology and Evolution of Communities,* ed. Martin L. Cody and Jared M. Diamond (Cambridge: Harvard University Press, 1975), 522–34; Edward F. Connor and Daniel S. Simberloff, "Species Number and Compositional Similarity of the Galápagos Flora and Avifauna," *Ecological Monographs* 48 (1977): 225. See Martin L. Cody and Jared M. Diamond, eds., *Ecology and Evolution of Communities* (Cambridge: Harvard University Press, 1975); Stephen D. Fretwell, "The Impact of Robert MacArthur on Ecology," *Annual Review of Ecology and Systematics* 6 (1975): 1–15.

25. See Edward F. Connor and Daniel S. Simberloff, "Rejoinders," in *Ecological Communities: Conceptual Issues and the Evidence,* ed. Donald R. Strong Jr., Daniel S. Simberloff, Lawrence G. Abele, and Anne B. Thistle, 341 (Princeton: Princeton University Press, 1984); and Michael E. Gilpin and Jared Diamond, "Rejoinders," in ibid., 336.

26. Jared Diamond, "Overview: Laboratory Experiments, Field Experiments, and Natural Experiments," in *Community Ecology,* ed. Jared Diamond and Ted J. Case, 3–22 (New York: Harper and Row, 1986); Jonathan Roughgarden, "Competition and Theory in Community Ecology," *American Naturalist* 122 (1983): 583–601; and Richard Levins and Richard Lewontin, "Dialectics and Reductionism in Ecology," *Synthese* 43 (1980): 47–78.

27. See Robert K. Colwell and David Winkler, "A Null Model for Null Models in Biogeography," in *Ecological Communities: Conceptual Issues and the Evidence,* ed. Donald R. Strong Jr., Daniel S. Simberloff, Lawrence G. Abele, and Anne B. Thistle (Princeton: Princeton University Press, 1984), 344–59.

28. Null models were not new to ecology. Donald R. Strong Jr., "Null Hypotheses in Ecology," *Synthese* 43 (1980): 271–85, points out earlier attempts that had failed to persuade most ecologists. The time was not right for them. Null models should not be confused with neutral models. Hal Caswell, "Community Structure: A Neutral Model Analysis," *Ecological Monographs* 46 (1976): 327–54, specifically distanced his approach from that of null models.

See also Stephen P. Hubbell, *The Unified Neutral Theory of Biodiversity and Biogeography* (Princeton: Princeton University Press, 2001) for a random colonization and extinction model that is meant to be explanatory.

29. Gotelli interview.

FURTHER READING

"A Round Table on Research in Ecology and Evolutionary Biology." *American Naturalist* 122 (1983).

Dritschilo, William. *Earth Days: Ecology Comes of Age as a Science* (iUniverse, 2004), http://www.ebookmall.com/ebooks/earth-days-dritschilo-ebooks.htm.

Gotelli, Nicholas J., and Gary R. Graves. *Null Models in Ecology* (Washington, D.C.: Smithsonian Institution Press, 1996).

Saarinen, Esa, ed. *Conceptual Issues in Ecology* (Boston: D. Reidel, 1980).

Strong, Donald R., Jr., Daniel S. Simberloff, Lawrence G. Abele, and Anne B. Thistle, eds. *Ecological Communities: Conceptual Issues and the Evidence.* (Princeton: Princeton University Press, 1984).

Epilogue:
Legitimation Is the Name of the Game

R. C. LEWONTIN

The stories of new ideas in biology that have been recounted in this book can only be understood as part of a historical process if the words *rebel* and *iconoclast* are taken seriously. To be a rebel or an iconoclast is not the same thing as being someone with a heterodox view of some matter. Several times a year I receive manuscripts from people, amateurs of science, who have a new general theory of almost everything, and for all I know one or another of these theories might turn out to be right. Yet none of these intellectual efforts can be thought of as rebellion or iconoclasm. They are idiosyncratic, that is, literally personal, private, peculiar. Acts of rebellion are struggles against a reigning social or political power and can only be understood in the context of the social structures that provide the power to individuals and groups that enable them to maintain and propagate their influence. Even the seemingly personal act of rebellion of child against parent is a struggle against institutionalized power. Breakers of idols are not smashing mere representations of others' gods but destroying potential rallying points for the collective activity of other sects. The Mosaic prohibition against graven images was precisely to prevent competing forms of worship from organizing. ("I am the Lord *thy* God. Thou shalt have no other gods *before me.*")

To understand the problem of establishing a new view of some natural phenomenon is to understand the problem of introducing that view into collective consideration and final acceptance by the social and political organization that constitutes Science. The operation of that social and political organization has four interlocking structural elements with which the rebel or iconoclast must cope, if there is any hope of incorporating a heterodox view into the corpus of accepted scientific knowledge. There is considerable interdependence of these elements by which they potentiate and support each other. The details of each of these elements of legitimation are well known by scientists, but it is their integration into a single coherent social and political system of power against which the rebel must contend that concerns us if we are to understand the individual histories recounted in this book.

The elements against which the rebel or iconoclast must contend are as follows:

1. *Public communication.* Communication of the heterodoxy to a relevant scientific community must take place by means that are already part of the communally accepted channels of information passage. Essentially these are the standard journals of technical communication and, to a much lesser extent, invited talks and lectures at universities and scientific meetings. Journals and lectures are arrayed along a scale of prestige so that, for example, publication in the notoriously selective journal *Nature* is worth more than publication in a less selective and less widely read journal, say, *Theoretical and Applied Genetics,* and an invited keynote lecture at an international congress confers much greater legitimacy than a presentation at a weekly seminar of a research group.

Books have a peculiar role in legitimation. Book publishers include university presses and commercial publishers with specialized technical series such as the Princeton University Press monographs or the Springer publications in mathematical biology, both of which confer considerable legitimacy on the work of their authors and are at least as difficult of access as to high-prestige journals. But there are also vanity presses, and there is the possibility of self-publication in print from computer files. These, in general, do not confer the needed entree for heterodoxy. In contrast, the inclusion of a scientist's findings in textbooks written by others establishes that work as integral to the fundamental corpus of knowledge.

2. *Employment and promotion.* To be a scientist is to be employed as one. Biologists are almost exclusively employed in universities and colleges or in specialized government or private research institutes. To carry out research requires time and, usually, access to research facilities, although theoretical research that does not require large-scale computing facilities could be carried out under less favorable circumstances. Whereas an insurance executive such as Charles Ives can be a successful composer, the scientist's ability to do creative work

and the legitimacy of that work are strongly dependent on employment by an appropriate institution. Not only initial employment, but promotion, shifts of employment to more prestigious and better-paying institutions, and lifetime tenure depend on research produced.

3. *Professional dependents.* The research scientist is surrounded to varying degrees by graduate students and postdoctoral fellows who are largely responsible for the day-to-day production of research results that are the career currency of the scientist. The future legitimacy of these novices depends on that of their sponsor and of the work that they themselves produce.

4. *Grants.* Experimental work and large-scale computer simulation projects depend on the receipt of government or, more rarely, foundation or commercial grants and contracts. These grants pay for laboratory expenses, partial salaries for principal investigators, and support for graduate students, postdoctoral fellows, and laboratory assistants. Grants are awarded on the basis of detailed justifications and research plans provided by the applicant investigator. The receipt of such support in the face of a very selective competitive process is a critical element in the reinforcement and legitimation of the hypotheses and research agenda of the applicant.

Although every professional is aware of these elements as essential to the building of a career and of acceptance of one's work into the body of agreed-upon science, it is not always understood that they are all under the control of a single powerful, socially determined source of power: the agreed-upon body of peers. For each of these four elements there must be acceptance and validation by scientists whose own work has marked them, by a mixture of formal and informal processes, as qualified representatives of "science." Manuscripts submitted to journals are given a first screening by an editor who, in the case of journals under the control of a scientific society, has been elected by members of the society. If the manuscript falls within certain broad limits of acceptability it is then sent to several reviewers for detailed scrutiny and eventual approval or disapproval; these reviewers have been chosen because their own work has already been validated by the peer community. In the case of books, many are produced because they have been solicited by publishers, based on the scientific reputation of the prospective author. If they have not been solicited, they may be sent to a publisher after prior conversations initiated by the author who is known by reputation. In any event, such book manuscripts are also submitted to a body of peer referees.

Employment and promotion depend, in the case of first appointment, on the recommendation of established scientists who have participated in the training of the candidate. For promotions and more senior appointments, the opinions of other senior scientists are solicited. The process of appointment has undergone an instructive evolution in the past thirty years in the United States. In the more remote past candidates for academic positions were identified by an informal process of communication between established scientists. On hearing through informal sources that a new position had become available, professors would recommend their students and postdoctoral fellows to their professional acquaintances. For appointments at the senior level, colleagues would approach colleagues and attempt to woo them away from their present positions. I got my first job, in 1954, because my professor was asked to recommend someone, and my three subsequent moves from one university to another, the last in 1973, were a consequence of direct personal contacts from colleagues. On the surface, all that has changed. Beginning in the 1970s in the United States, there was a rising consciousness that an "Old Boys Club" controlled the process of academic appointment to the detriment of women and minorities. This consciousness resulted in federal regulations requiring that all academic positions be formally advertised and that a justification be provided for the failure to appoint women or minority candidates. The advertisement of positions was already long established in the British Commonwealth, although in Australia the advertised pay scales for women were at one time lower than those for men. The formal open solicitation of applications for jobs at all levels has certainly increased the number of applicants, but it is not at all clear that it has changed the essential process of appointment. A heavy emphasis in the appointment process is still placed on the evaluations and recommendations of senior scientists, whose opinions have been solicited by the appointing committees. There has been little difficulty in providing unchallenged reasons for the failure to appoint women and minorities. The "Old Boys (and Girls) Network" still works.

Graduate students and postdoctoral fellows choose the institutions and sponsors to whom they apply based on their evaluation of the future career consequences of their choice. As novices their choices must be largely informed by the reputation of those sponsors, reputations that have reached the students through the medium of textbooks, lectures, scientific publications, and the advice of their teachers. The perceived status of their prospective mentors depends both on the importance given by the community of scientists to the subject matter and on the reputation of the individual. In both ways, communal views of what is important and who does it best are transmitted to the next generation of prospective community members.

The awarding of grants for research is, for the most part, the most overtly peer-dependent process of all. The National Institutes of Health and the National Science Foundation, for example, clearly state that grant applications are "peer reviewed." In practice this means that a rotating group of senior scientists who have themselves been recipients of grants at one time or another are asked by the scientific officers of the granting agencies to serve on so-called study sections or panels that review the proposals and judge their merits, establishing an order of priority for funding. Because only about 10 percent of proposals can be funded and because it is difficult, if not impossible, to carry out research without grant funds, the priority judgments of the peer reviewers have a determining effect on the problematic and the methodology of the science. Investigators, acutely conscious of the competition for funds, are constrained to propose research that not only fits clearly within the orthodoxy of the field but reflects current fashions in detail. It is widely understood that if one wants to do something innovative, even within the orthodox framework of problematic and theory, the best way is to divert funds from already approved research and then, if the experiments work, to use the results to justify a new research proposal. That is precisely what Jack Hubby and I did when we introduced gel electrophoresis of proteins into the problematic of population genetics, creating a fad that lasted twenty years.

The scientific officers of funding agencies can and do intervene to divert some funds to research areas that they judge to have been neglected, for example, by putting out requests for proposals to encourage research in such areas. These interventions are meant to alleviate an excessive narrowing of research agendas to what is most fashionable at the moment or to encourage research on specialized issues that are essential to a field but are not sufficiently glamorous. They are not designed to encourage iconoclasm and heterodoxy.

The four elements that enforce the scientific orthodoxy against which the rebel struggles are in a dialectical relation with the composition of the peer group that controls them. What is acceptable in science is decided by a group of peers with institutional power. But who are those peers? Why, they are the people who have successfully participated in building the science and its institutions. "Science" is what those who are admitted to be "scientists" do, and "scientists" are those who do what we all agree is "science." How, then, can there ever be a successful rebellion? How can what was once a heterodoxy be admitted into the corpus of orthodox knowledge and the iconoclast be incorporated into the body of peers? The general interest of the various histories recounted in this book is precisely in the illustration of the various dynamics that make such inclusion of the excluded possible. Although many histories illustrate the importance of persistence and of independent discoveries and confirmations made by other scientists, some of the most interesting cases involve the social structure of science itself.

A unique history is that of Barbara McClintock. As a consequence of her early work showing the cytological basis of Mendelian genetics, McClintock was in an extraordinary position of reputation and independence. When I was a high school student in the mid-1940s she was held up to me as an example of a famous woman scientist, among the first women to be elected to the National Academy. She belonged to the pantheon that included Thomas Hunt Morgan and Calvin Bridges. As a consequence of the general discrimination against women in universities and also because of her personal style of research and interaction she was, from the middle of her career, a staff member at the Cold Spring Harbor laboratories of the Carnegie Institution, where she was provided with the modest funds and infrastructure that she needed for her crossing experiments with mutations in maize. Thus she was freed of the necessity of moving in a university hierarchy, of constant publication, of recruiting and managing a group of students and postdoctoral fellows, and of obtaining grants from a peer-controlled system. She was free to pursue her heterodoxy. Moreover, the very high status she had attained on the basis of her earlier work made it impossible to casually dismiss her unorthodox claims about the mobility of genetic elements. I remember attending lectures she gave on the *Dt-Ds* system in the late 1950s and 1960s and hearing my colleagues as they left the lecture hall say things like, "That is the craziest thing I ever heard, but if Barbara says it there must be something in it." What the case of Barbara McClintock illustrates is that an acknowledged member of the scientific peerage can achieve credibility for a heterodox claim provided that the claim is within the domain of the scientist's already acknowledged credibility. The importance of this last proviso is illustrated by a case not included in this book.

Linus Pauling has an unchallenged place in the history of modern chemistry. He was awarded the Nobel Prize in chemistry in 1954 for his fundamental work on the nature of the chemical bond and was generally conceded to be the leading theoretical chemist of the time. Late in life, Pauling began to agitate for the use of massive doses of Vitamin C for treating the common cold and carried out a highly visible campaign for this therapy. This campaign never succeeded, and his views concerning the matter are seen as a kind of crackpot manifestation of advanced age in a scientist dizzy with previous success. After all, if you have won two Nobel Prizes, one in chemistry and one for promoting peace, you might be forgiven a certain hubris. What the case of Pauling illustrates, in contrast to that of McClintock, is that a high position in the scientific peerage confers a priori credibility for an unorthodox scientific claim provided that the domain of that claim is close to the previously acknowledged expertise. Successful scientists make deposits in an account in the Bank of Credibility. But every claim made outside their area of acknowledged expertise is a withdrawal from that account and even if, as in the case of Pauling, there is a very large balance, the account can be overdrawn. Pauling overdrew his account.

Another illuminating case is in the contrast between the careers of William Hamilton (Chapter 16) and someone to whom an essay in this book is not devoted: George Price.

It is not at all clear that Hamilton should be thought of as a rebel or an iconoclast in the sense of someone who was contradicting some well-established dogma of science such as the fixity of the position of genes on chromosomes. Rather, he devoted himself to explaining a puzzling aspect of evolution under natural selection that had been given only minor attention by the body of evolutionary theorists and population geneticists. The question was how to explain phenomena that involved altruism, the apparent sacrifice of individual fitness in the interest of increasing the fitness of others. The classic examples were the sacrifice of resources by parents to increase the resources available to offspring and the existence of sterile castes of ants whose activities serve the survivorship and reproduction of other colony members. Evolutionists had long recognized the problem raised by these phenomena for the standard theory of evolution by natural selection.

There was an attempt to deal with the issue by theories of group- and population-level selection, but the main reaction of evolutionary and population geneticists was to avoid the problem, treating it as puzzling but marginal to the main body of theory. Standard synthetic treatments of population genetics ignored it. Unlike the direct challenge to the central dogmas of genetics posed by McClintock's mobile genes or Temin's reverse transcription of RNA into DNA, the problem posed for evolutionary biology by observations of apparent altruism could always be handled by an appeal to higher-level population phenomena that in no way threatened accepted basic mechanisms of genetic evolution.

Hamilton's solution to the problems raised by apparent self-sacrifice was a theory of kin selection, which made the differential replication of genes, rather than the rate of reproduction of individuals, the central analytic device. In doing so he did not propose to overthrow the main apparatus of evolutionary genetics, but only to amplify it by reorienting its analysis. Indeed, he did not invent the principle of kin selection. J. B. S. Haldane is famously said to have remarked that, from the standpoint of natural selection, one could sacrifice one's life to save two of one's brothers. The cause of Hamilton's failure to thrive academically during the first half his career lay not in the scientific establishment's hostility to his ideas but in its indifference to them.

All that changed dramatically in 1975 with the appearance of E. O. Wilson's *Sociobiology*. Suddenly the problematic of the evolution of social interactions was pushed to the center of evolutionary consideration by a prominent figure in the peer structure, and a major reorientation of interest occurred. In the last quarter of the twentieth century, sociobiology and its successor, evolutionary psychology, formed a major theme in evolutionary biology, and with it, the theory of kin selection and "Hamilton's Principle" achieved iconic status.[1]

Hamilton's later academic and institutional recognition was linked to the new centrality conferred on his work by a reorientation of the evolutionary problematic.

Hamilton's work on kin selection intersected with that of another outsider, George Price, who, unlike Hamilton, never succeeded in becoming a member of the scientific peerage. Price came to evolutionary theory late in his life from outside of biology. He never held an academic position, was in his last years an associate of the Galton Laboratory in London, where he was given office space, and, after eight years there, committed suicide. During the period between 1970 and his death in 1975 he reformulated the standard equations of population genetics in such a way that all forms of selection including kin selection were subsumed under them. His approach, embodied in what became known as the "Price Equation," which was based on the covariation between fitness and phenotype, also clarified and amplified Fisher's Fundamental Theorem of Natural Selection. In 1970, by prearrangement, he and Hamilton published separate papers about kin selection in the same issue of *Nature.* In yet another aspect of his work, Price also inspired Maynard Smith's development of evolutionary game theory.[2]

Despite the recognition of Price's contributions to evolutionary theory by two prominent evolutionists and by colleagues such as Harry Harris and Cedric Smith at the Galton Laboratory, he remained until the end of his life an outsider and even now is someone whose name is virtually unknown in the evolutionary biology community at large. His failure to prosper, in contrast to the eventual success of Hamilton, raises interesting problems for the historian and sociologist of science. It would be easy to dismiss Price as a brilliant although psychologically unstable personality, but he seems to have been no more eccentric than a number of quite successful scientists.

Each of the histories discussed in this book presents idiosyncratic elements in the development of both individual careers and the body of accepted science. In considering these case histories it is important not to be so involved with historical idiosyncrasies that we lose sight of the significant structural properties of science as a socialized process carried out in a strongly structured environment.

NOTES

1. An analysis of the causes for the explosion of this major intellectual movement in biology, psychology, sociology, and anthropology has yet to be produced. Ullica Segerstrale's *Defenders of the Truth* (Oxford: Oxford University Press, 2000) deals with the intellectual and ideological controversies that arose concerning sociobiology but not with the intellectual and institutional forces that led to its rise.

2. An extensive treatment of Price's life and work is given in Steven A. Frank, "George Price's Contributions to Evolutionary Genetics," *Journal of Theoretical Biology* 175 (1998): 373–388.

Contributors

Garland E. Allen teaches biology and history of science at Washington University in St. Louis. His special interests are the history of genetics, embryology, and evolution and their interrelationships. He has also worked on the history of eugenics, particularly in the United States.

Mark Borrello is an assistant professor of history of science in the Department of Ecology, Evolution, and Behavior at the University of Minnesota. His current research focuses on the history of evolutionary theory in late nineteenth and twentieth centuries.

Nathaniel Comfort is associate professor of the history of medicine at Johns Hopkins University. He is the author of *The Tangled Field: Barbara McClintock and the Search for the Patterns of Genetic Control* and editor of *The Panda's Black Box: Opening Up the Intelligent Design Controversy.*

James F. Crow is professor emeritus of genetics at the University of Wisconsin. His research interests have been and continue to be experimental and theoretical population genetics.

Ute Deichmann is research professor at the Leo Baeck Institute London, research group leader in history of the biological and chemical sciences at the Institute of Genetics, University of Cologne, and research reader in intellectual history at the University of Sussex.

Vinciane Despret is an associate professor at the Department of Philosophy of the University of Liege, Belgium. She also teaches at the Free University of Brussels.

Michael R. Dietrich is an associate professor in the Department of Biological Sciences at Dartmouth College. In addition to his work on Richard Goldschmidt, he has also published works about the history of molecular evolution.

William Dritschilo has a Ph.D. in ecology and evolutionary biology from Cornell University and was an early faculty member of the Environmental Science and Engineering Program at UCLA. He currently resides in Proctor, Vermont.

Raphael Falk is retired professor at the Department of Genetics and the Program for the History and Philosophy of Science at the Hebrew University of Jerusalem. He is coeditor of *The Concept of the Gene in Development and Evolution* (with Peter Beurton and Hans-Jörg Rheinberger) and author of *Zionism and the Biology of the Jews* (in Hebrew).

Oren Harman is an assistant professor in the graduate program in science, technology, and society at Bar Ilan University. He is the author of *The Man Who Invented the Chromosome: A Life of Cyril Darlington.* His current research focuses on the history of evolutionary biology.

Tim Horder is a senior research fellow at the Department of Physiology, Anatomy, and Genetics at Oxford University. His research interests focus on laboratory-based experimental embryology together with the historical background of developmental biology. He previously worked on the development of the nervous system in the tradition initiated by Roger Sperry.

David L. Hull taught at the University of Wisconsin at Milwaukee and at Northwestern University. He is now emeritus.

Daniel J. Kevles teaches at Yale University, where he is the Stanley Woodward Professor of History and chair of the Program in the History of Science and Medicine. His works include *In the Name of Eugenics: Genetics and the Uses of Human Heredity* and *The Baltimore Case: A Trial of Politics, Science, and Character.*

R. C. Lewontin is Alexander Agassiz Research Professor emeritus at Harvard University. He is an experimental and theoretical evolutionary geneticist and has worked in and published in the history, philosophy, and sociology of biological research. He himself has never been a rebel in his research, but rather is continuing the research problematic of an earlier generation of evolutionists.

John Prebble is senior lecturer emeritus in biochemistry and former vice-principal of Royal Holloway, University of London. He carries out research in the history of bioenergetic biochemistry in the twentieth century.

Michael Ruse is a professor of philosophy at Florida State University. He used to teach in Canada.

Jan Sapp is professor of biology at York University, Toronto. His current research focuses on the history of microbial evolutionary biology.

Ullica Segerstrale is professor of sociology at Illinois Institute of Technology, Chicago. Her books include *Defenders of the Truth,* an empirical study of the sociobiology controversy, and *Nature's Oracle,* an intellectual biography of William Hamilton.

David Sepkoski is assistant professor of history at the University of North Carolina at Wilmington. His current research focuses on the relation between paleontology and evolutionary biology in the twentieth century.

Phillip V. Tobias is a professor of anatomy and human biology at the University of Witwatersrand in South Africa. His research interests are in paleoanthropology and the human biology of African people.

Bruce Weber is professor of biochemistry emeritus at California State University at Fullerton and Robert H. Woodworth Chair emeritus of science and natural philosophy at Bennington College. His current research focuses on the conceptual development of biochemistry in the twentieth century as well as on the application of complex systems dynamics to evolutionary theory.

Index

Note: Page numbers followed by
"*f*" indicate figures.

Abele, Lawrence, 362–63
Adaptation and Natural Selection
  (Williams), 223
adaptationism, 329–33
Adelberg, Edward, *The Microbial
  World,* 306
AIDS, 284, 294–95
Allen, Garland E., 4, 13, 73, 75
Allen, T. F. H., 356
Alloway, J. Lionel, 161
Altmann, Jeanne, 343
altruism, 223–25, 282, 286–89, 293,
  378–79
American Journal of Science, 324
American Museum of Natural
  History, 201–2
American Naturalist (journal), 119,
  141, 287
*Analytische Theorie der organi-
  schen Entwicklung* (Driesch),
  46–49
Anglican Church, 20, 22
*Animal Dispersion in Relation
  to Social Behavior* (Wynne-
  Edwards), 218, 221–23
animals, observers' effect on
  behavior of, 341–43
Annual Review of Biochemistry
  (journal), 241
Annual Review of Neuroscience
  (journal), 185

appointments, of scientists, 375
Aristotle, 52
Atwood, Kim, 309
Australopithecus africanus, 6,
  88–94, 97, 99
Avery, Oswald, 157*f;* and bacteria,
  164; conservatism of, 154, 166,
  168; contribution of, 154; and
  DNA, 5, 91, 154, 161–65, 167–68;
  early research of, 159–61; iconoclasm of, 15, 17, 94, 162–63, 168–
  69; and method, 5, 166–67; and
  microbiology, 159, 162; personality of, 96; as rebel, 5, 7; reception
  of, 163–65, 168; research approach of, 166–67; steadfastness
  of, 167, 169; temperament of, 7,
  168
Axelrod, Robert, 292

baboons, 338–42, 344–45, 347, 350
Bacon, Francis, 115
bacteria: Avery and, 160–64;
  classification of, 303–7, 310–16;
  early knowledge of, 158–59, 304;
  icons in, 10; virus versus, 304–5;
  Woese and, 311–16
Baer, Karl Ernst von, 40
Baltas, Aristides, 12–13
Baltimore, David, 248, 257–59
Bates, Henry Walter, 25, 26
Bateson, William, 11, 17, 74, 77,
  104–6, 113, 156
Bawden, Frederick, 158

Beadle, George, 148, 159; *An Intro-duction to Genetics,* 109
behavior: neuroscience and, 175; Sperry and, 182. *See also* social behavior
Belling, John, 109, 111–12, 114
Bergey's Manual of Determinative Bacteriology, 305, 313
Bernal, J. D., 57
Bigelow, R. S., 200
biochemistry, 232, 240, 243. *See also* bioenergetics
Biochemistry and Morphogenesis (Needham), 57
bioenergetics, 231, 234, 238, 241, 244–45
biogenetic law, 40, 41
biogeography: contemporary, 196–97; controversies in, 194–95; Croizat and, 194, 196–97, 199–209; explanation in, 365; island biogeography, 359–65; pan-biogeography, 196–97, 199–200, 204–7; scientific profession and, 207–8; vicariance, 196–97, 198*f,* 200–203, 205–6
Biological Reviews (journal), 238
Bishop, David, 311
Bogen, Joseph, 180
books, and legitimacy, 373–74
Born to Rebel (Sulloway), 296
Borrello, Mark, 4
Boveri, Theodor, 43, 55
Bowler, Peter, 92
Boyer, Paul, 240, 242, 245
brain research, 175–77, 180–84
Bridges, Calvin B., 156, 377
Bridgewater Treatises, 22
British Association for the Advancement of Science, 21
Broom, Robert, 89–90
Brown, W. L., 366
Brundin, Lars, 200, 207

Bruno, Giordano, 296
Bryan, William Jennings, 98

cancer research, 248, 251–56, 260–61
Carnap, Rudolf, 57
Carpenter, Frank, 359
Carson, Hampton, 108
Carson, Rachel, *Silent Spring,* 357
Caspersson, Torbjörn, 158, 159
Castle, William, 71–72
causality: experimentation and, 43; as scientific ideal, 22
The Cell (Wilson), 113
Chadwick, James, 307
Chambers, Robert, *The Vestiges of the Natural History of Creation,* 28–29, 32–33
Chance, Britton, 242
Chargaff, Erwin, 96, 164, 165
Chatton, Edouard, 305
chemiosmotic theory, 236–38, 237*f,* 240–45
chromosomal theory of inheri-tance, 78–79, 105, 113, 156
chromosomes, and evolution, 108, 114–15, 126–28. *See also* cytology
Churchill, Fred, 48, 70
cladism, 196–97, 206
Clarke, Bryan, 273
Clarke, Ronald J., 89
classification, principles of, 303–4
Climo, F., 205
Coats, Sadie, 203
cognitive neuroscience, 174, 185
Cohn, Ferdinand, 158
Cole, Rufus, 159
Comfort, Nathaniel, 6, 15
commitment, of scientists to their theories, 228, 231, 243–44, 255, 276. *See also* steadfastness
community ecology, 358–62, 365

competition, and evolution, 366
Competitive Exclusion Principle, 359
Connor, Edward F., 365
consciousness, 181–83
conservation, 362–64
contingency, in evolution, 323, 330–31
continuous evolution, 73–75
Coolidge, William David, 308
Copeland, Edwin, 304
Copeland, Herbert, 304
counterfactuals, 8
Cox, C. B., 206
cranks: and rebels, 4, 29–30; Wallace, 26–30, 33
Craw, Robin, 203–8; *Panbiogeography,* 204
creationism, 97–99, 132
Crick, Francis, 65, 79, 164, 165, 168, 307–9
Croizat, Leon, 195*f*; and biogeography, 194–209, 196–97, 199–209; life of, 197–99; *Panbiogeography,* 199, 205, 207; personality of, 208; as rebel, 5–7, 194, 197, 200; reception of, 7; as solo operator, 5, 194; status of, in scientific discipline, 3, 209; writing style of, 201
crossing over, 106–7, 107*f*, 113
Crow, James F., 4
cytogenetics, 115, 144
cytology: Darlington and, 106–16; disciplinary context of, 112–15; history of, 112–13; iconoclasm and, 14, 16; icons in, 10; McClintock and, 137; problems in, 106–7
cytoplasmic inheritance, 115

Dale, Henry, 163
Danielli, Jim, 235
Darlington, Cyril, 105*f*; and cytology, 106–16; and genetics, 104–6; iconoclasm of, 10, 11, 14, 16; journal founded by, 5; life of, 103–4; and method, 5, 111, 115; as rebel, 5–6, 104, 106, 112–16; *Recent Advances in Cytology,* 106, 107–8, 111, 114; reception of, 108–11; as solo operator, 5; temperament of, 6
Darlington, P. J., 195, 200, 206
Darrow, Clarence, 98
Dart, Raymond, 85*f*; and human origins, 84, 87–94, 98–100; iconoclasm of, 15, 18; life of, 86–87; as rebel, 6–7, 84–86, 99–100; reception of, 7, 89–91; temperament of, 6, 86, 96; unorthodox thought of, 12
Darwin, Charles: and altruism, 282; on classification, 303; Dart and, 18; *The Descent of Man,* 31, 97, 213–14; Driesch and, 40; and evolution, 27–29, 115; Gould and, 17, 321–23, 332–34; and group selection, 213–14; Hooker and, 194–95; and human origins, 31; Huxley's support of, 89; and natural selection, 4, 22–28, 30–31; not a rebel, 23; *On the Origin of Species,* 23–26, 40, 97, 209, 213, 285, 303; and species, 68; Wallace compared to, 26–28; Wynne-Edwards compared to, 215
Darwin, Erasmus, 23
Davies, Robert, 237
Davis, Bernard, 168
Davis, D. D., 95
Davis, W. Dwight, 127
Dawkins, Richard, 288, 289; *The Selfish Gene,* 224, 289
Dawson, Martin H., 160–61
de Beer, Gavin, 216
de Bruyn, M., 87

Deichmann, Ute, 7
Delbrück, Max, 2, 154, 156, 159, 165, 250–51, 254–55
Demerec, Milislav, 124, 125, 129–30
The Descent of Man (Darwin), 31, 97, 213–14
Despret, Vinciane, 7
determinism, 54, 331
Developmental Biology (Gilbert), 149
developmental macromutations, 126–28
DeVore, Irven, 340, 347
de Vries, Hugo, 17, 66, 69, 74
De Waal, Frans, 349
Dewey, John, 182
dialectical materialism, 57
Diamond, Jared, 362–67
disciplinary boundaries, 15–16, 115
discovery: acceptance of, factors delaying, 91–97; acceptance of, factors involved in, 146, 148, 372–73; and iconoclasm, 15; premature, 91–94, 99
The Discovery and Characterization of Transposable Elements (McClintock), 144
dissent, 1, 131. *See also* iconoclasm
DNA: Avery and, 5, 91, 154, 161–65, 167–68; discovery of, 96, 155; double helical structure of, 307; early knowledge of, 155, 158; neutral theory of molecular evolution and, 273; RNA and, 254, 256–60, 309; sexual reproduction and, 290
Dobzhansky, Theodosius, 123, 128, 218, 225; *Genetics and the Origin of Species*, 114–15, 126–27
Dohrn, Anton, 38
dominance, in primates, 340–41, 346–47
Doolittle, Ford, 314

Douderoff, Michael, *The Microbial World*, 306
Driesch, Hans, 39*f; Analytische Theorie der organischen Entwicklung*, 46–49; and embryology, 37, 43–53, 45*f*, 59; and experimental approach, 38–39, 46, 51; followers/collaborators of, 5; and holism, 54–58; iconoclasm of, 13, 16, 17; life of, 38–40; and mechanism, 46–54; method of, 5; and Nazism, 58–60; *The Philosophy of the Individual*, 56; as rebel, 4–7, 37, 41, 44–45, 58–60; reception of, 55–58; *The Science and Philosophy of the Organism*, 52, 56; status of, in scientific discipline, 4, 37, 55–57, 60; temperament of, 6; and vitalism, 37, 39, 51–58
Dritschilo, William, 5, 6, 14
Dubois, Eugene, 84, 97
Dubos, René, 160, 165, 166, 168
Duckworth, Wynfrid L. H., 93
Dulbecco, Renato, 248, 250–51, 254–55
Dunn, Leslie C., 70, 119–20, 123, 125, 131

East, Edward M., 71–72
Eccles, John, 183
ecology: approaches to, 357–58; community ecology, 358–62, 365; defined, 358; funding in, 367–68; iconoclasm and, 14; icons in, 10; and method, 361–68; re-emergence of, 357; Simberloff and, 356–68
Egelhardt, Vladimir, 232–33
Einstein, Albert, 274
Eldredge, Niles, 324, 326–28
Elemente der Exakten Erblichkeitslehre (Johannsen), 73, 74, 80

Elliot Smith, Grafton, 87, 89, 90
Elton, Charles, 214, 216, 220, 366
embryology: Driesch and, 37, 43–53, 59; and evolutionary theory, 37, 41; and ontogeny, 42–43
Emerson, Rollins, 139
employment, of scientists, 374–75
Engels, Friedrich, 57
entelechy, 52, 56, 58
Ernster, Lars, 241–42
Ethology (journal), 226
euchromatin, 129–30
eukaryotes, 302, 305–7, 312–16
Ever Since Darwin (Gould), 334
evolution: chromosomes and, 108, 114–15, 126–28; competition and, 366; constraints on, 332–33; contingency of, 323, 330–31; continuous, 73–75; Darwin and, 27–29; genetic code and, 308–11, 316; Goldschmidt and, 125–28, 327; Gould and, 321–35; gradual, 321–22, 324, 326, 332; Haeckel and, 40–41; heredity and, 108, 110; iconoclasm and, 15–18; macro- versus micro-, 126–27, 328, 334; punctuated equilibrium and, 326–28, 329*f;* recapitulation theory and, 40; religion and, 86, 97–99; saltational, 125–26; sexual reproduction and, 290, 292–93; social behavior and, 282, 286–89, 293; teaching of, 98–99; uniform, 321–22, 324; Wallace and, 27–29; Wynne-Edwards and, 213. *See also* human origins; natural selection
evolutionary medicine, 296
evolutionary psychology, 296, 379
evolutionary synthesis, 114–15, 128, 206–7. *See also* Modern Synthesis
Evolution (journal), 326

The Evolution of Primate Behavior (Jolly), 338
Evolution Through Group Selection (Wynne-Edwards), 225–26
exaptation, 333
experimental approach: Driesch and, 38–39, 46, 51; as iconoclastic, 15; importance of, 43; personal identity and, 150–51

Falk, Raphael, 17, 156
Fedigan, Linda, 347; *Primate Encounters,* 342
feminism: McClintock and, 145–46, 149; science and, 342–43, 347–48
Ferguson, Niall, 8
Ferris, Virginia, 202
Fisher, R. A., 5, 108–10, 218, 224, 265, 268–71; *The Genetical Theory of Natural Selection,* 286
Fisher's Fundamental Theorem of Natural Selection, 79, 271, 279, 286, 380
Fitch, Walter, 310
Fogel, R. A., 8
Ford, E. B., 216
Fossey, Diane, 342
Fox, George, 312–14
Franklin, Rosalind, 164
free will, 54, 182
Friess, Christine, 288
Froriep's ganglion, 84
frozen accident theory, 308

Gaia, 293
Galikas, Biruté, 342
Galileo Galilei, 96, 296
Galton, Francis, 66, 70, 73, 75, 78
gannetry, Cape St. Mary, Newfoundland, 219*f*
gender: academic discrimination based on, 344, 377; in primate behavior, 338–42, 345–46; in

gender (*continued*)
primatology, 342–45; in scientific
observation, 342–48, 351
gene regulation, 140–42, 144–45,
148, 149
*The Genetical Theory of Natural
Selection* (Fisher), 286
genetic code, 308–11, 316
The Genetic Code (Woese), 309–10
genetic hierarchy, 128–29
genetics: Avery and, 163–65; Dar-
lington and, 104–6; early history
of, 155–58; Goldschmidt and,
119–32; iconoclasm and, 14;
Johannsen and transformation
of, 71–80; major concepts of,
65; McClintock and, 137–51.
See also inheritance
Genetics (journal), 119, 125
Genetics and the Origin of Species
(Dobzhansky), 114–15, 126–27
genius, 27
genotypes, 68–71, 74–80, 156
Gestalt theory, 176, 177, 180, 183
Gibbs, G. W., 204
Gibson, J. J., 182
Gilbert, Scott, *Developmental
Biology*, 149
Gillespie, John, 274, 276
Gladstone, William, 26
Glynn Research Institute, 238
God: Darwin and, 24; science and,
22
Goldschmidt, Richard, 121*f;* and
evolution, 125–28, 327; icono-
clasm of, 11, 15, 17–18, 131–32,
169; *The Material Basis of Evolution,* 126–28, 132; and particulate
gene, 119–20, 123–24, 128, 130–
31; as rebel, 7, 119–20, 132; reception of, 7, 124, 127–28; and sex
determination, 120–22; and spontaneous mutation, 124–25; status

of, in scientific discipline, 120;
Theoretical Genetics, 129
Goldstein, Kurt, 176
Goodall, Jane, 342–43, 346
Goodrich, E. S., 216
Gotelli, Nicholas, 368
Gould, Stephen Jay, 322*f;* and adaptationism, 329–33; conservatism
of, 322–23, 334–35; and contingency, 323, 330–31; early life of,
323–25; *Ever Since Darwin,* 334;
and evolution, 321–35; followers/
collaborators of, 5; on Goldschmidt, 130, 132; iconoclasm of,
11, 15, 17, 132, 321–24, 327–28,
330, 332, 335; on Modern Synthesis, 218; *Ontogeny and Phylogeny,*
331–32; and paleontology, 323–
25, 327–28; and punctuated equilibrium, 326–27; as rebel, 5, 7,
323, 324, 334–35; reception of, 7,
321; Simberloff and, 360; *The
Structure of Evolutionary Theory,*
333–34
graduate students, 374–76
Grant, Robert, 23
grants, 374, 376
Gray, Asa, 24
Gray, Michael, 314
Gray, R., 205
Great Monkey Trial, 98
Greenberg, Rayla, 256
Grehan, John R., 204–8; *Panbiogeography,* 204
Grene, Marjorie, 356
Griffith, Fred, 160–61, 163
Gross, Paul, 331
group selection: in contemporary
biology, 215; Darwin and, 213–
14; Hamilton and, 285, 286, 288–
89; Wynne-Edwards and, 213,
217, 219–28
Groupé, Vincent, 252

Grundzüge einer Theorie von Phylogenetischen Systematik (Hennig), 94–95

Gurdon, John, 2

Haeckel, Ernst, 13, 38, 40–41, 43, 44, 56, 304

Haldane, J. B. S., 15, 56, 57, 108–10, 114, 218, 265, 268–70, 277, 286, 288, 379

Haldane, John Scott, 56

Hamburger, Viktor, 177

Hamilton, William D. (Bill), 283*f;* and altruism, 282, 286–89, 293, 378–79; contribution of, 4, 296; death of, 6, 295; early life of, 284–86; iconoclasm of, 15, 283–84, 289, 296–97; and kin selection, 30, 223–25, 282, 287–89, 379; as rebel, 3–4, 6–7, 289, 292, 294–96, 378; reception of, 7; and sexual reproduction, 290, 291*f,* 292–93; status of, in scientific discipline, 3, 282–83, 288; temperament of, 6, 284

Hamilton's Principle/Rule, 287, 379

Haraway, Donna, 341, 343, 350, 351

Harrington, Anne, 51, 55, 58

Harris, Harry, 380

Hartmann, Eduard von, 51

Heads, Michael, 204–8; *Panbiogeography,* 204

Heidegger, Martin, 55

Heidelberger, Michael, 160, 169

Hein, Hilde, 52

Heitz, Emil, 129

Henderson, T. M., 205

Hennig, Willi, 196, 200–202, 206; *Grundzüge einer Theorie von Phylogenetischen Systematik,* 94–95

Henslow, John, 23

Herbst, Curt, 38–39, 43

heredity. *See* inheritance

Heredity (journal), 5

heroes, 1

Herschel, John F. W., 20, 22, 23

Hertwig, Oskar, 43, 112

Hertwig, Richard, 43, 120

heterochromatin, 129–30

heterodoxy: communication of, 373; and credibility, 377–78; employment and financial hurdles for, 373–74, 376; iconoclasm versus, 372; McClintock and, 377

Hinde, Robert, 344

Hinkle, Peter, 241

His, Wilhelm, 84

H.M.S. *Beagle,* 23, 29

H.M.S. *Challenger,* 40

holism, 54–58, 175–77

Holthausen, C., 124

Homo erectus, 92, 97, 99

Homo habilis, 91–92, 94

Hooker, J. D., 194, 200

Hooper, Edward, *The River,* 294

Hopkins, Frederick Gowland, 235

Hotchkiss, Rollin, 164

Huang, Alice, 257, 258

Hubby, Jack, 376

Hubel, David, 175, 179

Hughes, Alun R., 89

Hull, David L., 3

Human Genome Project, 79

human origins: Dart and, 84, 87–94, 98–100; Darwin and, 30–31; paradigm of, in 1925, 92–94; Wallace and, 30–31, 33–34. *See also* evolution

Hume, David, 51

Huskins, Charles Leonard, 109

Hutchinson, G. E., 365, 366

Huxley, Julian, 216

Huxley, Thomas Henry, 12, 17, 22, 23, 26, 30, 32, 89, 218; *Man's Place in Nature,* 97
hypothetico-deductive systems, 22

iconoclasm: conceptual, 14–15; conservative, 16–17; disciplinary, 15–16, 115; discovery and, 15; function of, 8; heterodoxy versus, 372; methodological, 13–14; nature of, 12–13, 227–28; novel, 17; obstacles for, 373–77; reactionary, 16; rebelliousness versus, 6–7, 137, 147*f;* research characteristics and, 169. *See also* dissent
icons of science: challenges to, 2–3; defined, 2, 9–10; establishment of, 10–12. *See also* scientific paradigms
Imbrie, John, 323
immunochemistry, 159
inclusive fitness, 286–89
informal knowledge, 348
Ingram, Vernon, 148
inheritance: chromosomal theory of, 78–79, 105, 113, 156; cytoplasmic, 115; DNA and, 5, 91, 155; evolution and, 108, 110; Johannsen and, 65–80, 76*f;* Mendelian factors and, 71–72, 77; in species, 68–69; theories of, 66
intersexuality, 120–22
An Introduction to Genetics (Sturtevant and Beadle), 109
island biogeography, 359–65
Ives, Charles, 373

Jablonka, Eva, 2
Jacob, François, 141, 148, 309
Jagendorf, André, 238, 241
Janssens, Frans Alfons, 106, 113
Jay, Phyllis, 343, 346, 351

Jefferson, Thomas, 18
Johannesburg Star (newspaper), 98
Johannsen, Wilhelm, 67*f;* career of, 66; contribution of, 70–80, 156; *Elemente der Exakten Erblichkeitslehre,* 73, 74, 80; iconoclasm of, 11, 16–17; and inheritance theory, 65–80, 76*f;* as rebel, 3; status of, in scientific discipline, 3
John Innes Horticultural Institution, 104–5
Jolly, Alison, 343, 347, 350, 351; *The Evolution of Primate Behavior,* 338
Journal of Experimental Medicine, 162
Journal of Theoretical Biology, 227, 287
Jukes, Thomas H., 272–73

Kalckar, Herman, 233
Kandler, O., 315
Kant, Immanuel, 51
Keilin, David, 233, 234, 243
Keith, Arthur, 89, 91, 93–94
Keller, Evelyn Fox, 150
Kerr, Warwick, 287
Kevles, Daniel J., 4, 11
Kihara, Hitoshi, 267
Kimura, Motoo, 266*f;* advocacy by, 276; contribution of, 4; death of, 276; life of, 266–69; and neutral theory, 265–66, 272–77; and population genetics, 269–78; as rebel, 4, 266, 276–77; reception of, 277; status of, in scientific discipline, 265
King, Jack L., 272–73
Kingsley, Charles, 7, 60
kin selection, 223, 225, 282, 287–89, 379
Kiriakoff, S. G., 200

Knowles Lab, 308
Koch, Robert, 158
Kolmogorov equations, 270–71
Komai, Taku, 268
Kornberg, Arthur, 2
Krebs, Hans, 237
Kuhn, Thomas, 9, 323, 358; *The Structure of Scientific Revolutions,* 3
Kumazawa, M., 267
Kummel, Bernhard, 325

Lack, David, 223; *Natural Regulation of Animal Numbers,* 218–20
Lamarck, Jean-Baptiste, 23
Lamb, Marion, 2
Landsteiner, Karl, 169
Langmuir, Irving, 308
language, and transmission of science, 94–95
Lashley, Karl, 175–77, 180, 183, 184
Leakey, Louis S. B., 91, 94, 342
Lederberg, Joshua, 148, 159, 164
legitimacy: communication and, 373–74; employment and, 373–75; grants and, 374, 376; novice scientists and, 374–76; rebels and, 3–4
Lenton, Tim, 293
Levins, Richard, 335
Lewin, Kurt, 176
Lewontin, Richard, 18, 330, 332–33, 335
Lipmann, Fritz, 233
Locke, John, 51
Loeb, Jacques, 53, 155, 166–67, 169; *The Organism as a Whole,* 56, 70
Longwell, Chester, 324
Lovelock, James, 293
Løvtrup, Søren, 207
Luria, Salvador, 165
Lush, Jay L., 268–69

Lwoff, André, 305
Lyell, Charles, 23, 322, 324

MacArthur, Robert, 6, 357–58, 361–62, 365; *The Theory of Island Biogeography,* 358–59
MacFadyen, Amyen, 361
Machamer, Peter, 12–13
MacLeod, Colin M., 154, 161–64
Madison, James, 18
Malthus, Robert, 27
Manaker, Robert, 252
Manhattan Project, 357
Mann, Horace, 323
Man's Place in Nature (Huxley), 97
Margoliash, Emmanuel, 310
Margulis, Lynn, 2
Maruyama, Takeo, 271
Marx, Karl, 57
The Material Basis of Evolution (Goldschmidt), 126–28, 132
mathematics, and science, 275
Matsuda, Hirotsugu, 274
Matthew, W. D., 195, 206
May, Robert, 227, 364, 367
Mayden, Richard, 205
Maynard Smith, John, 223, 225, 289, 321, 380
Mayr, Ernst, 109–10, 195, 200, 204, 206, 209, 218, 316, 327
McCarty, Maclyn, 154, 160, 162–64, 168
McClintock, Barbara, 138*f;* and cytology, 137; *The Discovery and Characterization of Transposable Elements,* 144; feminism and, 145–46, 149; and gene regulation, 140–42, 144–45, 149; and genetics, 137–51; iconoclasm of, 11, 15, 17–18, 137–42, 147*f,* 377, 379; Nobel Prize won by, 7, 138–39, 142, 145; private iconoclasm of, 139–41; private rebellion of, 142–44,

McClintock, Barbara (*continued*)
150; public iconoclasm of, 141–
42; public rebellion of, 144–46,
149–50; as rebel, 5–7, 139, 142–
46, 147*f*, 149–50, 265; reception
of, 148–49; as solo operator, 5;
status of, in scientific discipline,
137–39, 145, 377; temperament
of, 6; and transposition, 137, 139–
42, 144–46, 148–49
McCulloch, Warren, 180
McDonald, Duncan, 268
McDowell, R. M., 203
mechanistic materialism, 42–43,
46–56
meiosis, 103, 106, 108
Mendel, Gregor, 65, 69, 71, 96, 112,
115, 156, 214
Mendelian factors, 71–72, 77
Merrill, E. D., 198–99
method: Avery and, 166–67; Darlington and, 5, 111, 115; ecology
and, 361–66; iconoclasm and,
13–14; rebels and, 4–5
Meyer, R. L., 180
Michaelis, Leonor, 169
The Microbial World (Stanier,
Douderoff, and Adelberg), 306
microbiology: Avery and, 159, 162;
early history of, 158–59; Woese
and, 315–16
Miescher, Friedrich, 96, 155
Mill, John Stuart, 32
Mirsky, Alfred, 159, 165
missing link, 88, 97, 326
Mitchell, Godfrey, 234
Mitchell, Peter, 232*f*; and chemiosmotic theory, 236–38, 237*f*,
240–45; Grey Books, 238, 242;
iconoclasm of, 16, 17; life and
early work of, 234–36; method of,
5; Nobel Prize won by, 7, 231,
242; and oxidative phosphoryla-

tion, 231, 234, 236–44; as rebel,
5–7, 242–43; research approach
of, 242; steadfastness of, 243–44;
temperament of, 234
Mizutani, Satoshi, 259
Modern Synthesis, 218, *220,* 321–
22. *See also* evolutionary
synthesis
molecular biology, 316
Molecular Biology of the Gene
(Watson), 309–10
molecular evolution: emergence
of field of, 310; neutral theory of,
4, 265–66, 272–78; Woese and,
315–16
molecular genetics, 169
Monod, Jacques, 141, 148, 309
Moore, P. D., 206
Morgan, Thomas Hunt, 38, 56, 65,
75–79, 105*f*, 106, 112, 113, 121,
156, 268, 377
morphology, 37, 40–41
Morris, Jeannie, 216
Morton, Newton, 268–69
mosaic theory of development,
42–43, 44*f*, 45–46
Moyle, Jennifer, 238–39
Muller, Hermann J., 71, 78, 123,
125, 127, 129, 131, 156, 163
Myers, R. E., 180
myths, of McClintock's life and
work, 139–51

Napier, John R., 91, 94
National Environmental Policy
Act, 357
National Geographic effect, 343
National Institutes of Health, 376
National Science Foundation,
376
*Natural Regulation of Animal
Numbers* (Lack), 218–20
natural selection: community ecol-

ogy and, 358–59; cytology and, 107–11; Darwin and, 4, 22–28, 30–31; Gould's questioning of, 322, 329–33; group versus individual, 213, 217, 219–28; and level of selection, 228; neutral theory of, 4, 265–66, 272–78; random processes and, 270; Wallace and, 4, 20, 25–31. *See also* evolution

Nature (journal), 88, 90, 93, 227, 236, 259, 260, 287, 288, 292, 294, 373, 380

Nazism, 58–60

Needham, Joseph, *Biochemistry and Morphogenesis,* 57

Nelson, Gareth, 196–97, 200–203, 206–9; *Systematics and Biogeography,* 196, 203, 204; *Vicariance Biogeography,* 200, 204

neo-Darwinism, 126–27, 132, 213, 218, 222–24, 284, 333–34

Neufeld, Fred, 161

neuroscience, 177–85. *See also* cognitive neuroscience

neutral theory of molecular evolution, 4, 265–66, 272–78

Newell, Norman, 323, 325

Newton, Frank, 105–6

Newton, Isaac, 21–22

New Zealand Journal of Zoology, 204

niche theory, 359

Nilsson, Nils Hjalmar, 66

Nobel Prize, 4, 7, 138–39, 142, 145, 148, 175, 184, 231, 242, 245, 248, 261, 267, 307–10, 378

normal science. *See* icons of science; scientific paradigms

null models, 366–68

Odum, Eugene P., 357–58

Ohta, Tomoko, 272, 275

On the Origin of Species (Darwin), 23–26, 40, 97, 209, 213, 285, 303

ontogeny, 40–43

Ontogeny and Phylogeny (Gould), 331–32

Oppenheimer, Jane, 46, 51

organicism, 175–77

The Organism as a Whole (Loeb), 56, 70

origin of organisms, 22–24, 302, 313, 315–16. *See also* human origins

orthodoxy, establishment of, 11–12. *See also* scientific paradigms

oxidative phosphorylation: initial concepts of, 231–34; Mitchell and, 231, 234, 236–44

Page, R., 205

Pagel, Mark, 205

paleoanthropology, 86–95

Paleobiology (journal), 132, 328

paleontology, 323–25, 327–28

Paley, William, 24

panbiogeography, 196–97, 199–200, 204–7

Panbiogeography (Craw, Grehan, and Heads), 204

Panbiogeography (Croizat), 199, 205, 207

paradigms. *See* scientific paradigms

Parasite Red Queen theory, 292, 293

parasites, 290, 291*f,* 292–93

Pareto, Vilfredo, 1

particulate gene, 119–20, 123–24, 128, 130–31

Pasteur, Louis, 158

Pauling, Linus, 272–73, 310, 378

Pavlov, Ivan, 184

Pearson, Karl, 73

peer review, 374, 376

Pera, Marcello, 12–13
personality: biogeography and, 208; and transmission of science, 95–96
persuasion, 12–13
Peters, Robert Henry, 361
phage group, 165, 250
phenotypes, 69–71, 74–75, 121–24, 156
Phylogenetic Systematics (Hennig), 95
phylogeny, 37, 40–42
Piltdown forgery, 92–93
Pirie, Norman, 158
Pithecanthropus, 92, 97
Planck, Max, 243
Platnick, Norman, 201–2; *Systematics and Biogeography,* 196, 203, 204
Plough, Harold, 124, 125
politics, and transmission of science, 95
Pollard, Ernest, 307–8
Popper, Karl, 32, 196, 201, 206, 244, 361
population dynamics, 214–17, 219–25
population genetics, 269–78
Portugal, Franklin, 143
position effects, 123–24, 129
postdoctoral fellows, 374–76
Powell, Baden, 21
Prebble, John, 5, 16
preformationism, 75–77
premature discovery, 91–94, 99
Prenant, Marcel, 57
Price, George, 288–89, 378–80
Price Equation, 380
Primate Encounters (Strum and Fedigan), 342
primates, social behavior of, 338–47, 350
primatology: dominance theory in, 340–41, 346–47; women in, 342–45, 347–48, 351
Principles of Geology (Lyell), 23
The Problem of Individuality (Driesch), 52
Proceedings of the National Academy of Sciences, 292
prokaryotes, 302, 305–7, 312–16
punctuated equilibrium, 326–28, 329*f*

Quammen, David, 364

Racker, Efraim, 241–42, 244
Randall, John, 164
Raup, David, 330–32
rebels: common features of, 7; diversity among, 3; heterodox thinkers versus, 372; historiographical category of, 8–9; iconoclasts versus, 6–7, 137, 147*f;* methods of, 4–5; non-academic paths pursued by, 5–6; obstacles for, 373–77; role of, 1, 59; solo versus collective efforts of, 5; status of, in scientific discipline, 3–4, 60; steadfastness of, 7; temperaments of, 6–7, 60, 85
recapitulation theory, 40
Recent Advances in Cytology (Darlington), 106, 107–8, 111, 114
Reich, Wilhelm, 54
Reichenbach, Hans, 57
Reifferscheidt, Margarete, 39
religion: and evolution, 86, 97–99; science and, 22, 33, 96–97
reputation, creation of, 174, 183–85
reverse transcriptase, 4, 15, 259–60
rhetoric, 12–13
Rhoades, Marcus, 139, 140
Richmond, Rollin, 273
Ridley, Mark, 226
Ripley, Suzanne, 343

The River (Hooper), 294
RNA: and bacteria classification,
 311–14; and genetic code, 309;
 Rous virus and, 254, 256–60
Robertson, Rutherford, 236
Robinson, John T., 89
Roll-Hansen, Nils, 69, 74, 78
Rosen, Donn, 201–3, 206; *Vicari-
 ance Biogeography*, 200, 204
Roughgarden, Jonathan, 366–67
Rous, Peyton, 138, 251
Rousseau, Jean-Jacques, 342
Rous virus, 251–61, 253*f*
Roux, Wilhelm, 41–43, 44*f*, 45–46,
 59
Rowell, Thelma, 339*f*; baboon stud-
 ies of, 338–39, 341–48; contribu-
 tion of, 350–51; early life of, 344;
 iconoclasm of, 347–51; method
 of, 5; and primatology, 343–45,
 347–48; as rebel, 5, 7; and sheep
 studies, 349–50; temperament
 of, 7
Rubin, Harry, 11, 251–55, 257
Ruse, Michael, 331
Russell, Bertrand, 341–42
Rutherford, Ernest, 307

Sakharov, Andrei, 260
saltationism, 17, 125–26
Sanger, Frederick, 310, 311
Sapp, Jan, 10, 14, 16, 75
Schaxel, Julius, 56
Schlick, Moritz, 57
Schopenhauer, Arthur, 51
Schopf, Tom, 325, 328, 330–32, 334
Schrödinger, Erwin, *What Is Life?*,
 158
science: disciplinary boundaries
 of, 15–16; fate of theories in,
 205–6, 209; gender and, 342–48,
 351, 377; ideal of, 21–22; informal
 knowledge versus, 348; mathe-

matics and, 275; religion and, 22,
 33, 96–97; reputation creation in,
 174, 183–85; social structure of,
 18, 146, 206–8, 372–77, 380; in
 Victorian England, 20–22
Science (journal), 260, 292, 294,
 314, 358
*The Science and Philosophy of the
 Organism* (Driesch), 52, 56
Scientific American (journal), 91,
 223
Scientific Controversies (Macha-
 mer, Pera, Baltas), 12
scientific paradigms: development
 of, 146, 148; discovery and, 91–
 94, 99; in paleoanthropology, 91–
 94, 99. *See also* icons of science;
 orthodoxy
Scopes, John Thomas, 98
Sedgwick, Adam, 23, 28
Segerstrale, Ullica, 15
The Selfish Gene (Dawkins), 224,
 289
Sepkoski, David, 5, 17
sexual reproduction, 290, 292–93
Shackleton, Ernest, 216
Shaw, George Bernard, 7
sheep, 349–50
Shellshear, Joseph, 84
Sherrington, Charles, 184, 185
Sia, Richard, 161
Silent Spring (Carson), 357
Simberloff, Daniel S. (Dan), 358*f*;
 and conservation, 362–64; and
 ecology, 356–68; and evolution,
 330; followers/collaborators of, 5,
 361; iconoclasm of, 10, 14, 360–
 61; and method, 361–68; and null
 models, 366–68; as rebel, 5, 7,
 357–58; temperament of, 7, 364
Simpson, George Gaylord, 3, 90,
 127–28, 195, 199–201, 206–9, 218,
 327

Slater, E. C. (Bill), 233, 239, 240, 242

Sloan Wilson, David, 226, 289

SLOSS (single large or several small refuges), 361–64

Smith, Cedric, 380

Smith-Woodward, Arthur, 89

Smuts, J. C., 89

Sober, Elliott, 289

social behavior: evolution and, 282, 286–89, 293; population dynamics and, 221–24; of primates, 338–47, 350; of sheep, 349–50

sociobiology, 289, 379

Sociobiology (Wilson), 289, 331, 379

Sogin, Mitchell, 311

Sollas, William Johnson, 90

Sonneborn, Tracy, 2

South Africa, 95, 98–99

Southwood, Richard, 290

Spearmann, Hans, 48

species-packing, 359, 365

Sperry, Roger, 176*f;* and brain research, 175, 180–84; iconoclasm of, 17; life of, 175; method of, 5; and mind-body problem, 174; and neuroscience, 177–85, 179*f;* Nobel Prize won by, 7, 175, 184; as rebel, 5, 7, 174, 177, 184–85; status of, in scientific discipline, 184–85; and values, 182–83, 185

Spiegelman, Sol, 309, 311

Spiers, A. E., 87

spiritualism, 26, 28–29, 31–34

split-brain research, 180–84

spontaneous mutation, 124–25

standpoint theories, 347–48

Stanier, Roger, 305, 314; *The Microbial World,* 306

Stanley, Steven, 328

Stanley, Wendell, 158

steadfastness: of Avery, 167, 169; of Mitchell, 243–44; of rebels, 7;

of Wallace, 29. *See also* commitment

Stebbins, G. Ledyard, 218

Stent, Gunther S., 91, 94, 96, 158

Stern, Curt, 119, 127

Stetson, R. H., 175

Stibbe, E. P., 87

Stoeckenius, Walther, 241

Strong, Donald R., 361

The Structure of Evolutionary Theory (Gould), 333–34

The Structure of Scientific Revolutions (Kuhn), 3

Strum, Shirley, 343, 347; *Primate Encounters,* 342

Sturtevant, Alfred H., 115, 156; *An Introduction to Genetics,* 109

Sugden, Bill, 257

Sulloway, Frank, *Born to Rebel,* 296

Sutton-Boveri hypothesis, 112–13

Systematics and Biogeography (Nelson and Platnick), 196, 203, 204

systemic mutations, 126–27

Tallahassee Mafia, 361, 364, 366–67

Tatum, Edward, 148, 159, 164

Taung child fossil, 15, 86–94, 97–98

teleology, 49–52. *See also* entelechy

Temin, Howard, 249*f;* and cancer research, 248, 251–56, 260–61; contribution of, 4; early life of, 248–50; iconoclasm of, 11, 15, 17, 379; method of, 5; Nobel Prize won by, 4, 7, 248, 261; as rebel, 4–7, 248, 250, 260–61; reception of, 257; status of, in scientific discipline, 4, 257–58, 260; temperament of, 6; and virology, 251–61

Tennessee Monkey Trial, 98

Thelma effect, 341, 351
Theoretical and Applied Genetics (journal), 373
Theoretical Genetics (Goldschmidt), 129
theories: fate of, 205–6, 209; simplicity of, 274
The Theory of Island Biogeography (Wilson and MacArthur), 358
Tobias, Phillip V., 6, 12, 89, 94
Tolstoy, Leo, 9
transposition, genetic, 137, 139–42, 144–46, 148–49
Travelyan, G. M., 8
Trends in Ecology and Evolution (journal), 227
Trivers, Robert, 223–25
Tuatara (journal), 204

Umwelt, 55
uniformitarianism, 321–22, 324
unit characters, 69–71, 77
Üxeküll, Jakob von, 55, 56

values, neuropsychology and, 182–83, 185
van Niel, C. B., 305, 314
Velikovsky, Immanuel, 201
The Vestiges of the Natural History of Creation (Chambers), 28–29
vicariance biogeography, 196–97, 198*f*, 200–203, 205–6
Vicariance Biogeography (Nelson and Rosen), 200, 204
Vilmorin, Louis, 66
virology, 250–51, 258
virus, versus bacteria, 304–5
vitalism, 34, 37, 39, 51–58
Vogel, Philip, 180
Vrba, Elizabeth, 333

Wade, Michael J., 226
Wald, George, 359

Wallace, Alfred Russel, 21*f*; and biogeography, 30; as crank, 26–30, 33; Darwin compared to, 26–28; and evolution, 27–29; iconoclasm of, 16; lack of training, 30, 32, 33; life and death of, 24–25; misfortunes of, 25–26; and natural selection, 4, 20, 25–31; as rebel, 3–4, 7, 30, 33–34; reception of, 7; and spiritualism, 26, 28–29, 31–34; status of, in scientific discipline, 3–4, 26, 28, 30, 33–34; steadfastness of, 29; unorthodox thought of, 12
Wallace's Line, 30
Washburn, Sherwood, 340
Watson, James, 65, 79, 164, 165, 168, 307; *Molecular Biology of the Gene*, 309–10
Weber, Bruce, 5, 16
Wedgwood, Josiah, the younger, 24
Weismann, August, 14, 17, 38, 42, 59, 75, 77, 78, 103, 113–14, 116
Weiss, Paul, 175–78, 184
Weldon, W. F. Raphael, 73
What Is Life? (Schrödinger), 158
Wheelis, Mark, 315
Whewell, William, 21, 22, 23, 28
Wiesel, Torsten, 175, 179
Wilkins, Maurice, 164, 165
Williams, George C., 30, 224; *Adaptation and Natural Selection*, 223
Williams, R. J. P., 236
Wilson, E. B., 112, 155; *The Cell*, 113
Wilson, E. O., 360–62, 365–66; *Sociobiology*, 289, 331, 379; *The Theory of Island Biogeography*, 358–59
Wilson, James T. "Jummy," 86
Wilson, Maurice, 307
Woese, Carl, 303*f*; and bacteria classification, 311–16; early life

Woese, Carl (*continued*)
  of, 307–8; and genetic code, 308–
  11, 316; *The Genetic Code,* 309–
  10; iconoclasm of, 10, 14, 16, 17,
  302, 314–17; and origin of life,
  302; reception of, 7
Wolfe, Ralph, 313
Woodger, Joseph, 235
Wright, Sewall, 108, 115, 127–28,
  218, 225, 265, 268–72, 286
Wynne-Edwards, Vero Copner,
  214*f; Animal Dispersion in Relation to Social Behavior,* 218, 221–
  23; Darwin compared to, 215;
  early life of, 215–16; and evolution, 213; *Evolution Through
  Group Selection,* 225–26; and
  group selection, 213, 217, 219–28;

iconoclasm of, 12, 17, 220–21,
  227–28; method of, 5; and population dynamics, 214–17, 219–25;
  as rebel, 4–5, 7, 215; reception of,
  7, 220, 223, 226

X-Club, 12

Yoon, M.-Y., 180
Young, Robert Burns, 87
Yukawa, Hideki, 267
Yule, Udny, 73

Zangerl, R., 95
Zuckerkandl, Emile, 272–73, 310
Zuckerman, Solly, 339–40, 342,
  346
Zuk, Marlene, 290, 293

Seattle Public Library
Green Lake Branch
(206) 684 7547

08/31/15 02:16PM

Borrower # 1148266

Rebels, mavericks, and heretics in biolo
0010064339137   Date Due: 09/21/15
acbk

A kid's guide to awesome duct tape proje
0010083072081   Date Due: 09/21/15
jcbk

TOTAL ITEMS: 2

Visit us on the Web at www.spl.org